MAR   2011

Phlox

Phlox

A Natural History and Gardener's Guide

James H. Locklear

Timber Press

Portland • London

Published in 2011 by Timber Press, Inc.

The Haseltine Building
133 S.W. Second Avenue, Suite 450
Portland, Oregon 97204-3527
www.timberpress.com

2 The Quadrant
135 Salusbury Road
London NW6 6RJ
www.timberpress.co.uk

Printed in China
Interior designed by Lila Braker

Library of Congress Cataloging-in-Publication Data

Locklear, James H., 1953–
    Phlox : a natural history and gardener's guide / James H. Locklear. — 1st ed.
        p. cm.
    Includes bibliographical references and index.
    ISBN 978-0-88192-934-8
  1. Phlox. I. Title.
    QK495.P77L63 2011
    635.9'3394—dc22

                                                          2010025877

A catalog record for this book is also available from the British Library.

In memory of
Amy Jo North

Contents

Color photographs follow page 160

Preface

There is a mountain, remote and reckless, that crests in Wyoming's Owl Creek Range. Hidden in the heart of the Wind River Indian Reservation, it bears no official place name today, and probably needs none. But on 17 July 1873 it received a christening both descriptive and poetic—Phlox Mountain (PLATE 1).

Captain William A. Jones, U.S. engineer, was leading a government reconnaissance and scientific mission to the Yellowstone country of northwestern Wyoming when he and a small party made a side trip to this peak. The men had traveled for a number of days across dry and desolate plains, so they must have experienced a refreshing jolt when they ascended to this flowery massif. Moved by the sight, Jones named the mountain for "the extensive fields of white phlox that its slopes display."

Sadly, the name didn't stick, even though formally published in the expedition's report to the United States House of Representatives in 1874. Phlox Mountain appears on no maps today, not even the most detailed topographic quadrangle, yet it endures in a few obscure references as the original collection locality of the celebrated Jones' columbine (*Aquilegia jonesii*)—an exquisite plant discovered here and named for Jones by the expedition's botanist, Charles Christopher Parry.

I'm not sure when I first came across a reference to Phlox Mountain, but I was compelled to make a pilgrimage. I found a copy of Jones's report in the University of Nebraska library, and its narrative, charts, and crumbling maps helped me zero-in on the target. Upon securing permission from the Shoshone and Arapaho tribes to visit their reservation lands, my buddy Brad Anderson and I retraced the expedition's route and climbed Phlox Mountain on 11 July 2000.

The Captain's inspiration was plain to see. The slopes and summit were cloaked with low mats of Yellowstone phlox (*Phlox multiflora*), past peak bloom but still a defining presence in the flora. Like Jones, we lingered over unguarded, backside views of the Bighorn, Absaroka, and Wind River mountains—ranges of renown, compassing the fringes of a secret universe visible only from this forgotten polestar. It was an exhilarating, privileged experience we tried hard to soak up, knowing full well we would never stand on this wild ground again.

There is a house, middle-aged and modest, that sits on a corner lot in the town of Warrensburg, Missouri. Nothing about it would turn the head of a passerby, but the sight of it stirs deep emotion in me. My wife, Lynn, grew up there and our children spent sweet, grandparent-indulged days in its embrace.

The house appears in a photograph, fastened by magnet to the refrigerator in our kitchen. The photo was taken a couple of days after the auction that closed thirty-seven years of residence by Lynn's family. Our niece Amy and her parents, Donnie and Linda, were actually the intended subjects. Gathered on the back porch, Donnie is standing behind Amy, hands on her shoulders, tipping her off balance. There are smiles, tired ones, all around.

It was high summer, when the cicadas are at full throttle. You can tell because waist-high wands of summer phlox (*Phlox paniculata*) are in bloom in the foreground, holding matronly sway over an undisciplined but well-loved garden. The color of their flowers, a muddy lilac-purple, reveals these were not fancy, named cultivars—just *ever-day* plants, as my *y*-dropping Hoosier-sprung in-laws would say.

Even so, these plants were not without pedigree. They were descendants, many summers removed, of clear-colored cultivars that bore splendid names like 'Starfire', 'Sir John Falstaff', and 'Dodo Hanbury Forbes'—bred a half century before by the likes of Ruys of Holland, Symons-Jeune of England, and Schollhammer of Germany. Now, purpled by the passing of many generations in this Missouri garden, they would win no flower show. But what is that, compared to cheering the world of a child?

Amy spent many days playing in this yard, but that phlox-season would be her last. We lost her in a car accident six months later—a feisty, beloved girl of sixteen summers. And now this photo, galleried in our kitchen, has become a cherished remembrance of a loved one, and the phlox plants, like old friends dear enough to be taken for granted, have gained an unspoken place in our family's memories.

I set out to write a book on a genus of plants . . .

What a genus! No other group of plants in North America can equal *Phlox* for its preeminence in the wild and in the garden. Its sixty-one species decorate and sometimes define plant communities up and down the continent, from Beringia to Chihuahua, the Siskiyous to Apalachicola. Yet no American genus has enjoyed a richer history in the gardens of the world. Borders without summer phlox were "unthinkable" to German plantsman Wilhelm Schacht. An over-the-top Reginald Farrer called for a yearly "horticultural festival" to commemorate the introduction of creeping phlox (*Phlox subulata*) into English gardens. Phloxes have even come to bear the weight of nostalgia, as when they showed up in the childhood dream-

land of English literary giant J. R. R. Tolkien—"There all the borders, trimmed with box, were filled with favourite flowers, with phlox."

What a journey! At the start of this project I knew only a handful of *Phlox* species, so I made it my goal to see as many in the wild as time and resources allowed. The ten-year pursuit placed me before some of America's most beautiful and profound scenery, and marked my senses with the tang of *krummholz* and *cove forest*, *shut-in* and *sky island*, *barren* and *fen*. I gained the acquaintance of wildings so rare few have seen them, and landscapes so powerful they show up in *my* dreams. I also met, through the library, great explorers and scientists and nature writers, who helped me to see better, and deeper.

Thomas à Kempis asked, "And what have we to do with *genus* and *species*, the dry notions of logicians?" I understand and agree with the caution of this fifteenth-century mystic, that the study of nature cannot by itself quench a searching soul. But my immersion in the genus *Phlox* these past few years has engaged both my mind and my heart, and has stirred ripples of wonder that will occupy me the rest of my life.

Acknowledgments

In the immortal words of Rocky Balboa, "She's got gaps. I got gaps. Together we fill gaps."

When Neal Maillet of Timber Press contacted me in 1996 about writing a book on the genus *Phlox*, I tried to help him understand how little I really knew about this group of plants. Sure, I had recently written an article on the subject for a gardening magazine, but it was little more than a superficial rehashing of things I gleaned from other articles. I even suggested names of people I thought could do better job. Neal would not go away.

So the contract was signed and I embarked on the project. It didn't take long to reach the boundary between my knowledge and my ignorance.

These acknowledgments recognize the people, institutions, and organizations that helped me fill the gaps.

I tapped the expertise of a number of plant taxonomists in the preparation of this book. Conversations with Carolyn Ferguson (Kansas State University), Ronald Hartman (University of Wyoming), Robert Kaul (University of Nebraska–Lincoln), Donald Levin (University of Texas at Austin), and Dieter Wilken (University of California at Santa Barbara) were especially helpful, but they should not be held accountable for errors of taxonomic judgment found within.

I am indebted to the many herbaria that gave me access to their collections, loaned specimens, or provided digital images so that I could observe as many type specimens as possible. Nearly fifty herbaria were consulted. Robert Kaul, curator of the Bessey Herbarium of the University of Nebraska State Museum was a tremendous help in securing loans of type specimens

I enjoyed the resources of many of America's great botanical and horticultural libraries, and was thrilled to have access to several rare book collections. I here admit to abusing my privilege as a staff member of the University of Nebraska–Lincoln in drawing old obscure books out of library storage and working the interlibrary loan system to its limits.

A great number of botanists, ecologists, and natural areas managers helped me to better understand the flora and ecological systems of their part of the country, and helped me locate populations of hard-to-find species. The list includes James Allison (Georgia), Robert Dorn (Wyoming), Walter Fertig (Utah and Wyoming),

David Hannah (Montana), Michael Homoya (Indiana), William Jennings (Colorado), Peter Lesica (Montana), Chris Ludwig (Virginia), Chris Oberholster (Alabama), J. Dan Pittillo (North Carolina), Michael Powell (Texas), Stanley Welsh (Utah), and George Yatskievych (Missouri).

There are many who have had much more horticultural experience with phlox than I, and conversations and correspondence with them has greatly enriched this book. Among them are Don Hackenbery, Richard Hawke, Don Jacobs, Panayoti Kelaidis, Rick Lupp, and Charles and Martha Oliver.

The North American Rock Garden Society, the Native Plant Society of Oregon, and the University of Nebraska–Lincoln provided grants that allowed me to travel across the country to do field research. This book is the far more complete thanks to their generous support.

My publisher Timber Press has exhibited great patience and support as this project became much bigger in scope and took far more time to complete than either party imagined at the outset. Thanks to Neal Maillet for contacting me in the first place, and to editor-in-chief Tom Fischer for keeping after me about this project.

Finally, while they added no botanical or horticultural expertise to the book, I owe the greatest debt of gratitude to my family. To my children, Kendra, Karen, and Greg, go thanks for putting up with my travels and distractedness for so many years. To my wife, Lynn, this book cost you more than anyone, and would never have been written without your amazing loving-kindness.

Part 1

·◆·

An Overview of the Genus *Phlox*

1

Naming the Flame:
Botanical History of the Genus *Phlox*

*"After the marvelous creation of the World, a work so unimaginable,
measureless and stupendous, the Earth was covered everywhere with Plants,
those machines of the supreme craftsmanship."*

—CARL LINNAEUS, writing in the dedication to his 1738 work, *Hortus Cliffortianus*

First Encounters

The earliest scientific encounters with the genus *Phlox* came at the hands of John Banister, an English missionary-naturalist who arrived in colonial Virginia in 1678. Banister was the first university-trained naturalist to study the flora and fauna of the British colonies, and he sent specimens of plants as well as insects, spiders, and mollusks to leading naturalists in England, also compiling descriptive Latin catalogs and preparing drawings of some species. Banister's 1680 plant catalog included precise drawings of *P. pilosa* and *P. subulata*, representing the first illustrations of any phlox species.

Leonard Plukenet, English physician, royal professor of botany, and gardener to Queen Mary, authored and published important early catalogs of newly discovered plants. The first edition of his *Phytographia* (1691) contains exact reproductions of Banister's drawings of *Phlox pilosa* and *P. subulata*, and his *Almagestum botanicum* (1696) provided complete Latin diagnoses for these. His *Almagesti botanici mantissa* (1700) added descriptions for *P. divaricata* and *P. paniculata*, with Virginia noted as the provenance. While not indicated by Plukenet, Banister was likely the source of the material from which the descriptions were made. Plukenet's works were important references for Swedish naturalist Carl Linnaeus in the preparation of his landmark *Species Plantarum*, published in 1753.

The next scientific collections of phloxes were made in 1698 by Dr. David Krieg and William Vernon in colonial Maryland. Krieg, a German-born English ship's surgeon and artist, collected for London apothecary James Petiver; and

Vernon, of Cambridge University, collected under the auspices of the Temple Coffee House Botany Club. Collections of *Phlox maculata* by Krieg and *P. pilosa* by Vernon were cited by English botanist John Ray in the third volume of his *Historia Plantarum* (1704), which was likewise referenced by Linnaeus in *Species Plantarum*.

English artist and naturalist Mark Catesby traveled widely in the American colonies, first (1712–1719) in Virginia and later (1722–1725) in the Carolinas and Georgia. Catesby provided seeds and specimens of American plants to a number of scientists and horticulturists in England and Europe, including *Phlox carolina* and *P. glaberrima*. "M. Catesby" was specifically cited in the Latin text accompanying the pre-Linnaean description and illustration of *P. glaberrima* in *Hortus Elthamensis*, an enumeration of the garden of James Sherard's Eltham Palace in London, authored by John James Dillenius in 1732. Through the agency of William Sherard (younger brother of James), some of Catesby's specimens reached Johan Frederik Gronovius, physician and senator in Leiden, Holland, and patron of Linnaeus. Among them was *P. glaberrima*.

Catesby never returned to the American colonies, but continued to work through acquaintances made during his travels to bring more plant material back to Europe. One of them was John Clayton, an English-born physician who immigrated to colonial Virginia in 1715, residing near Williamsburg and eventually becoming Clerk to the County Court of Gloucester County. Clayton was interested in natural history and with encouragement from Catesby began collecting seeds and plant specimens, sending his first shipment to Catesby at Oxford in 1734, some of which Catesby forwarded on to Gronovius, including at least one specimen of *Phlox glaberrima*.

European botanists would become much better acquainted with the genus *Phlox* through the duo of Bartram and Collinson. John Bartram, Pennsylvania-born Quaker farmer and self-taught naturalist, was America's first great botanist and plant explorer. In 1728, Bartram established a botanic garden at Gray's Ferry on the Schuylkill River, now in Philadelphia, from which he propagated and grew plants for shipment to Europe, eventually serving more than fifty subscribers. His main subscriber and patron was Peter Collinson, a London cloth merchant with a keen interest in new and rare plants. Details regarding the exchange of plants between the two men can be gleaned from their correspondence (Darlington 1849) and from *Hortus Collinsonianus*, an informal catalog of plants cultivated in Collinson's garden at Mill Hill (then a village, later a district of London), edited and published by Lewis Weston Dillwyn in 1843 from Collinson's own garden notes. From these sources it can be determined that Bartram sent Collinson stock of *P. divaricata*, *P. paniculata*, and *P. subulata* between 1739 and 1745. Linnaeus's descriptions of *P. divaricata* and *P. paniculata* in *Species Plantarum* were based in part on plants in cultivation in the botanic garden of Uppsala University in Sweden, provided to him by Collinson, almost certainly from material originally collected by Bartram.

The Founding of *Phlox*

The genus *Phlox* was established by Linnaeus in *Genera Plantarum* (1737). In *Hortus Cliffortianus* (1738), Linnaeus attributed the name to Theophrastus, an ancient Greek philosopher and student of Aristotle who authored an encyclopedia on the plant kingdom (*Historia Plantarum*) between the third and second centuries BC. Linnaeus defined *Phlox* in Latin as "floris flammeo igneoque colore"—*flowers the color of glowing flame*. In delineating *Phlox* in *Genera Plantarum*, Linnaeus cited the description of *P. glaberrima* in *Hortus Elthamensis*, making this species the "generitype" upon which the genus was founded.

Prior to Linnaeus, phloxes were considered "something like a *Lychnis*" (Miller 1731), a genus of European plants familiar to European botanists and horticulturists. The name "*Lychnidea*" was established by Belgian botanist Matthias de L'Obel ("Lobelius") in his *Plantarum, Seu, Stirpum Historia* (1576) for a species of *Silene* or *Lychnis*, and was used in connection with phloxes as early as 1696 by Leonard Plukenet, with Phillip Miller (1752) coining the compelling common name "Bastard Lychnis" in his *Gardeners Dictionary*. The name *Phlox* did not gain widespread use until after Linnaeus published his *Species Plantarum* in 1753.

Species Plantarum is internationally accepted as the starting point for botanical nomenclature and contains the original descriptions of nine *Phlox* taxa still recognized today. Two of these, *P. maculata* and *P. subulata*, were based on specimens collected by Linnaeus's student Pehr Kalm in colonial New Jersey in 1749. The second edition of *Species Plantarum*, published in 1762, described one additional species, *P. carolina*. In all, Linnaeus recognized ten phlox species, nine in North American and one, *P. sibirica*, from "Asia boreali."

In *Species Plantarum*, Linnaeus not only established his own unique name and description for a species, he cited names from other works that he considered synonymous. These "pre-Linnaean" names are termed phrase names or polynomials since they contained more than the simple genus and species names of the Linnaean binomial system. All of this information—species name, description, synonym phrase names, geographical information, collector or source—constitutes the Linnaean "protologue." Thus, the protologue for *Phlox paniculata* in *Species*

Plantarum includes citation of a synonym phrase name ("Lychnidea *Viriginiana* Blattariae accedens, umbellata, maxima, Lysimachiae luteae folliis amplioribus, binis ex adverso positis") from Plukenet's *Almagesti botanici mantissa* (1700), plus cites "Collinson" as the source of the material Linnaeus examined in preparing his description.

Siberian Sidebar

Botanists in eighteenth-century Europe were not only trying to sort out plants from North America, they were plowing through stacks of specimens from Africa, Asia, and South America. In 1750, Linnaeus authored *Plantae Rariores Camschatcenses* to account for plants collected during the Second (1733–1741) Kamchatka Expedition of the Academy of Sciences of Saint Petersburg, Russia. Most of the species enumerated in this work were collected by Georg Wilhelm Steller, a member of the Academic detachment of the expedition as it made its way east across Siberia and on to the Pacific Coast. Steller died in 1746 while traveling back to Saint Petersburg and his specimens were somehow acquired by Grigory Demidov, a wealthy Russian patron of science who in 1750 sent them to Linnaeus for identification. Among them was the original material of *Phlox sibirica*, which Steller likely collected in the vicinity of Irkutsk in southern Siberia. Linnaeus also exchanged correspondence and specimens with another member of the Academic detachment, Johann Georg Gmelin. The earliest known illustration of *P. sibirica* appeared in Gmelin's *Flora Sibirica* in 1769.

From Charleston to Cherokee Country

Following the publication of the second edition of Linnaeus's *Species Plantarum* in 1762, no additional phloxes came to light until Scottish nurseryman and plant hunter John Fraser began his rambles in the Southern Appalachians. Fraser made seven trips to eastern North America between 1785 and 1810, collecting plants for the Royal Botanic Gardens, Kew, the Linnean Society, and even the Czar of Russia. Fraser also sold his plants privately in partnership with his sons John Jr. and James Thomas through his "American Nursery" at Sloane Square in London, later called "Messrs. Frasers' Nursery for Curious American Plants." The descriptions of two phloxes, *Phlox amoena* and *P. stolonifera*, were based on English garden plants derived from stock originally collected by Fraser. The story of the latter illustrates the iron rule of priority in botanical nomenclature.

Fraser's first trip to the region began with his landing at Charleston, South Carolina, in September of 1786. The following spring (1787) he journeyed from the seacoast to "the Cherokee Nation" in the southern Appalachian Mountains, traveling briefly with French plant collector André Michaux and his son François, something the elder Michaux noted in his journal with a hint of exasperation.

Fraser made subsequent collecting trips into the southern Appalachians, but the only one for which there is much information was in the spring of 1799 when he and his son John Jr. reached the Roan Highlands on the border of North Carolina and Tennessee. This journey would have placed Fraser in the heart of the range of *Phlox stolonifera* at its peak season of bloom.

It is possible that Michaux actually discovered this species before Fraser. Like his English rival, Michaux arrived in Charleston in September of 1786, but while Fraser was traveling as a horticultural entrepreneur, Michaux came ashore under the auspices of King Louis XVI as his royal botanist in America for the purpose of collecting and shipping plants to France. The two traveled and collected together for a few weeks in April and May of 1787, parting ways on 29 May in the vicinity of Augusta, Georgia, and thereafter presumably taking different routes into the mountains. In Fraser's very sketchy narrative of his 1787 journey, published in 1789 as *A short history of the Agrostis cornucopiae . . . and also, some account of a journey to the Cherokee Nation, in search of new plants*, he described the course of his travels as "from the south boundaries of Georgia [in the vicinity of Savannah] to the northward of Carolina." From Michaux's much more detailed journal it can be determined that the Frenchman eventually reached the mountainous region of present-day Oconee County, South Carolina, in June, where he collected what would become one of the most celebrated plants in the flora of Appalachia— Oconee bells (*Shortia galacifolia*). Michaux journeyed back into the same area in the spring of 1795.

In his *Flora Boreali-Americana* (1803), Michaux described the species in question as *Phlox reptans*, a name that gained wide circulation and persisted for years in both the botanical and horticultural literature. The collection locality on Michaux's specimen in the herbarium of the Muséum National d'Histoire Naturelle in Paris states, "Lieux umbrages des hautes Montanges Carolina septentrionalis [high mountains of North Carolina]." Which of these legendary plant explorers was first to discover this beautiful species will probably never be known, but the name *P. stolonifera*, based on Fraser's material, was published one year earlier in the English gardening periodical *Curtis's Botanical Magazine*, giving Fraser the glory.

Race to the Pacific

The phloxes of the Atlantic seaboard and Appalachian piedmont were fairly well known as the eighteenth century passed into the nineteenth. Remarkably, the next species to be added to the genus would come, not from the other side of the Cumberland Gap, but from the other side of Lemhi Pass.

The Louisiana Purchase of 1803 is said to have added "800,000 square miles of ignorance" to the young United States of America. President Thomas Jefferson, anxious to learn what he actually acquired from the French and concerned about reports of British probings in Oregon Country, commissioned the Corps of Discovery to journey from Saint Louis to the Pacific Coast. The Lewis and Clark Expedition, as it would come to be known, headed up the Missouri on 14 May 1804, reached the Pacific in November of 1805, and headed back to Saint Louis on 23 March 1806. Prevented by heavy snows in the Bitterroot Mountains from crossing back over the Continental Divide, the party spent much of May and early June of 1806 along the Clearwater River in present-day Idaho. Meriwether Lewis, who served as both commander and naturalist on the expedition, collected several plant species during the delay that would prove new to science, including the beautiful *Phlox speciosa*, which was described by Frederick Traugott Pursh in his *Flora Americae Septentrionalis* (1813).

As the fledgling United States was trying to come to grips with its backcountry, the British government was sponsoring ambitious expeditions of its own holdings in "British America." These expeditions were launched primarily with military and economic motivations, particularly the search for a "Northwest Passage," a sea route through the Arctic Ocean along the northern coast of North America. Scottish physician and explorer-naturalist Sir John Richardson served on two legendary and perilous overland expeditions to the North American Arctic under the command of Captain Sir John Franklin, and was responsible for bringing the first of the western cushion-forming phloxes to the attention of science. Richardson collected *Phlox hoodii* in present-day Saskatchewan, Canada, on the first expedition (1819–1821), which he described as a new species in the expedition's botanical report in 1823. On the second expedition (1825–1827) he collected *P. richardsonii* on the Arctic Coast of Canada, which English botanist Sir William Jackson Hooker described and named in his honor in 1838.

Richardson named *Phlox hoodii* after Lieutenant Robert Hood, a junior officer on the first expedition who was murdered by a fellow member of the party in 1821. In his 1823 description of the species, Richardson noted:

> The specific name is a small tribute to the memory of my lamented friend and companion, whose genius, had his life been spared, would have raised him to a conspicuous station in his profession, and rendered him an ornament to any science to which he might have chosen to direct his attention.

The expedition's journal reveals that, at an opportune moment, Richardson took it upon himself to kill Hood's suspected murderer.

Dispatching the Scots

Though he never left Europe, Hooker had a profound influence on the botanical exploration of North America. Named Regius Professor of Botany at Glasgow University in 1820, Hooker was instrumental in the development of the Glasgow Botanic Garden and it was here that his path crossed that of a newly hired gardener named David Douglas. So impressed was Hooker with the young Scot that he recommended Douglas to the Royal Horticultural Society of London as a botanical collector. Douglas made his first collecting trip under the auspices of the Society in 1823, traveling to the northeastern United States and Canada. He was back in England in 1824 but later that same year set sail for the west coast of North America, arriving at the mouth of the Columbia River in April of 1825. For months Douglas scoured the watershed of the Columbia collecting seeds and plant specimens and in the process discovering scores of new species that today bear his name. Among them was *Phlox douglasii*, which he encountered in present-day Oregon and Washington in 1826 and which Hooker named in his honor.

Thomas Drummond was another Scot dispatched to America by Hooker. It was upon Hooker's recommendation that Drummond was appointed assistant naturalist to John Richardson on Franklin's Second Overland Expedition to the Arctic (1825–1827), where he demonstrated his "zeal and assiduity" as a collector. Hooker subsequently helped recruit a host of subscribers to facilitate Drummond's travels in the southern and western parts of the United States, for which they would obtain seeds and/or sets of specimens from his collections. These scientists and investors were particularly anxious to get Drummond into Texas, described by Hooker as "hitherto almost untrodden by the foot of a Botanist." Drummond arrived at Velasco, Texas, in March of 1833 and explored and collected between the Gulf Coast and the Edwards Plateau for the next twenty-one months. In the spring of 1834 he collected specimens and seeds of a new annual species of phlox, which Hooker would name *Phlox drummondii* and which would become a near-overnight garden sensation in Europe.

Hooker summarized the known flora of North America in his *Flora Boreali-Americana* (1829–1840), and in his treatment of the genus *Phlox* in 1838 noted, "[i]t must be allowed that Dr. Richardson, and Messrs. Drummond and Douglas, have been pre-eminently successful in discovering new and beautiful species of the Genus *Phlox* in N. America." Scotsmen all, each man had a species of phlox named for him. Sadly, two of the three "met with an untimely grave" in the New World, Douglas in 1834 in Hawaii and Drummond in 1835 in Cuba. The contributions of these remarkable men are memorialized herein in the name *Phlox* Scotia Alpines Group (see discussion in Appendix B).

Americans Take to the Field

The role of Europeans in the taxonomic history of the genus is well illustrated by the early botanical exploration of Texas, which engaged an international array of characters. Concerning *Phlox*, the notables include Swiss naturalist Jean Louis Berlandier who collected *P. drummondii* subsp. *glabriflora* in 1834, the aforementioned Scotsman Thomas Drummond and *P. drummondii*, and German geologist and naturalist Karl Ferdinand Roemer. Roemer collected in Texas for eighteen months during 1845–1847 and secured the types of the two other Texas annuals, *P. cuspidata* and *P. roemeriana*, which were described by German taxonomist Georg Henrick Adolph Scheele in the German botanical journal *Linnaea*.

On a broader scale, the first major enumerations of the genus were authored by Europeans. Treatments of *Phlox* in Michaux's *Flora Boreali-Americana* (1803), Pursh's *Flora Americae Septentrionalis* (1813), and Hooker's *Flora Boreali-Americana* (1838) have already been mentioned. The first real monograph of the genus was authored by English taxonomist George Bentham, long-time president of the Linnean Society. In 1845, Bentham published a treatment of the Polemoniaceae for Auguste Pyrame de Candolle's *Prodomus systematis naturalis regni vegetabilis* in which he listed twenty-five species and five varieties of *Phlox* including *P. floridana*, which he described from material collected by pioneer American botanist Alvin Wentworth Chapman in the Apalachicola region of Florida. In 1849, Bentham described *P. diffusa*, collected by Karl Theodore Hartweg for the Royal Horticultural Society of London on the other side of the continent in the Sierra Nevada of California.

It was Thomas Nuttall who initiated a transition to American authority over this very American plant genus. Nuttall was an English-born naturalist who arrived in Philadelphia in 1808 and spent most of his adult life in the United States. A biographical sketch published the year after his death called Nuttall "an Englishman by birth, but an American by his scientific labors and reputation." Nuttall collected and named thousands of new species but, unlike his European predecessors, published his findings in American scientific journals, notably those of the Philadelphia Academy of Natural Science.

Nuttall's primary contributions to the taxonomic history of *Phlox* came through his relationship with Nathaniel Jarvis Wyeth, with whom he became acquainted during a ten-year (1822–1833) association with Harvard University. A Cambridge, Massachusetts, ice merchant with aspirations to make his fortune in the western fur trade, "Captain" Wyeth made two overland expeditions to Oregon, the first 1832–1833. Said to be "[e]ndowed with an adventurous spirit and suited to bold enterprises," Wyeth collected plant specimens as circumstances allowed for "friend Nuttall" on the return (eastbound) leg of the expedition in 1833. In a paper published in 1834, Nuttall described many new species of plants from the specimens brought back by Wyeth "from the distant and perilous regions of the west," including three important western phloxes—*P. caespitosa*, *P. longifolia*, and *P. muscoides*. That same year Nuttall joined Wyeth on his second Oregon Expedition (1834–1836) and, at the less-than-prime age of forty-eight, traveled mostly on foot along the route that would later become the Oregon Trail, collecting scores of new species along his way west, including the type of *P. andicola*. Nuttall's last brush with the genus came in 1848 when he described the important southwestern species *P. nana* from the type collected by William Gambel near Santa Fe, New Mexico, in 1845.

Total break from European dominance in the taxonomic history of *Phlox* came in 1870 with publication of a monograph on the genus by Asa Gray. Professor of botany at Harvard University and best-known botanical figure in nineteenth century America, Gray authored "Revision of the North American Polemoniaceae" in the *Proceedings of the American Academy of Arts and Sciences*, listing twenty-seven species, eleven varieties, and one formae of *Phlox*.

Gray was a "museum botanist" and depended upon an array of loyal field collectors to send him specimens and potential new species from the western United States. Some of these hardy souls were associated with official government enterprises. This group included Samuel Washington Woodhouse, who collected the type of *Phlox woodhousei* in Arizona in 1851 while serving as surgeon-naturalist on one of the Pacific Railroad Surveys. In the same year, Charles Wright collected the type of *P. villosissima* in Texas while participating in a military-sponsored survey of the Mexico–United States boundary. Others were independent professional collectors who scratched out a living by selling sets of their specimens to herbaria in Europe and the eastern United States. Such was Charles Christopher Parry, who secured the type of *P. condensata* while exploring the flora of the Southern Rockies in 1861. In a happy bit of cosmic symmetry, *P. condensata*, which was described by Gray from Parry's collection, occurs in the alpine on Gray's Peak in Colorado, named by Parry for his friend and mentor.

Western Upstarts

Gray can be credited with wrested dominance from the Europeans, but he was subsequently guilty of attempting to concentrate botanical power and influence at Harvard. Toward the end of his career his authority was being challenged by upstart western botanists like Marcus Eugene Jones and Edward Lee Greene who began publishing their own discoveries rather than waiting for the blessings of Gray and other eastern botanists. Jones described *Phlox albomarginata* from Montana in 1894 and *P. longifolia* var. *gladiformis* (syn. *P. gladiformis*) from Utah in 1895. Greene described *P. alyssifolia* from Saskatchewan in 1896.

A few of the new western phloxes were discovered by frontier clergymen who pursued botany as an avocation. Such was Greene in the early days of his career. Although he would go on to become professor of botany at Catholic University, Greene served as Episcopal missionary to a church in Yreka in northern California from April 1876 to March 1877, confiding in correspondence to Gray that he had enough sermons in the hopper to free him up for an entire season of plant collecting. Among his several discoveries in the area was *Phlox hirsuta*, restricted to outcrops of serpentine rock near Yreka and now recognized as an endangered species. Then there was Francis Duncan Kelsey, a Congregational clergyman who held a pastorate in Helena, Montana, from 1885 to 1892. Kelsey discovered a number of new plant species in the vicinity of Helena including *P. albomarginata*, *P. kelseyi*, and his celebrated namesake *Kelseya uniflora*.

The next big bump in the number of phloxes would come through the influence of Aven Nelson. Professor of botany and president of the University of Wyoming, Nelson established the Rocky Mountain Herbarium which is today an important repository of *Phlox* type specimens. Between 1898 and 1938 Nelson described fifteen species, one subspecies, and one variety of *Phlox*, although only three of his species (*P. aculeata*, *P. cluteana*, and *P. multiflora*) have stood the test of taxonomic time. Nelson taught and mentored many students who went on to have significant careers in botany themselves, including James Francis Macbride, who, as "a boy just out of the Boise [Idaho] High School," collected the type of *P. aculeata* in Idaho in 1910, sending the specimen to Nelson for identification, who described it as a new species in 1911.

Nelson's first graduate student was Elias Emanuel Nelson, a Swedish-born American (of no relation) who received the first graduate degree awarded by the University of Wyoming. Elias's master's thesis was a monograph on western members of the genus, covering an area from the Great Plains and central Texas to the West Coast. Published in the university's Experiment Station annual report for 1899 as "Revision of the Western North American Phloxes," this important paper listed thirty-six species and ten varieties of *Phlox*, describing *P. hirsuta*, *P. tenuifolia*, and *P. viscida*, and provided the original descriptions and epithets upon which *P. hendersonii* and *P. condensata* subsp. *covillei* were founded.

While Aven Nelson and his students were helping bring the phloxes of the Rocky Mountains into focus, a series of botanists and collectors associated with Washington Agricultural College (now Washington State University) in Pullman, Washington, were sorting out the diverse and complicated phlox flora of the Columbia Plateau and Pacific Northwest. Chief among them was Charles Vancouver Piper, who founded the herbarium at the college, which is today an important repository of *Phlox* specimens and types from the Pacific Northwest. Piper collected the type of *P. viscida* in the Blue Mountains of southeast Washington in 1896, described by Elias Nelson in 1899. Piper's 1906 publication, *Flora of the State of Washington*, listed thirteen phloxes. William Cusick, Wilhelm Suksdorf, and Kirk Whited were other early collectors associated with the college, and their specimens have been important to understanding the phloxes of the Pacific Northwest.

Government-sponsored expeditions of the time were turning up other new phloxes in the western United States. Vernon Orlando Bailey collected the type of *Phlox austromontana* in Utah while serving as biologist on the Death Valley Expedition of 1890–1891, a project under the sponsorship of the United States Bureau of Biological Survey to explore and survey the region of Death Valley, California. Frederick Vernon Coville described the species in the expedition's botanical report in 1893. Percy Train collected the type of *P. griseola* in Nevada while serving with the "Indian Medicine Project" (1937–1941), an effort of the United States Department of Agriculture (Division of Plant Introduction, Bureau of Plant Industry) in cooperation with the Works Progress Administration of Nevada and the University of Nevada to study the uses of plants by the native peoples of Nevada. Two other associates of the project, Benjamin O. Moore and George E. Franklin Jr., collected the type of *P. griseola* subsp. *tumulosa* in Nevada. Both of these phloxes would be described in 1942 by Edgar Wherry.

The Wherry Era

The next major treatment of the genus came from German taxonomist August Brand. His 1907 monograph on the Polemoniaceae in Engler's *Das Pflanzenreich* listed forty-eight species, seventeen subspecies, thirty-nine varieties, and seven subvarieties of *Phlox*. Brand based his work solely on herbarium specimens in European institutions, apparently supplemented by those he obtained from the University of Wyoming Rocky Mountain Herbarium and the California Academy of Sciences. Brand described one new species that is still recognized—*P. amabilis* of Arizona—but his complicated system of categories below that of species actually increased confusion and muddied understanding of the genus. Thankfully, a new student of the genus would come on the scene to bring order from the chaos.

Edgar Theodore Wherry set out to be a mineral chemist, but the study of rocks drew him into the study of plants. Geological field work in the vicinity of his native Philadelphia brought him into contact with outcroppings of serpentine rock which supported open areas dominated by herbaceous plants quite different from that of the surrounding woodlands and forest. The most visible associate of these serpentine barrens is *Phlox subulata*, which in spring can sheet entire hills in pink. In a lecture presented to the Philadelphia Botanical Club in 1928, Wherry explained the inspiration behind his intention to study *Phlox* and the family Polemoniaceae: "Not only were the vast floral carpets exquisitely beautiful, but they also presented certain problems which aroused my scientific curiosity." What followed was a forty-one-year-long series of papers stretching from "The eastern subulate-leaved phloxes" in 1929 to "Notes on the phloxes in the Gulf Coast states" in 1970. Wherry's mutual interest in geology and botany engaged him in the study of the ecology and floristics of the serpentine barrens near Philadelphia and the shale barrens of the Middle Appalachian Mountains, about which he published several important early papers.

Unlike Bentham, Gray, Brand, and other earlier students of the genus who based their determinations solely on the examination of herbarium specimens, Wherry combined exhaustive and painstaking herbarium research with extensive field work all across the United States, often revisiting the locality where the type specimen of the species was originally collected. Wherry first tackled the eastern phloxes, publishing a series of papers on these species between 1929 and 1936, and then turned his attention to the western phloxes, publishing on them from 1938 to 1945. The culmination of his research was *The Genus Phlox* (1955), a 174-page monograph published as a book by the Morris Arboretum of the University of Pennsylvania in which Wherry recognized sixty-six species (including twenty-seven nominative subspecies) and fifty-seven subspecies. Wherry described numerous species and subspecies of *Phlox* over the course of his career, several of which he discovered himself including *P. caryophylla*, *P. glaberrima* subsp. *interior*, *P. idahonis*, *P. pilosa* subsp. *ozarkana*, *P. pulchra*, and *P. pulvinata*.

The only real challenge to Wherry's determinations came from Arthur John Cronquist, a plant taxonomist associated with the New York Botanical Garden. Cronquist greatly annoyed Wherry by his treatment of *Phlox* in *Vascular Plants of the Pacific Northwest* (part 4, published in 1959), in which he disregarded a number of Wherry's species and subspecies, tossing them into "subjective synonymy." Wherry contested many of these in a series of papers published in the 1960s in which he did little to hide his irritation. Cronquist repeated many of these plus did some more lumping in his treatment of *Phlox* in *Intermountain Flora* (volume 4), published in 1983, a year (mercifully) after Wherry's death. Both of these became highly regarded reference works and Cronquist's views on *Phlox* were generally accepted and repeated. But, as will be seen in this present treatment, my experience in the field has shown that in many cases it was Wherry who got it right.

Besides his considerable record as scientist, Wherry also was an early advocate for conservation and the preservation of natural areas, evidenced in his 1931 article "Save Our Wildflowers" in *The Scientific Monthly*. Wherry encouraged wildflower appreciation among lay audiences, particularly garden clubs. It should come as no surprise that when addressing the question of "Selecting a National Flower" in a lecture given at the New York Botanical Garden in 1928, Wherry argued on behalf of his favorite genus. In addition to pointing out the occurrence of phloxes all across the United States, a very rational consideration for a national emblem, Wherry went for the patriotic heartstrings—"Phloxes come in our national colors, red, white and blue, and the flowers of some kinds are five-pointed stars; they bloom, too, over the fourth of July."

Phlox Frontiers

Only four new species have been described since Wherry's monograph was published in 1955—*Phlox dispersa* (Sharsmith 1958) in the Sierra Nevada of California, *P. pungens* (Dorn 1988) and *P. opalensis* (Dorn 1992) in the Wyoming Basin, and *P. pattersonii* (Prather 1994) in the Sierra Madre Oriental of Mexico. If other new species are to be discovered they will likely turn up in the basins and plateaus of the Intermountain West, where *Phlox* and genera like *Astragalus*, *Eriogonum*, *Oxytropis*, *Penstemon*, and *Townsendia* are so highly diversified. More detailed study of the *Phlox* flora of Mexico may also yield additional species, or could result in recognition or affirmation of previously proposed entities like *P. mexicana*, which Wherry described in 1944 and which is still known from only two specimens collected in the vicinity of Durango, Mexico.

The genus *Phlox* has provided a field for discoveries of a different sort by Donald Levin of the University of Texas–Austin. Beginning in the 1960s, Levin and his students have authored a prolific series of publications dealing with evolution, hybridization, polyploidy, and breeding system in *Phlox*. Among the many contributions are elegant studies that demonstrate that butterflies can discern floral colors and outlines and thus discriminate between the flowers of different *Phlox* species.

The latest scientific research involving the genus *Phlox* is taking place mostly in the lab rather than the field, and is more about discovering relationships than about discovering new species. Studies by Carolyn Ferguson of Kansas State University and her students and colleagues involve the tools of molecular systematics, analyzing nuclear and chloroplast DNA sequences to discover patterns and relationships within the genus that help clarify the taxonomic boundaries of species. Ferguson and her students have also published valuable studies of pollination biology in *Phlox*.

Although the genus *Phlox* has been under the scrutiny of science for over 300 years, there are still plenty of problems yet to be resolved. In particular need of study are questions of taxonomic identity and ecological relationships within the *P. diffusa*, *P. longifolia*, *P. nana*, and *P. speciosa* complexes. It is hoped that new students of the genus will come along to build upon the foundation laid by this remarkable stream of people.

2

Taming the Flame:
Horticultural History of the Genus *Phlox*

"It is wonderful to see the fertility of your country, in Phlox."

—Peter Collinson to John Bartram, 19 September 1765

Early Records

The most reliable records of the first cultivation of phloxes can be found in the catalogs of Europe's early botanic gardens, and those of the gardens of wealthy enthusiasts. Linnaeus himself authored two such works—*Hortus Cliffortianus* (1738), his elaborate catalog of the garden and herbarium of George Clifford in Holland, and *Hortus Upsaliensis* (1748), his enumeration of plants in cultivation in the botanic garden of Uppsala University in Sweden. Both works record the cultivation of *Phlox glaberrima*.

The earliest published report of the cultivation of phlox is for *Phlox glaberrima*, which was being grown at the Society of Apothecaries' Physic Garden in the Chelsea district of London in 1725, as reported by Isaac Rand (1726), keeper of the garden, in the *Philosophical Transactions* of the Royal Society of London. Founded in 1673, the Chelsea Physic Garden was the pre-eminent botanic garden of the time in Europe. Listed under the phrase name "Lychnidea Virginiana," the source of the original material was not noted but likely came from Mark Catesby, who was a friend of Rand. This species may have been in cultivation even earlier, if the "Lichnoides" listed in the 1722–1723 nursery catalog of Thomas Fairchild was in fact *P. glaberrima*. Fairchild was a nurseryman in Hoxton (then a village, later a district of London) who also received seed and stock of American plants from Catesby.

The next species to be reported in cultivation was *Phlox carolina*, appearing under the phrase name "Lychnidea Caroliniana" in English botanist John Martyn's *Historia Plantarum Rariorum* (1728). It also was reported (under the same phrase name) in cultivation at Chelsea about the same time, both in the first edition of Philip Miller's *Gardeners Dictionary*, published in 1731, and in Rand's

(1735) annual report to the Royal Society. While not specified, Catesby was likely the original source in both cases.

John Bartram began shipping plants to European subscribers in the 1730s, most notably to his patron Peter Collinson, who is credited with the "introduction" of several phloxes into European horticulture. *Hortus Collinsonianus* provides dates of first cultivation for *Phlox divaricata* (1740), *P. maculata* (1740), *P. ovata* (1740), and *P. paniculata* (1744). The immensely popular *P. subulata* likely first reached European gardens by way of Collinson, to whom Bartram sent a shipment of "one sod of the fine creeping spring *Lychnis*" in 1745.

While Bartram was Collinson's main source of American plants, he was not his only one. Collinson identified the source of the *Phlox maculata* growing in his garden as "Dr. [Christopher] Witt, of Pennsylvania." Witt, a physician and friend of Bartram, is credited with establishing the first botanic garden in America in Germantown, Pennsylvania, in 1705. Collinson admired *P. maculata*, and spoke highly of it in a letter to Bartram dated 16 June 1742 (Darlington 1849). Bartram, perhaps a bit jealous, wrote Collinson on 11 June 1743 that he found "[t]he Doctor's famous *Lychnis*" to be "unworthy" of such esteem, adding, "[o]ur swamps and low grounds are full of them. I had so contemptible an opinion of it, as not to think it worth sending, nor afford any room in my garden."

Philip Miller, "Gardener to the Worshipful Company of the Apothecaries at their Botanic Garden at Chelsea," was superintendent of the Chelsea Physic Garden and one of Bartram's subscribers, first through the agency of Collinson but later (around 1755) striking up personal correspondence. Miller's *Gardeners Dictionary*, which went through eight editions (1731–1768), was an account of his own horticultural experience at Chelsea and contains early records of phlox cultivation. The genus made its debut (under the name "*Lychnidea*") in the first edition of *Gardeners Dictionary*, published 1731, with Miller listing *Phlox carolina* and *P. glaberrima* under pre-Linnaean names. In the sixth edition, Miller (1752) added *P. divaricata* and *P. maculata*. Miller first used the name *Phlox* in the seventh edition (issued in parts, 1756–1759), referencing Linnaeus's (1753) *Species Plantarum* and adding *P. ovata*, *P. paniculata*, and *P. pilosa*. Miller's (1759) account of *P. pilosa* represents the first published report of this species in cultivation.

In 1789, the Royal Botanic Gardens, Kew in England put out the first edition of *Hortus Kewensis*, in which royal gardener William Aiton listed seven phloxes in cultivation; these (allowing for synonyms) being *Phlox carolina*, *P. divaricata*, *P. maculata*, and *P. paniculata*. The same species were in cultivation at the Marburg Botanic Garden in Germany in 1794, as listed in Conrad Moench's *Methodus plantas horti botanici et agri Marburgensis*. In *Jardin de la Malmaison*, the 1804 account of the garden of Empress Josephine of France, author Étienne Pierre Ventenat recorded both *P. carolina* and *P. stolonifera* in cultivation. In *Enumeratio plantarum horti regii botanici Berolinensis*, Carl Ludwig Willdenow listed nine different phloxes in cultivation at the Berlin Botanical Garden by 1809.

Information on the horticultural history of phlox species can also be gleaned from early British gardening periodicals. One of the most important of these was the *Botanical Magazine*, founded in 1787 by its proprietor and editor William Curtis. Following Curtis's death in 1799, John Sims became editor and the name was changed to *Curtis's Botanical Magazine*. Both portrayed phlox species recently introduced into English horticulture. Curtis depicted *Phlox divaricata* in 1791 and *P. subulata* in 1798. The original descriptions of *P. stolonifera* and *P. amoena* were written by Sims and published (1802 and 1810, respectively) in *Curtis's Botanical Magazine*, both accompanied by beautiful full-color illustrations of cultivated plants. These now serve as the iconotype for each species—illustrations designated by a later authority as representing the type of a species. William Jackson Hooker followed Sims as editor in 1827, and used the magazine for publishing his original description of *P. drummondii* in 1835, also accompanied by an illustration that serves as the iconotype.

In 1823, Conrad Loddiges proposed *Phlox nivalis* as the name for a new species illustrated in his *Botanical Cabinet* (published 1817–1833), which depicted plants offered by his nursery, Conrad Loddiges & Sons, located in Hackney near London. Loddiges gave no technical account of the distinguishing characters of the new species, but in 1827 nurseryman Robert Sweet published a full description of the same horticultural material in his *British Flower Garden* (published 1823–1838) using the Loddiges name with a Latin diagnosis and a more detailed figure that included an illustration of a dissected flower. Wherry determined that Sweet validated the Loddiges epithet as the earliest-published independent name for the species and thus affirmed *P. nivalis*. Wherry later designated Sweet's illustration as the iconotype.

Even the plant lists of early nurseries yield information of historical value. Such are the extant catalogs of the nursery of John Fraser and his sons in London, which in some cases were the place of first publication for species names that are still recognized today. One such catalog appears to have been printed after the elder Fraser returned from his fourth trip to America in 1790 and promises "Living plants of the above . . . as well as Phloxes . . . will be ready for Inspection in a short Time." Fraser's catalog for the year 1796 listed *Phlox pilosa* and *P. subulata* among the offerings and the 1813 edition of the Fraser brothers catalog lists *P. amoena* among "rare North American Plants" offered for sale "at reduced Prices."

John Lyon was an enterprising American nurseryman and plant hunter who in 1812 brought to London a large collection of American plants for sale, renting space in Chelsea "at the Garden adjoining the Marlborough Tavern." His extensive catalog listed *Phlox divaricata*, *P. pilosa*, and *P. decussata*, the latter noted as a new species ("sp. nova") but actually a synonym for *P. paniculata*. Lyon's customers were a who's who of botanic garden curators, gardeners to various dukes and earls, the proprietors of nursery gardens, and other horticultural luminaries, which explains why the name *P. decussata* gained such wide circulation in the botanical and horticultural literature and persists even to this day.

On the Home Front

The early development of American horticulture took place in the environs of Boston, Charleston, New York, and Philadelphia. As in Europe, the horticultural history of phlox in America can be traced by studying the catalogs of the first American botanic gardens and the pages of early American gardening periodicals and nursery lists.

While John Bartram is best known for sending American plants to Europe, he also grew them himself in his garden on the Schuylkill River in present-day Philadelphia. His cultivation of *Phlox paniculata* is documented in a letter to Collinson dated 11 June 1743 in which Bartram described a phlox that "with me grows about five feet high, bearing large spikes of different coloured flowers, for three or four months in the year." Two other contemporaries of Bartram—Christopher Witt and Humphrey Marshall—were also shipping (and presumably cultivating) phloxes and other plants from the environs of Philadelphia to Europe. Coming along later, Scottish-born Philadelphia nurseryman Robert Buist was an early commercial grower of phloxes. The first edition of his *American Flower-Garden Directory* (1832) profiled several species including *P. stolonifera* and *P. subulata* and his last edition (1854) listed a number of additional "select varieties."

Several phloxes were in cultivation at the Elgin Botanic Garden in New York City as early as 1811 as reported by David Hosack in his *Hortus Elginensis* (second edition), including *Phlox maculata*, *P. paniculata*, *P. pilosa*, *P. setacea* (syn. *P. nivalis*), and *P. subulata*. Hosack's garden was located where Rockefeller Center now stands, in the heart of midtown Manhattan.

Boston was especially important in the early horticultural history of *Phlox*. In 1807, William E. Carter came to the United States from Yorkshire, England, to become the first gardener of the Massachusetts Botanic Garden in Cambridge, where he did considerable selection and breeding work with phlox, introducing a number of named varieties into the trade by the 1840s. During the same period, Boston nurserymen Charles Mason Hovey and Joseph Breck were engaged in cultivating and developing new varieties of phloxes. Both wrote about their experience with phlox as early as 1846: Hovey in *The Magazine of Horticulture, Botany and All Useful Discoveries and Improvements in Rural Affairs*, of which he was editor and proprietor, and Breck in *The Horticulturist and Journal of Rural Art and Rural Taste*. *Breck's Book of Flowers*, published in 1851, discusses the horticultural attributes of many phlox species, and describes a number of early American selections.

While some early American nurserymen listed phloxes as species, more often they offered "improved" selections imported from Europe. Hovey wrote of traveling to Paris in 1844 and selecting "thirty or forty of the finest new varieties" of *Phlox paniculata*, which he was confident American gardeners would prefer over "the *older* varieties" (italics added).

The Amazing Journey of Mr. Drummond's Phlox

The rise of *Phlox drummondii* in horticulture was breathtaking. In 1838, a mere four years after Thomas Drummond collected seed of the species in Texas, English botanist William Jackson Hooker declared it "the pride and ornament of our gardens." One-hundred-and-one-years later, Texas botanist Benjamin Carroll Tharp could call *P. drummondii* "the most loved of all garden annuals . . . cultivated in every civilized country on earth." In between are stories like that of a Mr. T. S. Gold of Cream Hill, Connecticut, who in 1850 chastised the editor of a new American horticultural magazine for getting too uppity and neglecting "the cottager's little plot of ground, as well as the precincts of the farm-house." Among the plants he suggested for gardeners lacking both know-how and means was *P. drummondii*, calling it "an *essential* in the *smallest* collection."

The global conquest of *Phlox drummondii* began in Scotland. The illustration that accompanied Hooker's description of *P. drummondii* was drawn from plants growing at Canonmill Cottage, the garden of Dr. Patrick Neill near Edinburgh in October of 1835. Neill was one of the subscribers secured by Hooker in support of Drummond's travels. Joseph Paxton reported "Mr. Drummond's phlox" in flower at the Glasgow Botanic Garden and the Botanical Garden of Manchester in 1835, and in June of 1836 it was on display at the show of the Devon and Exeter Botanical and Horticultural Society in an exhibit created by English nurseryman James Veitch. In the spring of 1836, the nursery of Buel & Wilson of Albany, New

York, imported seeds from London, which were distributed to commercial growers in New York and Philadelphia. Boston nurseryman Charles Hovey reported seeing *P. drummondii* in the exhibitions of the Massachusetts Horticultural Society in the fall of 1837, noting it "commanded universal admiration."

As soon as enough seed of *Phlox drummondii* became available, horticulturists began selecting unique forms from garden populations, and later began deliberate hybridization programs. As early as 1838, three years after the first seed was germinated in Europe, Hovey reported "[t]here are several varieties of this phlox, which have been produced from seed in Britain." Unique flower color was an early aim, and hues ranging from pale yellow and white through pinks, purples, and reds to almost black were achieved, along with varied coloration patterning such as "eyes" and picta. A scanning of 1915 gardening publications and seed catalogs from Germany, France, England, and America by James Kelly of the New York Botanical Garden revealed at least 200 named varieties of *P. drummondii* in existence at the time. Kelly, who published detailed studies of the genetics behind particular horticultural traits of *P. drummondii*, concluded that these varieties all arose from Drummond's original collection of seed. Kelly also presented a classification scheme for cultivated varieties that consisted of eleven different categories.

Eventually, *Phlox drummondii* made it back to Texas. Texas botanist Lloyd Shinners established the name *P. drummondii* var. *peregrina* in 1951 to describe "garden strains of *Phlox drummondii* [that] have escaped or been purposely planted along highways, about cemeteries, and in similar places," *peregrina* alluding to the pilgrim or wandering nature of these wayside occurrences. Shinners considered the highly variable plants of these occurrences to be distant descendents of *P. drummondii* and perhaps other species, repatriated "via English gardens."

Besting Nature

Beginning in the 1820s, a number of commercial growers began to make their own selections of phlox from seedlings, giving these names and marketing them to the gardening public. The initial aims were new and unique flower coloration, larger and fuller individual flowers, and a larger inflorescence. The earliest crosses, natural or human-mediated, appear to have involved *Phlox carolina* and *P. maculata*. While showy, these hybrids did not always perform well in the garden, causing growers to turn their attention to the more adaptable *P. paniculata*. Breeding shorter, more compact varieties of *P. paniculata* became a major focus.

Among the early notables in phlox selection and breeding was English nurseryman George Wheeler of Warminster, with the name *Phlox wheeleri* appearing as early as 1828 in *Hortus Epsomensis*, the catalog of the London nursery of Charles, James, and Peter Young. Other pioneering phlox growers and breeders in the early and mid 1800s were M. Rodigas of Belgium and M. Lierval of France. Rodigas was professor at the École d'Horticole de l'État à Gendburghe les Gand, and, in the view of Hovey, "one of the greatest cultivators of phlox in Europe." Rodigas is credited with developing over seventy named varieties of phlox. Lierval was equally dedicated to the genus, and published *Culture pratique des phlox* in 1866.

Rodigas was most famous for his cultivar 'Van Houttei', the first phlox with striped corolla lobes. "Van Houtte's Phlox" was written up and illustrated in *Edwards Botanical Register* in 1843 with the comment, "The appearance of the plant is beautiful, far beyond anything yet seen in the genus Phlox; and we were almost going to say, beyond any thing among the hardy perennials in cultivation." Belgium horticulturist Louis van Houtte, after whom the cultivar was named, purchased the original stock from Rodigas in 1840 for the then astounding sum of 1200 francs, a price that stimulated a burst of enthusiasm for phlox breeding in Europe that lasted until around 1860. While the exact parentage of Rodigas's creation is unknown, the distinctive elongate, columnar inflorescence of *Phlox maculata* is clearly apparent in the beautiful illustration that accompanied the article.

Interest in phlox breeding peaked again at the beginning of the twentieth century with important workers including Georg Arends, Wilhelm Pfitzer, Albert Schollhammer, and Goos & Koenemann Nursery, all of Germany; H. J. Jones of England, Victor Lemoine (famed lilac breeder) of France; and Bonne Ruys of the Royal Moerheim Nursery of Holland. Important American breeders of the time included William Schmeiske of Binghamton, New York; Frederick Rea of Norwood, Massachusetts; D. M. Andrews of Rockmont Nursery in Boulder, Colorado; and the Reverend Charles Simmons Harrison of York, Nebraska. Harrison authored and published *A Manual on the Phlox* (1906), a 31-page booklet packed with information on propagation, cultivation, and recommended cultivars, along with metaphysical musings. Many of these phlox breeders worked with other herbaceous garden plants, notably peonies and iris, developing cultivars of distinction within these genera. A scanning of American nursery catalogs in 1917 by C. L. Thayer revealed 584 named selections of *Phlox paniculata* in the trade.

Phlox breeding enjoyed another surge following the upheavals of the first and second World Wars. The most important European breeders were Captain Bertram Hanmer Bunbury (B. H. B.) Symons-Jeune of England and Karl Foerster of Germany. Symons-Jeune was both an expert in rock gardening and a prolific breeder of *Phlox paniculata* cultivars, his introductions entering the trade through the Bakers Nursery of Codsall. His *Phlox: A Flower Monograph* (1953) is an invaluable horticultural reference. Over 100 cultivars of summer phlox were introduced by various breeders in the former Soviet Union between 1935 and 1950.

Beginning in the 1960s, Alan Bloom of Bressingham Gardens in England emerged as the major force in phlox breeding, introducing a number of new cultivars that are still highly regarded, including *Phlox paniculata* 'Frans Schubert' and *P.* (Suffruticosa Group) 'Omega'. Breeders in the Netherlands continue to develop and market new phlox cultivars, notably the Jacob Th. De Vroomen nursery and Coen Jansen, working with the *P. ×arendsii* hybrids (see Appendix B), and Piet Oudolf working with *P. divaricata*.

No American horticulturist has been more active in phlox selection and breeding than Don Jacobs of Eco-Gardens in Decatur, Georgia. Jacobs has made selections and named cultivars from *Phlox amoena*, *P. divaricata*, *P. nivalis*, *P. paniculata*, and *P. pulchra*, his cultivar names all beginning with "Eco," such as *P. nivalis* 'Eco Flirty Eyes' and *P. pulchra* 'Eco Pale Moon'.

Rock Garden Stars

The emergence of rock gardening in the late 1800s further bolstered esteem for the genus *Phlox*. English horticulturist Reginald Farrer did much to stimulate interest in this somewhat obsessive subculture of the gardening world through his book *The English Rock-Garden* (1919). Farrer wrote in grandiose praise of phlox, calling the genus "incomparably the most important that America has yet evolved for the benefit of the rock-garden." Writing of phlox in his *Rock Garden and Alpine Plants* (1930), pioneer Swiss alpine gardener Henri Correvon declared, "with such beauties at their door our American friends have nothing to envy the Europeans."

Articles on rock gardening began to appear in American periodicals in the early 1900s, most notably the *Gardener's Chronicle of America*. The first important American book on the subject was Ira Noel Gabrielson's *Western American Alpines* (1932). Gabrielson, who would become a noted ornithologist and leader in wildlife conservation, wrote from experience gained while growing western native plants for sale during the Great Depression, and provides the first authoritative horticultural accounts of the western phloxes.

This growing enthusiasm coalesced into the founding of the American Rock Garden Society in 1934. The first issue of the *Bulletin of the American Rock Garden*

Society, published in 1943, featured articles by two unique but very different characters who shared a common passion for phlox.

Known most widely at the time by her married name of "Mrs. J. Norman Henry," Mary Gibson Henry was an internationally respected horticulturist and field botanist, who from her home near Philadelphia traveled North America from 1928 to 1967 collecting native plants. Some of these expeditions were conducted in a specially outfitted, chauffeur-driven Lincoln Continental, and on one foray into particularly remote country she used carrier pigeons to communicate with her banker husband. Henry had a strong interest in phlox, likely due to her association with fellow Pennsylvanian Edgar Wherry, and she made several selections from the wild which were cultivated in her garden at Gladwyne, some of which were eventually carried by Mayfair Nurseries of New Jersey. It was Henry who collected the original stock of the famous phlox cultivar 'Chattahoochee', sometime prior to 1946. Among the phloxes cultivated by Henry were the rare shale endemic *Phlox buckleyi* and a dwarf expression of *P. floridana* which she discovered near the Gulf Coast of Florida. Visiting her garden in 1939, Wherry observed a unique phlox that appeared to be a spontaneous hybrid between *P. bifida* and *P. subulata*, which he named *Phlox ×henryae* in her honor.

The American Rock Garden Society provided South Dakota rancher-turned-plantsman Claude Barr with much-needed connections to the eastern gardening establishment. From his isolated homestead near the Black Hills, Barr operated a mail-order nursery named Prairie Gem Ranch, issuing his first catalog in 1935. Barr introduced two cushion-forming phloxes from the Great Plains into horticulture—*Phlox alyssifolia* and *P. andicola*, a photograph of the latter featured on the cover of his 1936 catalog. Gardeners could obtain other phloxes from Barr, including westerners *P. hoodii* ("a lovely mite") and *P. longifolia*, and South Dakota strains of *P. divaricata* and *P. pilosa*. Like Henry, Barr became an acquaintance of Wherry, corresponding with him and even hosting him on his ranch in 1948. Barr honored his friend by naming a full-flowered selection of *P. andicola*, 'Dr. Wherry'. Barr's first-hand experience with the flora of the Great Plains, including *Phlox*, is captured wonderfully in his book *Jewels of the Plains* (1983).

While Wherry made invaluable contributions to the taxonomic understanding of the genus, he also was interested in the phloxes as garden subjects. His 1935 article "Our Native Phloxes and Their Horticultural Derivatives" in the *National Horticultural Magazine* provided a valuable summary of the garden attributes of the eastern members of the genus. Wherry was particularly taken with *Phlox pulchra*, which he discovered in Alabama in 1929 and succeeded in bringing into cultivation by 1935, noting its "considerable horticultural promise." Wherry was active in the American Rock Garden Society, and contributed many articles to the *Bulletin*, including "Rock Garden Phloxes" in 1946. The Society's Edgar T. Wherry Award recognizes "a person who has made an outstanding contribution to the dissemination of botanical or horticultural information about native North American plants." Claude Barr became the first recipient of the award in 1973.

Barr's award, now archived with his papers and herbarium at Chadron State College in Nebraska, consisted of a framed drawing of *Phlox andicola* by artist Laura Louise Foster, who with her husband, H. Lincoln, grew many eastern and western phloxes in their Connecticut rock garden, which they named Millstream. The Millstream series of phlox cultivars represent selections they made over the years from apparent spontaneous garden hybrids involving *P. bifida*, *P. nivalis*, and *P. subulata*, with one or more western cushion phloxes possibly contributing genes. The husband and wife team collaborated as author and illustrator on the book, *Rock Gardening: A Guide to Growing Alpines and other Wildflowers in the American Garden* (1968), which remains the standard American reference on the subject.

No discussion of phlox and horticulture would be complete without mention of Panayoti Kelaidis, long-time associate of the Denver Botanic Gardens and world-renowned authority on rock garden plants. Kelaidis has authored many articles and book chapters that provide valuable information on the use of phloxes as rock garden subjects. He also played a major role in bringing the "Mexican phloxes" into horticulture.

The Future of Phlox

Trends in horticulture come and go, and not even the genus *Phlox* is immune to changing fashion. This is apparent in the passing of the heyday of the traditional English perennial border, at least in America, and with it the immense popularity of tall border phloxes like *P. paniculata*. Borders may cycle back into style someday, but even if not, creative horticulturists will no doubt find new ways to weave these beautiful subjects into designs and plantings.

Almost every landscape has shade, which should ensure continuation of the two-and-a-half century love-affair with *Phlox divaricata*. Another shade-lover, *P. stolonifera*, seems to be emerging from woodland gardens to a greater variety of landscape settings. The genus also has candidates for the challenging condition of dry shade, such as *P. buckleyi* of the Appalachian shale barrens region and *P. villosissima* of the Edwards Plateau of Texas.

Landscapes sometimes have wet sites to deal with, and in some cases water is being intentionally retained on site through rain gardens. *Phlox glaberrima* and *P. maculata* are two species well-adapted to wet conditions and would make outstanding subjects for such gardens in eastern and central United States. While most often used in the rock garden, the mat-forming *P. kelseyi* is native to seasonally-moist habitat and could have similar landscape applications.

On the other end of the spectrum, water-conscious gardeners living in drier regions of the country should look to their native phloxes, some of which survive on less than 10 in. (25 cm) of precipitation per year. Very dwarf desert specialists like *Phlox muscoides* are challenging even to the most seasoned alpinist, but taller westerners like *P. speciosa* that occur in dry but less severe habitat should be easier to cultivate in more typical landscape situations.

Rock gardeners know the genus better than anyone and will continue to grow as many phloxes as they can get their hands on, doing whatever it takes to make them successful. While comprising a relatively small sector of the gardening world, these enthusiasts grow phlox all over the world, from Denver to the Czech Republic to Japan. Phloxes will always have a place of esteem and frustration in their hearts.

Phloxes have great potential as subjects for regional native plant horticulture. Interest in gardening and landscaping with native plants continues to grow, for aesthetic as well as environmental reasons. On the environmental side of the ledger, plants native to a particular region should be well adapted to cultivation in that region, requiring less water and other resources to grow. On the aesthetic side, plants from the local flora help bring a sense of place to cultivated landscapes that many today value. As an advocate for wildflowers of all kinds, Edgar Wherry would no doubt have appreciated this modern American movement, reminding us that there is no more American group of plants than the genus *Phlox*.

3

Keys to the Genus *Phlox*

The following keys to the genus *Phlox* Linnaeus (Polemoniaceae) build upon those of earlier workers, notably Wherry (1955) but also Cronquist (1959, 1984), Dorn (2001), Radford et al. (1964), Steyermark (1963), and Welsh (2003).

Keys to subspecies are provided in some of the individual species accounts in Part Two. Wherry (1955) recognized sixty-six species (including twenty-seven nominative subspecies) and fifty-seven subspecies in his monograph on the genus. The most strongly differentiated (morphologically and ecologically) of these are recognized in the species accounts. Further systematic and field research may prove others worthy of recognition.

The widespread western annual herb sometimes treated as *Phlox gracilis* (Douglas ex Hooker) Greene is properly *Microsteris gracilis* (Douglas ex Hooker) Greene (Ferguson and Jansen 2002; Ferguson et al. 1999; Wherry 1943a, 1955; Grant 1998; Porter and Johnson 2000).

Key to the Genus *Phlox*

1a. Plants of Asia and Eastern Europe 2
1b. Plants of North America 3
2a. Stems suffrutescent (woody at base), plant forming open tufts; widespread through middle Asia and into Ural Mountains of eastern Europe with rare disjunct occurrences above the Arctic Circle in northeastern Asia .... *Phlox sibirica*
2b. Stems herbaceous (not woody), plant forming condensed cushions; confined to the Chukotka Peninsula of the Russian Far East in the vicinity of the Bering Sea, above the Arctic Circle (disjunct from Alaska) .... *Phlox richardsonii*
3a. Plants from the eastern Great Plains, central Texas, and northeastern Mexico eastward to the Atlantic Coast .... Go to **Plants of Eastern North America**
3b. Plants from the western Great Plains, Trans-Pecos Texas, and northwestern Mexico westward to the Pacific Coast .... Go to **Plants of Western North America**

Key to Plants of Eastern North America

1a. Plants annual, completing life cycle in one year or less. . . . . Go to **Annual Phloxes**

1b. Plants perennial, normally persisting more than two years, with no definite limit to life span 2

2a. Plants stoloniferous, the flowering shoots arising from elongate modified stems (stolons) that spread horizontally along the surface of the ground . . . .*Phlox stolonifera*

2b. Plants not stoloniferous 3

3a. Plants rosulate, the flowering shoots arising from short rosettes of narrow persistent leaves clustered and crowned around a common basal point of attachment, the rosettes terminating slender modified stems (rhizomes) that spread laterally just below the surface of the ground; endemic to Virginia and West Virginia . . . . *Phlox buckleyi*

3b. Plants not rosulate, the flowering shoots not arising from rosettes 4

4a. Plant growth habit caespitose, with numerous short stems arising in discrete tuftlike clusters; leaves small, linear (narrow with sides parallel over most of length) to subulate (narrowly triangular) . . . . Go to **Eastern Mat-Forming Phloxes**

4b. Plant growth habit upstanding, the erect-ascending flowering shoots simple or sparingly branched and relatively tall 5

5a. Leaves with lateral veins plainly visible on underside, conspicuously reticulate (like a network) or areolate (divided into small angular spaces), the margin serrulate (with fine sharp teeth pointing forward) and ciliate (fringed with minute hairs); anthers pale yellow or white; flowering season mid- to late summer . . . . Go to **Eastern Veiny Leaved Phloxes**

5b. Leaves with lateral veins obscure and margin smooth or barely roughened; anthers deep yellow; flowering season spring to early summer 6

6a. Styles short, 1.5–3(–4) mm long, included in calyx, united for only ¼–½ (rarely ⅝) their total length; anthers placed below corolla orifice . . . . Go to **Eastern Upstanding Short-Styled Phloxes**

6b. Styles elongate, (10–)14–19(–26) mm long, exceeding the length of the calyx, united well over ½ and usually nearly throughout their length; at least some anthers placed at or exceeding corolla orifice . . . . . . Go to **Eastern Upstanding Long-Styled Phloxes**

Key to Annual Phloxes

1a. Calyx lobes equaling or exceeding the length of the corolla tube; corolla tube constricted at orifice; corolla lobes averaging 14 mm long; corolla eye with white laterals, inner rim of yellow, and throat nearly black; ovary/capsule with multiple (3–5) ovules/seeds per locule (chamber); primarily associated with limestone on the Edwards Plateau of Texas . . . . ***Phlox roemeriana***

1b. Corolla tube exceeding length of calyx lobes; corolla lobes mostly less than 12 mm long; corolla lacking conspicuous yellow eye; ovary/capsule with 1 ovule/seed per locule 2

2a. Plants relatively robust, pubescent with relatively coarse hairs; leaves broad to narrow, acute (tapering evenly along straight sides to a sharp tip) to acuminate (tapering gradually along slightly concave sides to a slender sharp tip), often short-aristate (tipped by a stiff slender bristle); corolla dimensions larger, tube (10–)12–17(–25) mm long, lobes averaging 12 × 9 mm; habitat dry, soils sandy . . . . ***Phlox drummondii***

2b. Plants relatively delicate, pubescent with relatively fine hairs; leaves narrow, cuspidate (bearing an abrupt, small, sharp point); corolla dimensions smaller, tube 8–13 mm long, lobes to 11 × 8 mm; habitat mesic, soils mostly clay-loam . . . . ***Phlox cuspidata***

Key to Eastern Mat-Forming Phloxes

1a. Styles short, 1.5–4 mm long, included in calyx, stigmas comprising up to ⅔ of style length; anthers placed below corolla orifice 2

1b. Styles elongate, mostly 5–12 mm, exserted from calyx, stigmas comprising less than ⅓ of style length; at least some anthers placed at or exceeding corolla orifice 3

2a. Leaves linear (narrow with sides parallel over most of length) to lanceolate (lance-shaped); corolla tube 8–12 mm long, corolla lobe dimensions averaging 8 × 6 mm; ovary/capsule with 2 ovules/seeds per locule (chamber); endemic to Great Plains (Kansas and Oklahoma) . . . . ***Phlox oklahomensis***

2b. Leaves subulate (narrowly triangular) to linear-subulate; corolla tube 10–13(–17) mm long, corolla lobe dimensions averaging 11 × 7 mm; ovary/capsule with 1 (rarely 2) ovule/seed per locule (chamber); Atlantic and Gulf Coastal Plain and Appalachian Piedmont . . . . ***Phlox nivalis***

3a. Plants forming carpetlike mats from a dense network of prostrate (lying flat upon the substrate) nonflowering stems; leaf nodes numerous, internodes (space along stem between leaf nodes) mostly crowded, fascicles (clusters) of smaller, nonflowering shoots present in leaf axils (region between the upper side of leaf and the stem); leaves and bracts (reduced leaves below the inflorescence) linear-subulate, similar in size; maximum leaf length (8–)10–20(–25) mm; corolla lobe notched 1–1.5(–3) mm deep; corolla hue predominantly purple; occurs eastern United States, primarily Appalachian . . . . ***Phlox subulata***

3b. Plant sparingly and diffusely branched, forming open tufts or festoons from a taproot; leaf nodes few, internodes mostly well-spaced, shoots lacking in leaf axils; leaves on nonflowering shoots linear-lanceolate, much smaller than lanceolate bracts; maximum leaf length 30–60 mm; corolla lobe notched (1.5–)3–5(–5.5) mm deep, corolla hue pale lavender to white; occurs central United States, primarily Ozarkian . . . . ***Phlox bifida***

Key to Eastern Veiny Leaved Phloxes

1a. Height 75–200 cm; leaf nodes numerous (15–40); leaves tending to be near opposite (arranged in pairs along the stem), narrow to moderately broad, their surface mostly glabrous (lacking hairs); inflorescence-herbage glabrous to fine-pubescent, the hairs only rarely gland-tipped; corolla tube usually pubescent; 1 or more anthers exserted from corolla tube . . . . ***Phlox paniculata***

1b. Height 50–150 cm; leaf nodes fewer (7–15); leaves opposite, relatively broad, their surface bristly; inflorescence-herbage glandular-pubescent (with gland-tipped hairs); corolla tube glabrous; all anthers included within corolla tube . . . . . . ***Phlox amplifolia***

Key to Eastern Upstanding Short-Styled Phloxes

1a. Stems suffrutescent (woody at base); flowering season late summer; endemic to temperate forest/woodland in Sierra Madre Oriental in Mexico . . . . . . ***Phlox pattersonii***

1b. Stems herbaceous (not woody); flowering season mostly spring to early summer . . . 2

2a. Flowering shoots accompanied by nonflowering shoots, the nonflowering ones becoming decumbent (reclining on the ground with stem tip ascending) and rooting at nodes, their leaves broad-elliptic; leaves of flowering shoots short-lanceolate (lance-shaped); corolla hue mostly violet . . . . *Phlox divaricata*

2b. Nonflowering shoots, if present, decumbent to ascending but not normally rooting at nodes; leaves of flowering shoots linear-attenuate (tapering gradually to a slender tip), lance-linear (narrow with sides parallel over most of length), or lanceolate; corolla hue mostly purple to pink 3

3a. Leaves oblong-elliptic to lanceolate; inflorescence compact and subtended by a hirsute (beset with coarse rough elongate, more or less straight and erect hairs) involucre (whorl) of bracts (reduced leaves); corolla tube glabrous (lacking hairs) . . . . .
. . . . *Phlox amoena*

3b. Leaves mostly linear to lanceolate; inflorescence rather open, bracts scattered throughout; corolla tube various 4

4a. Leaf size greatest below middle of stem; leaves glabrous (lacking hairs), the upper ones reduced in size, appressed (lying flat against the stem) and passing gradually into glandular-pubescent (with gland-tipped hairs) bracts; corolla tube glabrous; East Gulf Coastal Plain . . . . *Phlox floridana*

4b. Leaf size greatest above middle of stem; leaves glabrous or pubescent, the upper ones spreading and distinctly longer than bracts; parts varying from glabrous to pubescent 5

5a. Stems tending to be branched; pubescence copious, glandular at least upward; primarily Edwards Plateau in C Texas. . . . . *Phlox villosissima*

5b. Stems tending to be simple; pubescence sparse 6

6a. Hairs fine, glandular upward; calyx lobe subulate (narrowly triangular); corolla tube up to 16 mm (averaging 8–12 mm) long, the lobes up to 12 mm (averaging 10 mm) long; widely distributed throughout eastern United States . . . . *Phlox pilosa*

6b. Hairs coarse, eglandular (lacking glands); calyx lobe linear; corolla relatively large, the tube up to 21 mm (averaging 17 mm) long, the lobes up to 16 mm (averaging 12 mm) long; endemic to E Texas . . . . *Phlox pulcherrima*

Key to Eastern Upstanding Long-Styled Phloxes

1a. Flowering shoots mostly arising from decumbent (reclining on the ground with tip ascending or erect) nonflowering shoots or accompanied by a few ascending but shorter nonflowering ones; leaf nodes below inflorescence 6 or fewer; calyx averaging 10 mm long; plants of relatively dry habitat 2

1b. Flowering shoots mostly arising from modified stems (rhizomes) that spread laterally just below the surface of the ground, with nonflowering shoots rare; leaf nodes below inflorescence 7 or more; calyx averaging less than 10 mm long; plants of relatively mesic to wet habitat 3

2a. Nonflowering shoots short; flowering shoots with ca. 4 leaf nodes, lower leaves with long petiole (stalk) . . . . ***Phlox ovata***

2b. Nonflowering shoots elongate; flowering shoots with ca. 6 leaf nodes, lower leaves with short petiole; endemic to Alabama . . . . ***Phlox pulchra***

3a. Rhizome slender, curving up into a flowering shoot; stems often maculate (spotted) with conspicuous purplish red or dark wine-colored spots; inflorescence comprised of cymes in a narrow-conical or cylindrical panicle (compound inflorescence) that is at least twice as long as broad . . . . ***Phlox maculata***

3b. Rhizome thick, short, sending up flowering shoots from irregularly spaced nodes; stems rarely maculate; inflorescence comprised of cymes in a panicle that is about as broad as long 4

4a. Calyx subcampanulate (nearly bell-shaped), the lobes narrow with conspicuous midrib and with membranes broad, firm, flat to somewhat carinate (raised lengthwise into a keel-like ridge) . . . . ***Phlox glaberrima***

4b. Calyx subcylindrical (nearly cylinder-shaped), the lobes broader with moderate midrib and with membranes narrow, thinnish, early becoming conspicuously plicate (raised lengthwise into a flattened fold) . . . . ***Phlox carolina***

Key to Plants of Western North America

1a. Plants stoloniferous, the flowering shoots arising from elongate modified stems (stolons) that spread horizontally along the surface of the ground; endemic to N California and S Oregon . . . . ***Phlox adsurgens***

1b. Plants not stoloniferous 2

2a. Plants rosulate, the flowering shoots arising from short rosettes of narrow persistent leaves clustered and crowned around a common basal point of attachment, the rosettes terminating slender modified stems (rhizomes) that spread laterally just below the surface of the ground; endemic to Colorado Plateau (Arizona, New Mexico, and Utah) . . . . ***Phlox cluteana***

2b. Plants not rosulate, the flowering shoots not arising from rosettes 3

3a. Plant growth habit caespitose, with numerous short flowering shoots arising in discrete tuftlike clusters to form loose to dense mats, cushions, or mounds; leaf nodes more or less crowded; leaves crowded, narrow, often rigid; flowers solitary or 2 or 3(–6) together at the stem-tip . . . . Go to **Western Cushion-Forming Phloxes**

3b. Plant growth habit upstanding, the flowering shoots erect-ascending, simple or sparingly branched, and relatively tall; most (or all) of the principal leaves well-spaced and borne well above the ground rather than in a basal cluster or mat; flowers borne on a conspicuous stalk (pedicel) in loosely branched 3- to many-flowered terminal cymes 4

4a. Corolla tube somewhat funnelform (gradually expanding upward in the shape of a funnel); nearly endemic to Arizona . . . . ***Phlox tenuifolia***

4b. Corolla tube salverform, the tube scarcely expanding upward then abruptly spreading to a flat limb 5

5a. Ovary/capsule with multiple (2 or 3) ovules/seeds per locule (chamber); corolla tube somewhat constricted at orifice; distribution centered in southwestern United States and northern Mexico . . . . ***Phlox nana***

5b. Ovary/capsule with 1 or rarely 2 ovules/seeds per locule; corolla tube not constricted; distribution more northern 6

6a. Styles short, 1.5–3(–4) mm long, united for only ¼–½ (rarely ⅝) their total length, which is less than that of the calyx; anthers placed well below corolla orifice . . . .
. . . . Go to **Western Upstanding Short-Styled Phloxes**

6b. Styles elongate, (5–)9–40 mm long, united well over ½ and usually nearly throughout their length, which often exceeds that of the calyx; at least some anthers placed at corolla orifice . . . . Go to **Western Upstanding Long-Styled Phloxes**

Key to Western Cushion-Forming Phloxes

1a. Tissues of stems and leaves succulent (fleshy and juicy); habitat seasonally moist with poorly drained soils, often saline . . . . ***Phlox kelseyi***

1b. Tissues of stems and leaves firm and not succulent or, if succulent, plant occurs in dry to xeric habitat with mostly well-drained soils 2

2a. Distribution above or near Arctic Circle (Alaska and northwest Canada) . . . .
. . . . **Phlox richardsonii**

2b. Distribution far south of Arctic Circle (except rare disjunct occurrences of *P. hoodii*) . . .
. . . . 3

3a. Plants arising from a diffuse caudex (root crown) with numerous branches that
develop as long, very slender rhizomes that spread laterally just below the surface of
the substrate and terminate in densely caespitose, small leafy tufts; endemic to Sierra
Nevada in California . . . . **Phlox dispersa**

3b. Plants arising from an apparent taproot or from short rhizomes 4

4a. Plant a subshrub, branched from the base like a shrub and annually producing
herbaceous flowering growth that dies back to woody tissue at the end of the
growing season; growth habit generally open, but can be more condensed under
harsh environmental conditions 5

4b. Plant a perennial herb (with no woody tissue in shoots) forming more or less compact
mats, cushions, or mounds 8

5a. Calyx membranes conspicuously carinate (raised lengthwise into a keel-like ridge)
. . . . **Phlox austromontana**

5b. Calyx membranes flat to somewhat plicate (raised lengthwise into a flattened fold)
. . . . 6

6a. Leaves pale green, relatively broad (linear-oblong) and flat, thickish, with coarse
glandless cilia (hairs confined to leaf margin); Northern and Middle Rocky Mountains
. . . . **Phlox caespitosa**

6b. Leaves dark or bright green, relatively narrow and slender, cilia lacking or fine; mostly
west of the Rocky Mountains 7

7a. Pubescent with glandular hairs throughout; leaves somewhat acicular (needle-
shaped) and stiff, their surfaces pilose (beset with relatively sparse, soft, slender, more
or less erect hairs) to glabrate (very sparsely hairy) with fine gland-tipped hairs;
primarily dry to xeric habitat on the Columbia Plateau . . . . **Phlox douglasii**

7b. Pubescence not glandular; leaves linear (narrow with sides parallel over most of
length) to linear-subulate (narrowly triangular) and only moderately rigid; primarily
moderate to high elevations along the Cascade–Sierra axis . . . . **Phlox diffusa**

8a. Calyx membranes conspicuously carinate (raised lengthwise into a keel-like ridge)
and strongly bulged toward the base . . . . Go to **Phlox aculeata** Group

8b. Calyx membranes flat to weakly carinate 9

9a. Leaves relatively broad, oblong to elliptic-ovate to lanceolate, margins thickened and
white . . . . Go to **Phlox albomarginata** Group

9b. Leaves relatively narrow, linear or subulate, leaf margin thin or slightly thickened
. . . . 10

10a. Leaf margins bearing coarse, conspicuous cilia 11

10b. Leaf cilia fine or lacking altogether 13

11a. Pubescence rather copious; leaves covered with gray wax, their surfaces pilose (beset with relatively sparse, soft, slender, more or less erect hairs), their margin markedly thickened; Great Basin conifer woodland and desert shrubland .... *Phlox griseola*

11b. Pubescence rather sparse; leaf surfaces glabrate (very sparsely hairy); inflorescence-herbage glandular-pubescent; mostly alpine and subalpine communities 12

12a. Plants forming dense, tight cushions, shoots closely packed together; leaves appressed (lying flat against the stem), erect, 5–10 mm long; corolla tube 6–10 mm long, lobes 3–5 mm long; styles 1.5–3 mm long .... *Phlox condensata*

12b. Plants forming loose, open cushions, shoots more spreading; leaves not appressed to stem, 6–12 mm long; corolla dimensions larger, the tube 7–14 mm long, lobes 6–7 mm long; styles 2.5–5.5 mm long .... *Phlox pulvinata*

13a. Inflorescence-herbage glandular-pubescent (with gland-tipped hairs) nearly through-out; calyx membranes tending to be narrow 14

13b. Inflorescence-herbage pubescent but with glandular hairs sparse or lacking altogether; calyx membranes tending to be broad 15

14a. Plant forming diffuse mats to more condensed cushions 5–10 cm tall; larger leaves 15–25 mm long; calyx mostly united less than ½ its length; endemic to Northern Rocky Mountains (Montana) .... *Phlox missoulensis*

14b. Plant forming compact mounds 2–6 cm tall; leaves 6–12 mm long, appressed to stems; calyx mostly united more than ½ its length; endemic to Pacific Northwest in alpine and subalpine communities .... *Phlox hendersonii*

15a. Plant mostly glabrous (lacking hairs) to sparsely pilose (beset with relatively sparse, soft, slender, more or less erect hairs); leaves broadly linear (narrow with sides parallel over most of length); calyx lobes with a conspicuous elevated midrib, joined by depressed, deep-seated calyx membranes .... Go to *Phlox multiflora* **Group**

15b. Plant more or less canescent (beset with dense vesture of fine relatively short gray hairs) on portions of the leaf and calyx; leaves tending to be subulate; calyx lobe midrib only moderately elevated and membranes not markedly depressed ....
.... Go to *Phlox hoodii* **Group**

Key to *Phlox aculeata* Group

1a. Leaves linear (narrow with sides parallel over most of length), up to 30 mm long, only the uppermost glandular-pubescent (with gland-tipped hairs); corolla tube glabrous (lacking hairs); endemic to Columbia Plateau (Idaho, Oregon, and Washington) . . . . *Phlox aculeata*
1b. Leaves linear-lanceolate (lance-shaped), not over 22 mm long, mostly glandular-pubescent; corolla tube glandular-pubescent; endemic to Colorado Plateau (Nevada and Utah) . . . . *Phlox gladiformis*

Key to *Phlox albomarginata* Group

1a. Leaves mostly lanceolate (lance-shaped) or lance-linear (narrow with sides parallel over most of length), very stiff and pungent (sharp) with prominent, strongly thickened midrib; endemic to Wyoming Basin in Wyoming . . . . *Phlox pungens*
1b. Leaves oblong to elliptic-lanceolate or elliptic-ovate, only moderately rigid, midrib obscure; absent from Wyoming Basin 2
2a. Largest leaves 3–6 mm long; calyx 5–8 mm long . . . . *Phlox albomarginata*
2b. Largest leaves 8–30 mm long; calyx 8–13 mm long . . . . *Phlox alyssifolia*

Key to *Phlox multiflora* Group

1a. Plant forming diffuse to condensed mats of decumbent (reclining on the ground with tip ascending or erect) and crowded erect stems; leaves linear (narrow with sides parallel over most of length), lax; inflorescence-herbage glabrous (lacking hairs) to pubescent with mostly short-appressed hairs; glands if any tipping short hairs; primarily Rocky Mountains . . . . *Phlox multiflora*
1b. Plant forming colonies of discrete treelike shoots; leaves linear-subulate (narrowly triangular), firm; inflorescence-herbage pubescent with long kinky hairs; glands if any tipping long hairs; primarily Great Plains . . . . *Phlox andicola*

Key to *Phlox hoodii* Group

1a. Growth habit loosely matted, internodes (space between leaf nodes along the stem) elongate and apparent; corolla tube 10–12 mm long, lobes 6–7 mm long, limb 12–16 mm in diameter . . . . ***Phlox opalensis***

1b. Growth habit densely matted, internodes more crowded; corolla dimensions smaller, limb diameter under 12 mm 2

2a. Leaves densely crowded and overlapping, closely appressed (lying flat against the stem) and often 4-ranked with stems quadrate (appearing square in cross-section); leaves 3–5 mm long, broad-subulate (narrowly triangular) to narrowly lanceolate (lance-shaped), covered with dense woolly hairs at least at base. . . . . ***Phlox muscoides***

2b. Leaves less crowded, stems appearing round in cross-section; leaves 6–12 mm long, often linear, usually stiff and pungent (very sharp), often loosely pubescent with somewhat cobwebby hairs . . . . ***Phlox hoodii***

Key to Western Upstanding Short-Styled Phloxes

1a. Calyx membranes carinate (raised lengthwise into a keel-like ridge); corolla lobe much longer than wide and apiculate (ending in a short, abrupt point); endemic to Hells Canyon region of Columbia Plateau (Idaho, Oregon, and Washington) . . . . ***Phlox colubrina***

1b. Calyx membranes flat; corolla lobes little longer than wide, apex entire or notched (retuse to emarginate) 2

2a. Plant relatively robust, mostly over 15 cm tall, tending to become shrubby below; leaves mostly long-acuminate (tapering gradually along slightly concave sides to a slender sharp tip); corolla tube usually evidently less than twice as long as the calyx; corolla lobes 10–15 mm long; Pacific Northwest and Sierra Nevada, remote to Northern Rocky Mountains . . . . ***Phlox speciosa***

2b. Plant smaller, mostly under 15 cm tall, woody (if at all) only at the very base of stems; leaves mostly oblong, obtuse (blunt) to acute (tapering evenly along straight sides to a sharp tip); corolla tube usually nearly or fully twice as long as the calyx; corolla lobes smaller, mostly 6–11 mm long; endemic to Colorado Plateau (Arizona and New Mexico) . . . . ***Phlox woodhousei***

Key to Western Upstanding Long-Styled Phloxes

1a. Plants of seasonally moist to wet habitat; leaves distinctively broader than linear; endemic to Clearwater Basin of N Idaho . . . . ***Phlox idahonis***

1b. Plants of dry to xeric habitat; leaves mostly linear (narrow with sides parallel over most of length) to lanceolate (lance-shaped) 2

2a. Styles 5–8 mm long; plant a subshrub, branched from the base like a shrub with branches producing herbaceous (not woody) flowering growth that dies back to the wood at the end of the growing season; endemic to N California . . . . ***Phlox hirsuta***

2b. Styles (7–)9–40 mm long; plant a perennial herb (lacking woody tissue) or suffrutescent perennial with stems only woody at the base 3

3a. Calyx membranes conspicuously carinate (raised lengthwise into a keel-like ridge) . . . . ***Phlox longifolia***

3b. Calyx membranes flat or weakly carinate 4

4a. Corolla tube under 20 mm long; calyx membranes not at all carinate 5

4b. Corolla tube 20 mm long or more; calyx membranes wrinkly to near-carinate 7

5a. Herbage pubescence eglandular (lacking glands); endemic to Southern Rocky Mountains (Colorado and New Mexico) . . . . ***Phlox caryophylla***

5b. Herbage pubescence glandular, at least in the inflorescence 6

6a. Leaf texture thinnish, leaf tip acuminate (tapering gradually along slightly concave sides to a slender sharp tip); corolla lobe entire to emarginate (with a shallow notch); endemic to Columbia Plateau (Idaho, Oregon, and Washington) . . . . ***Phlox viscida***

6b. Leaf texture thickish, leaf tip obtusish (bluntish) to short-acuminate; corolla lobe conspicuously notched; endemic to Arizona . . . . ***Phlox amabilis***

7a. Corolla tube length 20–33 mm . . . . ***Phlox stansburyi***

7b. Corolla tube length ca. 40 mm; endemic to San Bernardino Mountains of S California . . . . ***Phlox dolichantha***

Part 2

Species Accounts

Phlox aculeata
Payette phlox

As you head across the sage flats south of Boise, Idaho, the levelness of the land is broken by the sudden slash of the Snake River Canyon. The stony austerity of the scene belies the fact that this chasm is a raptor factory, hosting the greatest concentration of breeding eagles, hawks, falcons, and other birds of prey in North America, perhaps the world. The nest-bearing ledges and crevices are particularly important for the propagation of prairie falcons, and they dive and cliff-race and scream around the canyon like they own the place.

Payette phlox grows along the canyon's brink, blooming as eggs are hatching in the aeries below. The distribution of this small, tufted plant reflects the extent of the Snake River Plain, a massive plateau built up from ancient lava flows. The soil that weathers from this volcanic parent material nourishes vegetation that nourishes ground squirrels that nourish falcon hatchlings—a vortex of energy exchange centered over the canyon each spring that is both elegant and awful, gentled somewhat by the unfurled white linen of phlox blossoms.

Phlox aculeata A. Nelson, Bot. Gaz. 52: 270. 1911 [prickly, of the leaves; "leaves . . . rather rigid and aculeate"] Caespitose perennial herb forming crowded tufts 5–10 cm tall from a short-branched caudex, the numerous erect-ascending flowering shoots 3–8 cm long; *leaves* firm, narrowly linear-subulate, mostly 1–3.5 cm long and 0.5–2 mm wide, sharp-pointed, basally ciliate, the surfaces glabrous to sparse-pilose; *inflorescence* (1-) 3- to 5-flowered, its herbage copiously glandular-pubescent, pedicels (1–)3–6(–9) mm long; *calyx* 7–12 mm long, united ⅜–⅝ its length, lobes subulate with apex sharp-cuspidate, membranes conspicuously carinate, the carina (keel) strongly bulged toward the base; *corolla* tube 10–15 mm long, less than twice as long as the calyx, glabrous, lobes elliptic-obovate (average dimensions 7 × 4 mm), hue purple to deep pink, shading to lighter or even white; *stamens* borne on upper corolla tube, with some anthers exserted; *style* 5–11(–13) mm long, united to tip which is free for 1 mm, with stigma placed among the anthers near corolla orifice; *flowering season* spring, April–May.

Geography
Chiefly Columbia Plateau, concentrated in the Payette and Snake River Plain sections (S Idaho and SE Oregon), extending north into the southern foothills of the Boise Mountains of the Northern Rocky Mountains (SW Idaho), with outlying

occurrences west in the Walla Walla Plateau section of the Columbia Plateau (NC Oregon) and south in the Steens Mountain region of the northern Great Basin (SE Oregon); occurs in the Craters of the Moon National Monument and the Snake River Birds of Prey National Conservation Area in Idaho and in the John Day Fossil Beds National Monument in Oregon.

Environment
Dissected plateaus and low mountains, on terraces, ledges, bluffs, slopes, ridges, and rocky pavement, 3000–5000 ft. (900–1500 m); *MAP* (mean annual precipitation) 8–14 in. (20–36 cm) in SW Idaho, with most falling as rain or snow between November and April; *habitat* xeric to dry; *soils* sandy to stony, sometimes alkaline, shallow with bedrock at or near the surface; *parent material* in Idaho chiefly basalt of the Snake River group, occurs on outwash derived from the highly eroded deposits of volcanic ash and tuff in the John Day River basin in NC Oregon.

Associations
Primarily shrub-steppe dominated by Wyoming big sagebrush (*Artemisia tridentata* subsp. *wyomingensis*) or antelope bitterbrush (*Purshia tridentata*) in association with perennial bunchgrasses including Indian mountain-ricegrass (*Achnatherum hymenoides*), bottlebrush squirrel-tail (*Elymus elmyoides*), needle-and-thread (*Hesperostipa comata*), curly bluegrass (*Poa secunda*), and bluebunch wheatgrass (*Pseudoroegneria spicata*). On rocky bluffs and other more xeric habitat, Payette phlox occurs in shrubland dominated by dwarf sagebrush (*Artemisia arbuscula*), rubber rabbitbrush (*Ericameria nauseosa*), spiny hop-sage (*Grayia spinosa*), gray ball sage (*Salvia dorrii*), littleleaf horsebrush (*Tetradymia glabrata*), and winter-fat (*Krascheninnikovia lanata*). At higher elevations in the Owyhee Mountains of SW Idaho it occurs in shrubland dominated by Wyoming big sagebrush, antelope bitterbrush, spiny greasebush (*Glossopetalon spinescens*), and granite prickly-phlox (*Leptodactylon pungens*), with Indian mountain-ricegrass a major associate. Payette phlox occurs to a lesser extent in bunchgrass steppe, with bluebunch wheatgrass and bottlebrush squirrel-tail being important graminoids. Disjunct populations of Payette phlox occur on sparsely vegetated badlands associated with the volcanic John Day and Clarno formations in NC Oregon.

Frequently associated forbs include arrow-leaf balsam-root (*Balsamorhiza sagittata*), northwestern Indian paintbrush (*Castilleja angustifolia*), hoary pincushion (*Chaenactis douglasii*), shaggy fleabane (*Erigeron pumilus*), silvery lupine (*Lupinus argenteus*), Hood's phlox (*Phlox hoodii*), and white-stem globemallow (*Sphaeralcea munroana*).

Natural history notes

Payette phlox is consumed by Townsend's ground squirrel (*Spermophilus townsendii* subsp. *idahoensis*) in the Snake River Birds of Prey National Conservation Area (Van Horne et al. 1998). These burrowing rodents are a key prey source supporting the largest concentrations of breeding eagles, hawks, falcons, owls, and other raptors in North America.

Cultivation

Payette phlox has not been reported in cultivation, although seed has been available. It is an attractive species with potential as a drought-tolerant landscape plant for drier portions of the Pacific Northwest. Wherry (1955) noted some of the more strongly fragrant forms of Payette phlox were "worthy of trial as rock garden subjects." Grow in full sun and dry, well-drained soil.

Phlox adsurgens
Siskiyou phlox
PLATE 2

The Siskiyous swallowed up David Rains Wallace. He wandered into this ancient, knurled mountain range, buckling northern California to southern Oregon, on a personal search for origins and meanings. The tangled geology, dense biology, and "unimaginably venerable forest" provided ample depth for his soundings. Sifting Bigfoot folklore and Darwinian orthodoxy swept him into the shadowlands where myth and truth entwine. The Siskiyous set him reeling—they gave him metaphysical vertigo.

It is fitting that this deep, mysterious range, global center of biodiversity and stomping ground of Sasquatch, should harbor a phlox all its own. Siskiyou phlox is endemic to the montane cosmos plumbed by Wallace in *The Klamath Knot*, gracing a succession of forest types from the dry Douglas-fir zone up into the subalpine where spires of noble fir and mountain hemlock sway. With soft pink, candy-striped flowers over mats of glossy evergreen foliage, it is one of the glories of the Siskiyou flora, and of the genus *Phlox*.

༄ ༅

Phlox adsurgens Torrey ex A. Gray, Proc. Amer. Acad. Arts 8: 256. 1870 [up-standing, of the flowering stems from prostrate shoots; "caulibus . . . adscendentibus"] Stoloniferous perennial herb with flowering shoots arising from elongate modified

stems (stolons) that spread horizontally along the surface of the ground, these rooting sparingly at the nodes and giving rise to loosely curved-ascending flowering
shoots 20–30(–40) cm tall with 4–7 well-spaced leaf nodes below the inflorescence,
glabrous; *leaves* firm, glabrous, shining, the upper sparse-ciliate, those on stolons
partly persistent, elliptic, averaging 23 mm long and 12 mm wide, on flowering
shoots deciduous, ovate, smaller; *inflorescence* compact, 6- to 12-flowered in open,
terminal cymes, its herbage glandular-pubescent, pedicels 15–25 mm long; *calyx*
10–12 mm long, united ⅜–½ its length, lobes subulate with apex bluntish, membranes somewhat plicate; *corolla* tube 14–20 mm long, nearly or fully twice as long
as the calyx, sparsely pilose to glabrate, lobes obovate (average dimensions 11 × 6
mm), hue purple, lilac, or pink, the eye usually paled and bearing a deep-hued stripe
toward the base of each corolla blade; *stamens* borne on upper corolla tube, with
some anthers exserted; style 13–15 mm long, united to tip which is free for 1 mm,
with stigma placed among the anthers at or near corolla orifice; *flowering season* late
spring, June–August.

∽∾∾

Geography
Pacific Mountain System, chiefly in the Klamath Mountains (NW California and
SW Oregon), extending south into the North Coast Ranges of NW California.

Environment
Mountains, primarily mid-elevation slopes and ridges, 1400–7000 ft. (400–2100
m); *MAP* 40–120 in. (102–305 cm); *habitat* dry to mesic; *soils* coarse-textured,
typically shallow with bedrock at or near the surface; *parent material* diorite and
quartz diorite, an igneous rock similar to granite. In his paper on the vegetation of
the Siskiyou Mountains, Whittaker (1960) noted Siskiyou phlox occurs across a
wide range of elevation from 1500–7000 ft. (460–2140 m), being most common at
middle elevations of 4500–5500 ft. (1370–1680 m).

Associations
Montane conifer forest and woodland, the composition of which varies with elevation and geology. Siskiyou phlox is often locally abundant and a prominent element in the herbaceous layer. Lower elevation forests are dominated by Douglas-
fir (*Pseudotsuga menziesii*), with Port Orford-cedar (*Chamaecyparis lawsoniana*),
Pacific yew (*Taxus brevifolia*), and sugar pine (*Pinus lambertiana*) associated. Evergreen hardwood trees are also typical of these drier forests, including Pacific madrone (*Arbutus menziesii*), golden chinquapin (*Chrysolepis chrysophylla*), tanoak
(*Lithocarpus densiflorus*), and canyon live oak (*Quercus chrysolepis*). Siskiyou phlox
also occurs in middle- to high-elevation forests dominated by white fir (*Abies con-*

color), with Douglas-fir and incense cedar (*Calocedrus decurrens*) associated, and in subalpine forests dominated by noble fir (*Abies procera*), California red fir (*A. magnifica*), and mountain hemlock (*Tsuga mertensiana*).

Siskiyou phlox occurs in the herbaceous layer of these forests and woodlands, along with a rich diversity of forbs including deerfoot (*Achlys triphylla*), red baneberry (*Actaea rubra*), Scouler's bellflower (*Campanula scouleri*), Menzies' wintergreen (*Chimaphila menziesii*), white-flower hawkweed (*Hieracium albiflorum*), white-vein wintergreen (*Pyrola picta*), northern starflower (*Trientalis borealis* subsp. *latifolia*), western trillium (*Trillium ovatum*), and white inside-out-flower (*Vancouveria hexandra*).

Cultivation

Siskiyou phlox is not easily grown in gardens outside of its native region, unless its cool woodland habitat is replicated. But gardeners in North America and Europe have been more than willing to accommodate the needs of this plant, regarded by many as the most beautiful of all the phloxes. Siskiyou phlox is valued for its shiny, nearly evergreen leaves and its large, attractive flowers. Often colored by darker rings and markings in a candy-striped effect, the flowers combine white and pink, in the opinion of Ira Gabrielson (1932), "more exquisitely than any other flower I have ever seen." Wherry (1946) added his praise, stating the flowers of Siskiyou phlox possess "perhaps the most entrancing coloring in the genus." The prostrate, colonizing habit of Siskiyou phlox makes it an effective groundcover in woodland gardens or shade gardens. Grow in shade to partial shade in evenly moist, well-drained, humus-rich, neutral to acid soil. Provide winter protection in regions with little or no snow cover. Cultivars are available. Horticultural hybrids are documented (see Appendix B).

Phlox albomarginata
Helena phlox
PLATES 3, 4

The Madison limestone is humble custodian to the crown jewels of the Rocky Mountain flora. Pearl-gray and calcium-rich, it is the veneer and sometimes the substance of a scattering of mountain ranges in Montana and Wyoming. Its outcrops, ledges, and hogbacks provide footing for some of the Rockies' most celebrated plants, including Kelsey-moss, a legendary, cliff-plastering shrublet, and

Jones' columbine, darling of rock gardeners the world over. To this roster of renown must be added another of the Madison's treasures—Helena phlox.

The distribution of Helena phlox is centered in southwest Montana, the driest part of the state, where it occurs on gravelly prairie uplands and on the lower slopes of mountains. A small, tufted plant when growing among prairie bunchgrasses, it assumes the character of an alpine when growing directly on the Madison, welding to the contours of rocky ledges in flowing, close-packed cushions. It also occurs in the empty environs of Idaho's Lost River Range, a raw-boned region of sage-bound mountains with their own secret wealth of limestone endemics.

❧ ❦

Phlox albomarginata M. E. Jones, Zoe 4: 367. 1894 [white margined, of the leaves; "margins . . . cartilaginous, thick, white"] Caespitose perennial herb from a short-branched caudex, ranging in habit from the nominate expression (subsp. *albomarginata*) which forms tufts 5–8 cm tall with erect-ascending flowering shoots up to 6 cm long to more strongly condensed forms in more xeric situations, the most extreme expression (subsp. *diapensioides*) forming dense hard cushions little more than 3 cm tall with numerous close-packed erect flowering shoots 0.5–1.5 cm long; *leaves* oblong to elliptic-lanceolate or elliptic-ovate, the margin thickened, cartilaginous, and tending to be whitish, cuspidate, coarse-ciliate and pilose, 3–6 mm long and 1.5–3 mm wide; *inflorescence* 1- to 3-flowered, its herbage glandular-pubescent, pedicels 2–15 mm long; *calyx* 6–8 mm long, united ½–⅝ its length, lobes oblong-triangular with obscure midrib and thick margin and with apex sharp-cuspidate, membranes flat; *corolla* tube 8–13 mm long, often twice as long as the calyx, lobes elliptic (dimensions varying from 4 × 3 mm for subsp. *diapensioides* to 6 × 4 mm for subsp. *albomarginata*), hue pink to purplish or white; *stamens* borne on upper corolla tube, with some anthers exserted; *style* 4–8(–10) mm long (2.5–5 mm in subsp. *diapensioides*), united to tip which is free for ca. 1 mm, with stigma placed among the anthers near corolla orifice; *flowering season* spring, May–July.

❧ ❦

Taxonomic notes

Wherry (1955) recognized *Phlox albomarginata* subsp. *diapensioides*, which he separated from the nominate expression (subsp. *albomarginata*) by differences in habit, stature, and size of floral parts, with the more reduced subsp. *diapensioides* forming low cushions that in exposed rock outcrop habitat can be as dramatically condensed as any in the genus. As presently understood, the morphological variation represented by these subspecies does not show clear correlation with geographic/ecological segregation.

Geography

Chiefly Rocky Mountain System, concentrated in the Northern Rocky Mountains with occurrences in the Beaverhead, Big Belt, and Madison ranges of SW and WC Montana, and in the Lemhi and Lost River ranges of EC Idaho, with outlying occurrences to the south in mountain ranges (Albion Mountains and Mount Putnam) at the northern edge of the Great Basin in S Idaho and in the Middle Rocky Mountains in SW Wyoming.

Environment

Plains, hilly piedmont, and mountains at low to middle elevations, on upper slopes of rolling to dissected terrain, escarpments, and rock ledges, 4400–7500 ft. (1350–2300 m); *MAP* 11–13 in. (28–33 cm); *habitat* primarily dry, sometimes xeric; *soils* stony, shallow with bedrock at or near the surface; *parent material* typically calcareous, notably limestone of the Madison Formation.

Associations

Shrubland, grassland, and rock outcrop communities. On foothills and the lower slopes of mountains, Helena phlox occurs in shrubland dominated by open stands of curl-leaf mountain-mahogany (*Cercocarpus ledifolius*) or skunkbush (*Rhus trilobata*), typically in association with bluebunch wheatgrass (*Pseudoroegneria spicata*). On more exposed, xeric sites, Helena phlox occurs in shrubland dominated by dwarf sagebrush (*Artemisia arbuscula*). Scattered limber pine (*Pinus flexilis*) and Rocky Mountain juniper (*Juniperus scopulorum*) may be associated with these shrublands.

On plains adjacent to the east slope of the Rocky Mountain Front, and in intermountain valleys, Helena phlox occurs in mixedgrass prairie dominated by rough fescue (*Festuca campestris*), Idaho fescue (*F. idahoensis*), bluebunch wheatgrass, and needle-and-thread (*Hesperostipa comata*), with bluegrama (*Bouteloua gracilis*) associated. Scattered big sagebrush (*Artemisia tridentata*) shrubs may also be associated. Prairie sagebrush (*A. frigida*) and forbs such as three-leaf milkvetch (*Astragalus gilviflorus*), oval-leaf wild buckwheat (*Eriogonum ovalifolium*), and Hood's phlox (*Phlox hoodii*) are common associates.

In rocky, shallow soil habitat within these larger ecological systems, Helena phlox (subsp. *diapensioides*) occurs in sparsely vegetated barrens and openings dominated by low-growing forbs, of which it is sometimes a prominent element. Associates include caespitose rockmat (*Petrophyton caespitosum*), a mat-forming shrub, along with forbs Tweedy's fleabane (*Erigeron tweedyi*), alpine bladderpod

(*Lesquerella alpina*), low nailwort (*Paronychia sessiliflora*), stiff-leaf beardtongue (*Penstemon aridus*), and silvery ragwort (*Packera cana*).

Cultivation
Helena phlox has been cultivated to a limited extent as a rock garden plant. The typical expression (subsp. *albomarginata*) forms loose cushions that attractively display flowers against a foil of relatively wide, white margined leaves. The dwarf expression (subsp. *diapensioides*) forms mounds as remarkably dense and dramatic as the much-admired Kelsey-moss (*Kelseya uniflora*), both of which were discovered in Montana by Francis Duncan Kelsey. Grow in full sun in dry, well-drained soil.

Phlox alyssifolia
alyssum-leaf phlox
PLATES 5, 6

It was 1910 when the young man claimed his South Dakota prairie homestead, barely twenty years from those unspeakable hours on nearby Wounded Knee Creek. He came to raise crops and cattle, but his sensitive heart was drawn away to other things. He would start a nursery and sell wildflowers by mail, naming his enterprise *Prairie Gem Ranch*. And he would write, wonderfully, about his intimate acquaintance with the flora of the plains and the neighboring Black Hills. He would even find himself engaging in a most unusual pastime—"phlox hunting."

Claude Barr's quarry was alyssum-leaf phlox and the Black Hills were his favorite hunting ground. Denizen of the northern Great Plains, dweller of bluff and butte, the mat-forming alyssum-leaf phlox makes its best showing in and about this pine-purpled mountain island. A large-flowered variant of this phlox is especially abundant in the Black Hills region on limestone formations honored with ringing, remembering Lakota names—*Minnekahta* and *Pahasapa*—tinting outcrops and entire ridges with a fragrant pink haze in spring.

Phlox alyssifolia Greene, Pittonia 3: 27. 1896 [foliage like that of sweet alyssum, *Alyssum maritimum* (syn. *Lobularia maritima*)] Caespitose perennial herb forming diffuse mats 3–11 cm tall from a branched, spreading caudex, the numerous decumbent to ascending flowering shoots with 3 or 4 nodes; *leaves* firm, elliptic-lanceolate to oblong, the margin thickened, cartilaginous, and tending to be whitish, tip cuspidate, coarse-ciliate and often glandular-pubescent, max. length 10–20 mm and width 2–5 mm; *inflorescence* 1- to 3- (5-) flowered, its herbage densely glandular-pubescent, sometimes pilose with glandless hairs, pedicels 2.5–12.5 mm long (up to 25 mm in subsp. *abdita*); *calyx* 8–13 mm long, united ⅜–⅝ its length, lobes narrow-triangular to linear-subulate with obscure midrib and thick margin and with apex sharp-cuspidate, membranes flat; *corolla* tube 10–18 mm long, less than twice as long as the calyx, lobes narrowly to broadly obovate (dimensions ranging from 7 × 5 mm for subsp. *collina* to 13 × 8 mm for subsp. *abdita*) with apex acutish to obtuse, hue purple to pink to rarely white, with a pale eye bearing 1–3 striae; *stamens* borne on upper corolla tube, with some anthers exserted; *style* 6–12 mm long, united to tip which is free for ca. 1 mm, with stigma placed among the anthers near or exserted beyond corolla orifice; *flowering season* late spring to early summer, May–June, depending on latitude and altitude.

Taxonomic notes

Wherry (1955) recognized *Phlox alyssifolia* subsp. *abdita* and subsp. *collina*, which he separated from the nominate expression (subsp. *alyssifolia*) by differences in stature, size of floral parts, and pubescence, the larger subsp. *abdita* occurring at the eastern limits of the species range and the more reduced subsp. *collina* at the drier, western limits. As presently understood, the morphological variation represented by these subspecies does not show clear correlation with geographic/ecological segregation.

Geography

Chiefly Great Plains, concentrated in the northwestern sector from the eastern slope of the Northern Rocky Mountains (SE British Columbia and NW Montana) to the Black Hills region (SW South Dakota and NE Wyoming), extending locally up from the plains to higher elevations in the Northern Rocky Mountains (NW Montana) and the Southern Rocky Mountains (C Wyoming).

Environment

Plains, hilly piedmont, and mountains at low to middle elevations, on upper slopes of rolling to dissected terrain, badlands, escarpments, bluffs, and ridges, 3300–8000 ft. (1000–2450 m), reaching highest elevations in the Laramie Range of Wyoming and the Rocky Mountain Front in Glacier National Park of Montana; *MAP* 12–17 in. (29–43 cm); *habitat* primarily dry, sometimes xeric; *soils* stony, shallow with bedrock at or near the surface; *parent material* frequently calcareous such as the Greenhorn, Minnekahta, and Pahasapa formations in the Black Hills of South Dakota and Wyoming, but also associated with sandstone, clay badlands, and glacial outwash. Alyssum-leaf phlox occurs on a variety of substrates, the more unusual of which are buttes comprised of scoria (iron-oxide porcelainite produced by burning lignite) in the coal-rich Powder River Basin of E Wyoming and quartzite rubble associated with prehistoric quarrying activities at the so-called Spanish Diggings in SE Wyoming.

Associations

Conifer woodland, shrubland, grassland, and rock outcrop communities. In the Black Hills region of Wyoming and South Dakota, alyssum-leaf phlox is common in ponderosa pine (*Pinus ponderosa*) woodland and savanna. It occurs to a lesser extent at higher elevations in white spruce (*Picea glauca*) forest. Alyssum-leaf phlox is a component of a diverse herbaceous layer in ponderosa pine woodland, along with few-flower shooting star (*Dodecatheon pulchellum*), prairie bluebells (*Mertensia lanceolata*), silvery ragwort (*Packera cana*), and American pasqueflower (*Pulsatilla patens*). In NW Montana and adjacent Alberta, Canada, alyssum-leaf phlox occurs in limber pine (*Pinus flexilis*) woodland and savanna associated with escarpments, foothills, lower mountain slopes, and rocky glacial moraine.

In NE Wyoming and SE Montana, alyssum-leaf phlox occurs in sagebrush steppe dominated by Wyoming big sagebrush (*Artemisia tridentata* subsp. *wyomingensis*) in association with western wheatgrass (*Pascopyrum smithii*). On the steep sides of buttes in this region, it occurs in a shrub community dominated by creeping juniper (*Juniperus horizontalis*). In the Black Hills region it occurs in birchleaf mountain-mahogany (*Cercocarpus montanus*) shrubland.

Alyssum-leaf phlox is associated with mixedgrass prairie throughout its range, the composition of which is influenced by local soils and topography. Characteristic graminoids of these upland grasslands include Idaho fescue (*Festuca idahoensis*), western wheatgrass, bluebunch wheatgrass (*Pseudoroegneria spicata*), and thread-leaved sedge (*Carex filifolia*).

In rocky, shallow soil habitat within these larger ecological systems, alyssum-leaf phlox occurs in sparsely vegetated barrens and openings dominated by low-growing forbs, of which it is often an abundant and conspicuous element. In the vicinity of the Black Hills, this community includes mountain cat's-eye (*Cryptantha cana*), alpine bladderpod (*Lesquerella alpina*), slender parsley (*Musineon tenuifolium*), plains phlox (*Phlox andicola*), and Hood's phlox (*P. hoodii*). Alyssum-leaf phlox occurs with the rare plant Wyoming fever-few (*Parthenium alpinum*) in barrens habitat in SE Wyoming. In NW Montana, alyssum-leaf phlox is a dominant plant in sparsely vegetated shallow soil communities that include oval-leaf wild buckwheat (*Eriogonum ovalifolium*), stemless four-nerve-daisy (*Tetraneuris acaulis*), alpine bladderpod, Hood's phlox, Howard's forget-me-not (*Eritrichium howardii*), low nailwort (*Paronychia sessiliflora*), and other cushion-forming plants. In this same area it reaches up into the mountains and occurs in alpine tundra on exposed ridges along the east edge of Glacier National Park. In the Prairie Provinces of Canada, alyssum-leaf phlox occurs in rock outcrop communities dominated by bent-flowered milkvetch (*Astragalus vexilliflexus*).

Cultivation
South Dakota rancher-turned-nurseryman Claude Barr introduced alyssum-leaf phlox into horticulture in 1935 through the inaugural catalog of his mail-order nursery, Prairie Gem Ranch. The large-flowered expression from the Black Hills (subsp. *abdita*) with its ground-covering habit and tolerance of part-shade conditions merits attention as a landscape plant for the Great Plains and Mountain West. Grow in full sun to light shade in dry, well-drained soil.

Phlox amabilis
Yavapai phlox
PLATE 7

In Arizona's Verde Valley, at the northern fringes of the Sonoran Desert, occurs a plant community that is surly as they come. Its signature species is a scary shrub-tree called crucifixion thorn. Densely branched, unkempt, pungent, crucifixion thorn crowds the crests of rocky hills and breaks where it is joined by other twiggy shrubs in a menacing chaparral few would be inclined to explore. But if you brave it in spring you could be rewarded with a glimpse of Yavapai phlox, tempering the demeanor of the place with starry, soft pink flowers.

The distribution of this low phlox is centered in the "Transition Zone," a sliver of central Arizona shoehorned in between the Colorado Plateau to the north and desert basins to the south. Similar to Mogollon phlox of the Plateau's ponderosa pine forests, Yavapai phlox lives in rougher neighborhoods, from desert scrub to juniper woodland. It seems to prefer outcroppings of limestone like the Verde Formation, which, in addition to growing mean stands of crucifixion thorn, supports a band of rare plants found nowhere else in the world.

❧❧

Phlox amabilis Brand, Pflanzenr. (Engler) 4 (250): 74. 1907 [worthy of love] Up-standing suffrutescent perennial with 1–5 erect-ascending flowering shoots from a short, deep-seated rhizome, 10–30 cm tall with ca. 6 leaf nodes below the inflorescence; *leaves* linear-elliptic to oblong, thick-textured, obtusish to short-acuminate, the upper ciliate and pilose, max. 30–45 mm long and 2.5–5(–6) mm wide; *inflorescence* compact, 6- to 12- (25-) flowered, its herbage copiously glandular-pubescent, max. pedicel length 8–20 mm; *calyx* 7–10 mm long, united ½–⅝ its length, lobes broadly subulate with apex cuspidate, membranes flat; *corolla* tube 12–17 mm long, lobes obovate (average dimensions 7 × 5 mm) with apex conspicuously notched 1–3 mm deep, hue purple to bright pink, the eye often paler and bearing deep-hued striae; *stamens* borne mid- to upper corolla tube, with some anthers placed near orifice; *style* 7–15 mm long, united to tip which is free for 1.5 mm, with stigma placed among the anthers near corolla orifice; *flowering season* late spring, March–June.

Taxonomic notes

The identity of *Phlox amabilis* has at times been entangled with that of *P. woodhousei*. These two taxa, the former endemic to Arizona and the latter near-endemic, are similar in growth habitat and other traits, notably relatively thick-textured leaves and notched corolla lobes, but differ markedly in style length: *P. amabilis* has styles 7–15 mm, nearly as long as the corolla tube, with the stigma placed among the anthers; *P. woodhousei* has styles 2–5 mm, much shorter than the corolla tube, with the stigma placed below the anthers. Cronquist (1984) acknowledged these differences, but incorporated *P. amabilis* into *P. woodhousei*, noting the two occur in similar habitats and have virtually the same geographic distribution. The two species in fact occur in quite different habitats and have distributions that are parallel due to their relative relationship to the Mogollon Rim, the southern escarpment of the Colorado Plateau and a significant physiographic feature of Arizona—*P. woodhousei* occurs along the top and north of the Mogollon Rim in conifer forest and woodland, while *P. amabilis* occurs below and south of the escarpment in desert scrub and chaparral.

Geography

Chiefly Basin and Range Physiographic Province concentrated at the northern limits of the Mexican Highlands section in C Arizona, with outlying occurrences on the southern edge of the Colorado Plateau in N Arizona. Yavapai phlox is endemic to Arizona, with most populations occurring below the southern escarpment (the Mogollon Rim) of the Colorado Plateau in a region of C Arizona known variously as the "Central Highlands," "Sub-Mogollon," or "Transition Zone." Yavapai phlox is particularly abundant in the Verde Valley of C Arizona, and the common name commemorates the Yavapai people whose history is tied to the valley. The few remote occurrences on the Colorado Plateau are known from the lower San Francisco Volcanic Field south of the Grand Canyon and from the Kaibab, Shivwits, and Uinkaret plateaus north of the Grand Canyon. Yavapai phlox occurs within the boundaries of Prescott National Forest and Montezuma Castle National Monument. The distribution of Yavapai phlox very closely approximates that of crucifixion thorn (*Canotia holacantha*) in Arizona, a dominant shrub at the northern limits of the Mexican Highlands.

Environment

Dissected plateaus, hilly piedmont, and mountains at low to middle elevations, on upper slopes of divides, mesas, and bluffs, 3500–7800 ft. (1000–2400 m); *MAP* 16–20 in. (41–51 cm); *habitat* xeric; *soils* sandy to somewhat fine-textured and clay-based with rock interspersed, shallow with bedrock at or near the surface; *parent material* limestone and other calcareous substrates, notably the Verde and Redwall formations, also basalt. A number of rare plants are restricted to the Verde Formation in C Arizona.

Associations

Primarily occurs at the ecotone between conifer woodland and the upper elevation limits of the Mogollon Chaparral and Sonoran desert scrub. Redberry juniper (*Juniperus coahuilensis*) dominates these woodlands where they develop on limestone, but Utah juniper (*J. osteosperma*) may codominate or replace it on other substrates. Crucifixion thorn, a large shrub, is often present in the understory of these very open woodlands, along with other species typical of the chaparral described below. Yavapai phlox occurs to a lesser extent in pinyon-juniper woodland codominated by two-needle pinyon pine (*Pinus edulis*) and Utah juniper and, at higher elevations, in pine-oak-juniper woodland with ponderosa pine (*P. ponderosa*) and Gambel oak (*Quercus gambelii*) dominating.

On limestone soils at the northern limits of the Sonoran Desert, Yavapai phlox occurs in the creosotebush (*Larrea tridentata*) / crucifixion thorn association, a unique shrubland limited to the Arizona Upland subdivision of the Sonoran desert scrub. A number of chaparral shrubs are often associated with this plant community including three-leaf Oregon-grape (*Mahonia trifoliolata*), desert-sweet (*Chamaebatiaria millefolium*), feather-plume dalea (*Dalea formosa*), green Mormon-tea (*Ephedra viridis*), Stansbury cliffrose (*Purshia stansburiana*), and fleshy-fruit yucca (*Yucca baccata*), as well as cacti (primarily *Opuntia* spp.).

Within these associations, Yavapai phlox occurs as a component of a sparse herbaceous layer dominated by graminoids including purple three-awn (*Aristida purpurea*), sideoats grama (*Bouteloua curtipendula*), black grama (*B. eriopoda*), common curly-mesquite (*Hilaria belangeri*), and Porter's muhly (*Muhlenbergia porteri*).

Cultivation

Yavapai phlox has not been reported in cultivation, but it is a showy species with potential as a drought-tolerant landscape plant for the southwestern United States. It would also be a striking subject for the rock garden. Grow in full sun in dry, well-drained soil, although soils with higher clay content would probably be tolerated.

Phlox amoena
chalice phlox

Barrenness is in the eye of the beholder. To the early settlers of the southeastern United States, with their *back-East* mindsets, the grassy openings they found scattered through the otherwise wooded country were *barrens*—they had no trees. But to a point of view framed on the plains, they are shining, glad places of relief from the press of the forest. To poke around the Coosa Valley Barrens of Georgia and Alabama is to stand upon an Iowa prairie hilltop, awash in bluestems, blazing stars, silphiums, and the like, with breeze and sunshine on your face.

While the prairie influence on the barrens is evident, a number of plants are local specialties of this habitat, including chalice phlox, a handsome species named for its cuplike clusters of flower. Chalice phlox is typical of barrens and other open habitat in the Southeast, including the flood-scoured cobble bars of the Obed and other still-wild streams of the Cumberland Plateau. It also occurs on the extraordinary Ketona dolomite glades of Alabama, a botanical "lost world" where eight new plants were discovered in the 1990s.

Phlox amoena Sims, Curtis's Bot. Mag. 32: plate 1308. 1810 [charming] Upstanding perennial herb with both decumbent nonflowering shoots and simple or sparingly branched erect-ascending flowering shoots, the latter 5–30(–45) cm tall with 5–9 leaf nodes below the inflorescence; *leaves* oblong and obtusish (subsp. *amoena*) to linear and acuminate (subsp. *lighthipei*), their max. length 20–40(–50) and width 3–8(–12) mm, tending to be erect or ascending, the margin coarse-ciliate and surface pilose, those on nonflowering shoots partly persistent; *inflorescence* compact, subtended by a cuplike compact (subsp. *amoena*) to lax (subsp. *lighthipei*) involucre of leaves and bracts, (3-) 6- to 18-flowered, its herbage pubescent with coarse glandless hairs, max. pedicel length 1–5(–6) mm; *calyx* 7–12 mm long, united ⅜–½ its length, lobes linear-subulate with apex bearing an awn 0.5–1 mm long, membranes flat to somewhat plicate; *corolla* tube (11–)12–16(–19) mm long, glabrous, lobes obovate (average dimensions 10 × 7 mm) with apex obtusish or apiculate, hue purple to pink, rarely white or lavender, the eye often paled bearing deep-hued striae, sometimes coalescing into a deep-hued ring; *stamens* with anthers included; *style* 1–3 mm long, united ¼–½ its length, with stigma included; *flowering season* spring, early April through May in lowlands, May into June at higher elevations.

◈◈◈

Taxonomic notes

Wherry (1955) recognized *Phlox amoena* subsp. *lighthipei*, which he separated from the nominate expression (subsp. *amoena*) by "minor divergences" in height and morphology, subsp. *lighthipei* being taller with narrower leaves and bracts and with an involucre not entirely hiding the calyces. Wherry also cited the "distinctive range" of subsp. *lighthipei*, which his map showed concentrated in NE Florida and SE Georgia at the southeastern limits of the distribution of chalice phlox. Levin and co-workers upheld recognition of subsp. *lighthipei* in their evolutionary studies of the *P. pilosa* complex (Levin 1966; Levin and Smith 1966; Levin and Schaal 1970), but Ferguson (1998) did not in her dissertation on the systematics of eastern *Phlox* complexes, noting traits used to separate the subspecies were not consistent across the range of chalice phlox. As presently understood, the morphological variation represented by these subspecies does not show clear correlation with geographic/ecological segregation.

Geography

Southeastern United States in the southern Appalachian Highlands and adjacent Interior Low Plateaus, south onto the Atlantic and East Gulf coastal plains.

Environment

Coastal plains, karst plains, dissected plateaus, hilly piedmont, and mountains, on upper slopes of rolling terrain and divides, escarpments, bluffs, ridges, and on gravel bars and road cuts, elevations from sea level to 1000 ft. (300 m); *habitat* dry to dry-mesic; *soils* sandy to stony, shallow with bedrock at or near the surface; *parent material* granite, gneiss, sandstone, limestone, dolomite. Wherry (1931b) described typical habitat of chalice phlox as "thin woods in rather sterile soil, although it sometimes extends into deeper woods where the soil is richer, and again pushes out into swamp thickets or even boggy meadows."

Associations

Conifer woodland, oak woodland, riverwash, and rock outcrop communities. On the Atlantic and Gulf coastal plains, chalice phlox is associated with longleaf pine (*Pinus palustris*) woodland and savanna. Mohr (1901) described chalice phlox as frequent in Alabama "in the grassy pine barrens of the Lower Pine Region," a description that corresponds to longleaf pine savanna. Wiregrass (*Aristida stricta*) typically dominates the herbaceous layer of these communities, which occur in dry, often very sandy soils. Harper (1906) listed chalice phlox among the plants occurring in "dry pine-barrens" in the Altamaha Grit region of the Coastal Plain in Georgia. Chalice phlox occurs in oak and oak-hickory forest and woodland on the Cumberland Plateau and other portions of the Appalachian Highlands, and in "hammock" forests on the East Gulf Coastal Plain. These forests and woodlands occur in relatively dry, upland habitat and have an open canopy structure.

Chalice phlox is associated with a variety of herbaceous plant systems in the southeastern United States, most of which occur as small patch communities on shallow, rocky soils within a matrix of forest or woodland. In the Interior Low Plateaus and the Ridge and Valley Appalachian regions, chalice phlox occurs in prairie barrens characterized by grasses and forbs typical of Midwest tallgrass prairie, with little bluestem (*Schizachyrium scoparium*) almost always dominant. Chalice phlox occurs to a lesser extent in more sparsely vegetated rock outcrop communities referred to as glades or cedar glades, including the Ketona dolomite glades of C Alabama where eight plants new to science were discovered in the 1990s (Allison and Stevens 2001). Chalice phlox occurs here in a zone at the margin of the glades where deeper soil has accumulated over bedrock, along with rare species including Alabama croton (*Croton alabamensis* var. *alabamensis*), Mohr's Barbara's-buttons (*Marshallia mohrii*), Alabama skullcap (*Scutellaria alabamensis*), and Kral's Indian paintbrush (*Castilleja kraliana*), an endemic of the Ketona glades. Chalice phlox also occurs in riverwash habitat associated with streams on the Cumberland Plateau in plant communities dominated by perennial grasses typical of the Midwest tallgrass prairie.

Cultivation

Horticultural respect has been slow in coming for chalice phlox. Not long after its description as a new species in 1810, its name and identity became entangled with that of a novel but hard-to-grow hybrid phlox developed in Germany. Later, when the true chalice phlox was actually being cultivated in the United States, reports of its performance by the opinion-making gardeners of the Northeast were unenthusiastic, one reference describing it as a "doughty Spartan from dry southern wastelands" that "soon degenerates in fertile loams." Thankfully, chalice phlox has seen more use in the southeastern United States in recent years, where its refined good looks and drought tolerance are earning it a place in native plant horticulture and rock gardens. Grow in full sun to partial shade in light, well-drained soil. Cultivars are available.

Phlox amplifolia
broadleaf phlox

"The whole country leaned toward the river," and so did his soul. Jayber Crow, bachelor barber of Wendell Berry's imagined village of Port William, Kentucky, felt the sway of the Kentucky River on the land and on himself, calling him back as a young man and rooting him there the rest of his days. He appreciated how the mellow roll of the Bluegrass gives way near the river to stronger topography, ending at last in slope and cliff, woods and forest, providing a timbered sanctuary where a lovelorn old man could go "for a certain quietness of mind."

These forested folds also provide harbor for broadleaf phlox, a summerblooming species that seeks sunshine in canopy gaps and along creek beds. The distribution of this tall phlox is centered in Kentucky, Tennessee, and southern Indiana, a stony region of *karst* geology where caves, sinkholes, and springs hint at underlying limestone formations. Close relative of the widespread and widely traveled summer phlox, broadleaf phlox, like Mr. Crow and Mr. Berry, has an affinity for smaller places.

❧❧

Phlox amplifolia Britton, Man. Fl. N. States 757. 1901 [ample-leaved; "Leaves large and broad"] Upstanding perennial herb from a thick short rhizome which from irregularly spaced nodes sends up a solitary or few flowering shoots 50–150 cm tall with 8–15 leaf nodes below the inflorescence; *leaves* opposite, with a short, margined petiole and rhombic-ovate blade, thinnish with conspicuously areolate veins, max. length 85–170 mm, max. width (20–)40–80 mm, the margin serrulate and ciliolate, lower surface fine-pilose and upper beset with papillose-based bristles or locally (as in S Indiana) glabrate; *inflorescence* a panicle of small cymes with ca. 100–150 flowers, its herbage glandular-pubescent, max. pedicel length 5–10 mm; *calyx* 6–9 mm long, united ½ its length, lobes long-subulate with apex sharp-cuspidate to subaristate, membranes broad, subplicate; *corolla* tube 18–27 mm long, glabrous, lobes elliptic to obovate (dimensions varying from 9 × 5 mm to 14 × 10 mm), hue pink to white; *stamens* with anthers cream-colored, 1 or 2 often exserted; *style* 16–24 mm long, united to tip which is free for 1 mm; *flowering season* summer, July–August.

Geography
Eastern and central United States in the southern Appalachian Highlands, Interior Low Plateaus, and Ozark Plateaus.

Environment
Karst plains, hilly piedmont, dissected plateaus, and mountains, on hillsides, talus slopes, and in hollows, ravines, along streams, and on rocky floodplains; *habitat* dry-mesic to mesic; *soils* stony but humus-rich; *parent material* mostly limestone, also meta-volcanic mafic rock (rich in magnesium).

Associations
Hardwood forest. In the eastern portion of its range, broadleaf phlox occurs in dry-mesic upland forests comprised of white oak (*Quercus alba*), sugar maple (*Acer saccharum*), American beech (*Fagus grandifolia*), and tuliptree (*Liriodendron tulipifera*). American chestnut (*Castanea dentata*) was historically important in these forests. In the western portion of its range it occurs in dry-mesic upland forest and woodland dominated by oaks and hickories including white oak, black oak (*Q. velutina*), and chinquapin oak (*Q. muhlenbergii*), along with sweet pignut hickory (*Carya glabra*), shagbark hickory (*C. ovata*), and white ash (*Fraxinus americana*). It also occurs in association with riparian forest dominated by silver maple (*A. saccharinum*), river birch (*Betula nigra*), green ash (*Fraxinus pennsylvanica*), sycamore (*Platanus occidentalis*), and American elm (*Ulmus americana*). Herbaceous associates include black bugbane (*Actaea racemosa*), American hog-peanut (*Amphicarpaea bracteata*), tall bellflower (*Campanulastrum americanum*), and starry catchfly (*Silene stellata*). Broadleaf phlox appears to be a canopy gap species, occurring in areas where the forest canopy has been opened by windfall, drought-related dieback, disease-related dieback, fire, or other disturbances.

Cultivation
Broadleaf phlox is essentially unknown in horticulture, although Indiana botanist Charles Deam reported in 1940, "We have had this species under cultivation for eleven years and it is very thrifty, has a long blooming period, and is one of the best phloxes for cultivation." It shares many attributes with wild forms of *Phlox paniculata*, which, by the time broadleaf phlox was discovered and described, had been in cultivation for over 150 years and had yielded hundreds of cultivars. Broadleaf phlox should find a place in native plant horticulture in the eastern United States. Grow in light to partial shade in moist, humus-rich, well-drained soil.

Phlox andicola
plains phlox
PLATE 8

From high ground in the heart of the Nebraska Sandhills, the land looks liquid, heaving and crashing in grassy waves all about you. Little wonder this prairie-anchored dune field, twenty thousand square miles in extent, calls up oceanic analogies. "A great sea, caught and held forever in a spell," was how pioneer-era novelist Mari Sandoz saw her native country. Little bluestem rides the swells, with modest tufts of plains phlox in tow, white flowers flickering through the wind-parted blades of the bunchgrass canopy.

A *psammophyte* if ever there was, plains phlox is a recurring thread in the weave of sandy Great Plains plant communities from Colorado north to Montana. My favorite encounters have come while walking the low sandstone bluffs that frame streams like the Arikaree, the Niobrara, and the Little Powder, where plains phlox is a citizen of rock plant communities. To find it in flower as tumbling flocks of lark buntings return to the shortgrass prairie is to know the essence and fleeting kindness of a High Plains spring.

‿⁀‿

Phlox andicola Nuttall ex A. Gray, Proc. Amer. Acad. Arts 8: 254. 1870 [dweller of the Andes, in reference to the Rocky Mountains which Nuttall also called the "Northern Andes"] Caespitose perennial herb forming colonies of discrete treelike tufts from a rhizomatous caudex, the erect-ascending flowering shoots 5–12 cm tall; *leaves* linear-subulate, firm and sub-acerose, pilose to glabrate, max. length 10–20 mm and width 0.75–2 mm; *inflorescence* 1- to 5-flowered, its herbage pubescent with long kinky hairs, usually glandless but in a rare variant (formerly treated as *P. planitiarum*) gland-tipped, max. pedicel length 1–5 mm; *calyx* 6.5–11 mm long, united ⅜–¾ its length, lobes subulate with moderately prominent midrib and with apex sharp-cuspidate, membranes broad and flat; *corolla* tube 6–13(–17) mm long, lobes obovate (dimensions ranging from 7 × 5 mm to 8 × 6 mm), hue pale lavender to white, sometimes yellowish or purplish, sometimes with yellowish eye; *style* 5–9 mm long, united to tip which is free for ca. 1 mm; *flowering season* spring and early summer, May–July.

‿⁀‿

Taxonomic notes
Wherry (1955) described *Phlox andicola* subsp. *parvula*, which he separated from the nominate expression (subsp. *andicola*) by differences in stature, subsp. *parvula*

being the more reduced expression found mostly in rocky (versus sandy) habitat. Wilken (1986) noted these expressions are mostly sympatric throughout the range of the species. As presently understood, the morphological variation represented by these subspecies does not show clear correlation with geographic/ecological segregation.

Geography

Chiefly Great Plains, concentrated in the central and northern sectors, extending into the Wyoming Basin of the Rocky Mountain System. Plains phlox is strongly associated with the distribution of sand habitat in the Great Plains, the largest area of which is found in the Nebraska Sandhills, a massive, 20,000 square mile (52,000 square km) dune field in NC Nebraska. Significant dune areas also occur in association with the South Platte River in NE Colorado, and with other streams in the Great Plains and in the Wyoming Basin in C Wyoming. In NE Montana, plains phlox occurs in the sand dune habitat associated with Medicine Lake. Plains phlox occurs in association with some of the more prominent escarpments of the Great Plains, including the Pawnee Buttes region of NE Colorado, the Wildcat Hills and Pine Ridge regions of W Nebraska, and the Hat Creek Breaks region of E Wyoming.

Environment

Plains and hilly piedmont, on escarpments, buttes, and upper slopes of sandhills and sand dunes, 3200–6000 ft. (960–1800 m); *MAP* 10–17 in. (25–43 cm); *habitat* primarily dry, sometimes xeric; *soils* sandy or, in rocky habitat, stony, shallow with bedrock at or near the surface; *parent material* typically sandstone, notably the Arikaree, Ogallala, and White River formations on the plains, but also Greenhorn and Minnekahta limestone formations in the Black Hills.

Associations

Conifer woodland, shrubland, grassland, and rock outcrop communities. Plains phlox is associated with ponderosa pine (*Pinus ponderosa*) woodland and savanna, particularly on escarpments and buttes in the Black Hills and Pine Ridge regions. It occurs to a lesser extent in limber pine (*P. flexilis*) woodland.

In NE Wyoming and SE Montana, plains phlox occurs in sagebrush steppe dominated by Wyoming big sagebrush (*Artemisia tridentata* subsp. *wyomingensis*) in association with bluebunch wheatgrass (*Pseudoroegneria spicata*) or western wheatgrass (*Pascopyrum smithii*). These stands are typically open with much soil surface exposed. Plains silver sagebrush (*A. cana*) frequently replaces big sagebrush on sandier sites in which plains phlox occurs. In this same region, plains phlox occurs on the slopes of buttes in a shrub community dominated by skunkbush (*Rhus aromatica*) in association with Idaho fescue (*Festuca idahoensis*). In NE Colorado and adjacent NW Kansas and SW Nebraska, plains phlox occurs in sandsage prairie comprised of grasses and forbs similar to sandhills prairie but differing in the presence and dominance of sand sagebrush (*A. filifolia*). Plains phlox also occurs in sandhills and dune habitat in C and SW Wyoming dominated by silver sage, with rubber rabbitbrush (*Ericameria nauseosa*) and small soapweed yucca (*Yucca glauca*) associated.

In NE Colorado and adjacent NW Kansas, plains phlox is associated with shortgrass prairie dominated by blue grama (*Bouteloua gracilis*) and buffalograss (*Buchloe dactyloides*). From the Colorado-Wyoming line northward, plains phlox occurs in mixedgrass prairie where needle-and-thread (*Hesperostipa comata*) and Idaho fescue dominate, often in association with blue grama and thread-leaved sedge (*Carex filifolia*). Plains phlox occurs in upland sites within these grasslands, where the vegetation is sparse and open and a diversity of forbs are associated including purple prairie-clover (*Dalea purpurea*), scarlet gaura (*Gaura coccinea*), dotted gayfeather (*Liatris punctata*), stemless point-vetch (*Oxytropis lambertii*), silvery scurfpea (*Pediomelum argophyllum*), white-flowered beardtongue (*Penstemon albidus*), dense spike-moss (*Selaginella densa*), and red globemallow (*Sphaeralcea coccinea*).

In the Nebraska Sandhills, plains phlox occurs where prairie vegetation has partially stabilized the sand. These communities are dominated by bunchgrasses, most notably little bluestem (*Schizachyrium scoparium*) and sand bluestem (*Andropogon hallii*), with sand reedgrass (*Calamovilfa longifolia*), needle-and-thread, and sand lovegrass (*Eragrostis trichodes*) also important. Frequently associated forbs include fragrant white sand-verbena (*Abronia fragrans*), sand milkweed (*Asclepias arenaria*), painted milkvetch (*Astragalus ceramicus*), silky prairie-clover (*Dalea villosa*), bush morning-glory (*Ipomoea leptophylla*), narrow-leaved puccoon (*Lithospermum incisum*), large-bracted scurfpea (*Pediomelum cuspidatum*), lance-leaf scurfpea (*Psoralidium lanceolatum*), various beardtongues (*Penstemon* spp.), and prairie spiderwort (*Tradescantia occidentalis*).

In rocky, shallow soil habitat within these larger ecological systems, plains phlox occurs in sparsely vegetated barrens and openings dominated by low-growing forbs. Frequently associated forbs include Hooker's sandwort (*Arenaria hookeri*), tufted milkvetch (*Astragalus spatulatus*), lavender-leaf evening primrose (*Calylophus lavandulifolius*), few-flower wild buckwheat (*Eriogonum pauciflorum*), breadroot scurfpea (*Pediomelum esculentum*), and Hood's phlox (*Phlox hoodii*). On escarpments in the Nebraska panhandle and adjacent Colorado and Wyoming, plains phlox occurs with regional endemics silky orophaca (*Astragalus sericoleucus*), mountain cat's-eye (*Cryptantha cana*), slender parsley (*Musineon tenuifolium*), and plains nailwort (*Paronychia depressa*).

Cultivation

Plains phlox once enjoyed fairly wide circulation among rock gardeners through the agency of South Dakota rancher-nurseryman Claude Barr, who introduced it into horticulture around 1936. Barr offered "*Phlox andicola parvula* 'Dr. Wherry'," a selection with wide and overlapping petal blades in his 1963 catalog, but this apparently disappeared from the trade. Because it spreads readily by underground rhizomes, plains phlox is a potential hazard in the rock garden, but the same attribute could make it an effective groundcover for colonizing open areas. Grow in full sun in dry, light, well-drained soil.

Phlox austromontana
desert phlox
PLATES 9, 10

Nevada's Wassuks are tough enough to have two timberlines. Rain-shadowed, first by the Sierra Nevada, then by the Sweetwater Range, these water-robbed mountains shoulder only a strap of tree cover. Desert scrub claims the footslopes, and alpine rockland mantles the heights. The timber, a *pygmy woodland* of chunky pinyon pine and juniper, is limited to the middle ground. It is daunting habitat, yet ramping upslope and down, in and out of the thin, aromatic needle-shade, are ivory-flowered mats of desert phlox.

Desert phlox ranges throughout the Intermountain West, most often in the *P-J*. A beautiful variant of this species occurs in Utah's Zion Canyon, gracing sandstone slickrock with piney festoons studded in spring with bright pink flowers. It is a worthy plant for this word-wrecking landscape of soaring walls, colossal monoliths, and plunging-from-the-rimrock cloudburst waterfalls, just as *Zion* is a worthy name, radiant in celestial imagery as the tabernacle of *The Uncontainable*— The City of God.

༄༅

Phlox austromontana Coville, Contr. U.S. Natl. Herb. 4:151. 1893 [of southern mountains] Subshrub, freely and diffusely branched from a woody base, ranging in habit from sprawling festoons of decumbent stems up to 30 cm long with ascending annual flowering branches 5–10 cm long (subsp. *jonesii* and subsp. *prostrata*) with internodes apparent to more strongly condensed (subsp. *austromontana*) or cushion-like (subsp. *lutescens*) expressions 5–15 cm high with flowering shoots 2–10 cm long; *leaves* linear-subulate, firm and acerose, pilose to glabrate, max. length 10–25 mm (up to 35 mm in subsp. *jonesii*), max. width 1–2 mm; *inflorescence* 1- to 5-flowered, its herbage appressed-pubescent to glabrate, max. pedicel length 0.5–2 mm; *calyx* 7–12 mm long, united ½–¾ its length, lobes subulate with apex sharp-cuspidate, membranes broad and conspicuously carinate; *corolla* tube 8–18 mm long, lobes obovate or orbicular (dimensions ranging from 6 × 4 mm for subsp. *austromontana* to 10 × 7 mm for subsp. *jonesii*), hue pink (bright pink in subsp. *jonesii*) or lavender to white (yellowish in subsp. *lutescens*); *stamens* borne mid- to upper corolla tube, with some anthers exserted; *style* 2–6 mm long (5–10 mm in subsp. *jonesii*), united to tip which is free for ca. 1 mm, with stigma included and placed just below the anthers; *flowering season* spring, April–June.

Key to the Subspecies of *Phlox austromontana*

1a. Plant growth habit more or less open, internodes typically apparent; plants of SW Utah in W Kane and much of Washington counties 2
1b. Plant growth habit variously open to compact; distribution various 3
2a. Calyx usually glabrous; plants lax, flowering stems arising above basal mat of leaves, or festooning on cliffs and ledges; leaves (some of them) 20–35 mm long and up to 3 mm wide, green and glabrous or nearly so; corolla hue usually bright pink; styles 5–10 mm long; morphology transitional to the next .... **subsp. *jonesii*** (Wherry) Locklear, J. Bot. Res. Inst. Texas 3: 645. 2009 | Zion desert phlox
2b. Calyx usually at least moderately villous; leaves typically 10–22 mm long; corolla hue commonly white; styles 4–6 mm long ....
.... **subsp. *prostrata*** (E. E. Nelson) Wherry, Wash. Acad. Sci. 29: 518. 1939 | Silver Reef desert phlox
3a. Corolla hue yellowish (fading to lemon yellow); plants forming cushions; leaves 10–25 mm long; calyx campanulate ....
.... **subsp. *lutescens*** (S. L. Welsh) Locklear, J. Bot. Res. Inst. Texas 3: 646. 2009 | Canyonlands desert phlox
3b. Corolla hue white, pink, or lavender (sometimes fading to cream in color); plant generally mat-forming; leaves mostly less than 15 mm long; calyx turbinate (inversely conical) to subcylindrical; most widely distributed expression ....
.... **subsp. *austromontana*** | desert phlox

Geography

Western interior of the United States. The nominate expression of desert phlox (subsp. *austromontana*) has the largest distribution, chiefly Colorado Plateau and Great Basin, extending north onto the Columbia Plateau and Northern Rocky Mountains and east into the Middle Rocky Mountains, with outlying occurrences to the south in mountain ranges in the Sonoran Desert (S Arizona) and Mexican Highland (S New Mexico) sections of the Basin and Range Physiographic Province, the San Gabriel, San Bernardino, and Peninsular Range mountains of S California, and the Sierras Juarez and San Pedro Mártir in Baja California, Mexico. Subspecies *jonesii* is endemic to the Zion Canyon region of SW Utah, and is rather common in Zion National Park. Subspecies *lutescens* is limited to the Canyonlands section of the Colorado Plateau in SE Utah (Garfield, Grand, Kane, and San Juan counties) and adjacent Arizona (Apache County). Subspecies *prostrata* is apparently limited to the Silver Reef mining district of SW Utah (Washington County), or at least is most common there.

Environment

Intermountain basins, dissected plateaus, hilly piedmont, and mountains at low to middle elevations, on upper slopes of rolling terrain and divides, mesas, escarpments, bluffs, ridges, and in canyons, 2500–9000 ft. (800–2800 m); *MAP* 8–14 in. (20–36 cm); *habitat* xeric to dry; *soils* medium- to coarse-textured, stony, shallow with bedrock at or near the surface; *parent material* primarily sandstone.

The three subspecies of desert phlox occurring in S Utah are associated with sandstone formations, typically occurring in cracks and crevices in the bedrock. Exposed, sparsely vegetated expanses of these smooth, tight-grained sandstones are termed "slickrock." These varieties may also occur in sandy alluvium derived from the bedrock. Subspecies *jonesii* is associated with Navajo Sandstone, the rock formation that dominates the geology of the Zion Canyon region. Subspecies *lutescens* is associated with the Cedar Mesa Sandstone formation at the type locality along Cataract Canyon. Subspecies *prostrata* is associated with geological formations exposed in the vicinity of the Silver Reef mining district, perhaps the Silver Reef Sandstone formation which forms rocky ledges in the area.

Associations

Conifer woodland, shrubland, chaparral, grassland, and rock outcrop communities. Desert phlox is a relatively constant associate of juniper woodland and pinyon-juniper woodland across the Intermountain West. Juniper woodland is comprised of open stands of Utah juniper (*Juniperus osteosperma*). The composition of pinyon-juniper woodland varies, with two-needle pinyon pine (*Pinus edulis*) and Utah juniper codominant on the Colorado Plateau, and single-leaf pinyon pine (*P. monophylla*) and Utah juniper codominant in the Great Basin. Desert phlox is often a major species in the sparse herbaceous layer of these pygmy woodlands of dwarf, stunted, widely spaced trees. At its upper altitudinal limits, pinyon-juniper woodland grades into ponderosa pine (*P. ponderosa*) woodland and savanna, and desert phlox makes the transition into these associations. Desert phlox ascends into Intermountain bristlecone pine (*P. longaeva*) woodland on badlands of the Claron limestone formation in Bryce Canyon National Park.

In the Blue Mountains region of NE Oregon, specifically the Elkhorn, Greenhorn, and Strawberry mountains, desert phlox occurs in xeric, subalpine woodland dominated by whitebark pine (*Pinus albicaulis*) and Engelmann spruce (*Picea engelmannii*). In the nearby Seven Devils Mountains of Idaho, desert phlox occurs in bedrock crevices in ridge tops within open forests of whitebark pine and Rocky Mountain subalpine fir (*Abies bifolia*).

In the San Bernardino Mountains of S California, desert phlox occurs in open areas within Jeffrey pine (*Pinus jeffreyi*) forest, in rare plant communities associated with limestone outcroppings and in a unique habitat referred to as "pebble plains." The dense clay soils of pebble plains are armored by a pavement of Saragosa quartzite pebbles pushed to the surface by frost heaving. While devoid of the surrounding Jeffrey pine forest vegetation, these flats have a distinctive flora of low growing cushion- and mat-forming herbaceous species, a number of which are endemic to the pebble plains.

Desert phlox occurs in several shrubland systems throughout its range, often where these intersect the lower reaches of pinyon-juniper woodland. Big sagebrush (*Artemisia tridentata*) or blackbrush (*Coleogyne ramosissima*) typically dominate these shrublands. At the upper altitudinal reaches of pinyon-juniper woodland in Utah, desert phlox extends into mountain brush shrubland dominated by Gambel oak (*Quercus gambelii*) and curl-leaf mountain-mahogany (*Cercocarpus ledifolius*). In the Elkhorn Mountains of NE Oregon, desert phlox occurs in subalpine shrubland dominated by mountain big sagebrush (*A. tridentata* subsp. *vasseyana*).

Subspecies *jonesii* occurs in sparsely vegetated slickrock habitat within the context of ponderosa pine woodland, pinyon-juniper woodland, and mountain brush shrubland. The latter is comprised of small trees and shrubs including Utah serviceberry (*Amelanchier utahensis*), Gambel oak, Turbinella live oak (*Quercus turbinella*), and single-leaf ash (*Fraxinus anomala*). This subspecies is typically associated with other crevice plants including slickrock Indian paintbrush (*Castilleja scabrida*), littleleaf mountain-mahogany (*Cercocarpus intricatus*), Nuttall's linan-thastrum (*Linanthastrum nuttallii*), caespitose rockmat (*Petrophyton caespitosum*), and the endemic Canaan daisy (*Erigeron canaani*). It also is an occasional associate of the rare Zion tansy (*Sphaeromeria ruthiae*), which only occurs on Navajo Sandstone in and near Zion Canyon. Subspecies *jonesii* is sometimes associated with plants in seasonally moist seeps and wet walls, including Zion shooting star (*Dodecatheon pulchellum* var. *zionense*) and King's clover (*Trifolium kingii* subsp. *macilentum*).

Subspecies *lutescens* occurs in rimrock shrubland comprised of shrubs and small trees with blackbrush often the dominant species accompanied by Utah serviceberry, rubber rabbitbrush (*Ericameria nauseosa*), single-leaf ash, Stansbury cliffrose (*Purshia stansburiana*), and fragrant sumac (*Rhus aromatica*).

In his thesis on the Pine Valley Mountains of SW Utah, Robert Warrick (1987) found subsp. *prostrata* to be "[l]ocally common in chaparral to mountain brush communities." The Pine Valley Mountains region is noteworthy for the disjunct occurrence of true evergreen chaparral, floristically similar to the Mogollon Chaparral of C Arizona, and containing many of the dominant genera of the California chaparral. This community is comprised of evergreen broadleaf shrubs including Turbinella live oak, Mojave desert whitethorn (*Ceanothus greggii*), Mexican manzanita (*Arctostaphylos pungens*), ashy silktassel (*Garrya flavescens*), curl-leaf mountain-mahogany, and narrowleaf yerba santa (*Eriodictyon angustifolium*). In Utah, this association is mainly developed in the Pine Valley and Bull Valley mountains of Washington County, a region which corresponds to the known distribution of subsp. *prostrata*.

Desert phlox occurs to a lesser extent in grassland systems. In NW Colorado, desert phlox occurs in Great Basin grassland dominated by bluebunch wheatgrass (*Pseudoroegneria spicata*) and Indian mountain-ricegrass (*Achnatherum hymenoides*). On rocky subalpine slopes in the Elkhorn Mountains of NE Oregon, desert phlox is a dominant forb in a graminoid community dominated by Geyer's sedge (*Carex geyeri*). On the Kaibab Plateau of N Arizona, on the north side of the Grand Canyon, desert phlox occurs in montane grassland.

Cultivation

Desert phlox has been cultivated to a limited extent as a rock garden plant. While the typical expression of the species (subsp. *austromontana*) does not possess any special attributes that would set it apart from other rock garden phloxes, subspp. *jonesii* and *lutescens* from the canyon country of S Utah have exciting horticultural potential, the former a festooning plant with bright pink flowers, the latter a cushion-forming expression with yellowish flowers. Both of these subspecies would be attractive subjects for rock garden or wall gardens, and merit development as drought-tolerant landscape plants for the southwestern United States. Grow in full sun to light shade in dry, well-drained soil.

Phlox bifida
cleft phlox

PLATE 11

Cleft phlox stars on many stages—sand prairies in Iowa, river bluffs in Illinois, cedar glades in Tennessee, the dunes of Lake Michigan. But to my eye its most beautiful venue is the rumpled Saint Francois Mountains of the Missouri Ozarks, where cleft phlox is a woodlander, blooming in stride with Dutchman's britches, bloodroot, and other spring ephemerals, all recently arisen through the winter-worn leather of last year's oak leaves.

Tucked away in these old, milled-down mountains is a local feature Ozark folks call a "shut-in"—a narrow gorge with a streambed choked by a jumble of boulders and talus. On a misty overcast day in early April, all this exposed rock seems to emanate its own light—a rose-gray, before-time smokiness that echoes through the surrounding uplands in the bough of the redbud tree, and settles to the forest floor in the petal of the cleft phlox.

∽∾∾

Phlox bifida L. C. Beck, Amer. J. Sci. 11: 170. 1826 [split in two, of the cleft corolla blades; "segments . . . deeply cleft, sometimes nearly to the base"] Caespitose suffrutescent perennial, sparingly and diffusely branched, forming open tufts or festoons from a taproot, the numerous erect-ascending flowering shoots 8–20(–25) cm long with 4 or 5 leaf nodes below the inflorescence; *leaves* linear to narrowly elliptic or lanceolate, ciliate, the upper pilose, max. length 30–60 mm, max. width 2–4 mm; *inflorescence* (3-) 6- to 9- (12-) flowered, lax, its herbage pubescent (rarely glabrous) with hairs mostly gland-tipped for nominate expression (subsp. *bifida*), glandless for subsp. *stellaria*, and mostly glandless (except adaxial surface of calyx lobes) for subsp. *arkansana*, max. pedicel length (8–)12–25(–40) mm; *calyx* 6.5–9.5 mm long, united ½–⅝ its length, lobes linear-subulate with apex cuspidate, membranes flat to subplicate; *corolla* tube 9–14 mm long, lobes obovate (dimensions ranging from 9.5 × 6.5 mm for subsp. *stellaria* to 10 × 7.5 mm for subsp. *bifida*) with apex conspicuously notched, degree of notching ranging from 1.5–3 mm deep (⅛–½ lobe length) for subsp. *stellaria*) to deep (¼–⅔ lobe length) in subsp. *arkansana* and subsp. *bifida*, hue lavender or rarely reddish purple or white, the base of the lobes sometimes bearing paired violet striae; *style* 5–12 mm long (mostly 2–4 mm in subsp. *arkansana*), united to tip which is free for 1 mm (stigma comprising more of style length for subsp. *arkansana*); *flowering season* spring to early summer, late March into June (depending on latitude), peaking late April to early May.

Key to the Subspecies of *Phlox bifida*

1a. Styles usually shorter than 6 mm (mostly 2–4 mm long), included in calyx; stigmas comprising less than ⅓ of style length; longer corolla lobes, corolla hue reddish purple or bright violet; restricted to NW Arkansas ....
...................**subsp. *arkansana*** D. L. Marsh, Trans. Kansas Acad. Sci. 63: 16. 1960 |
Arkansas cleft phlox

1b. Styles usually longer than 6 mm (5–14 mm long), exserted from calyx; stigmas comprising less than ⅕ of style length; corolla hue bright violet or white............ 2

2a. Inflorescence-herbage (pedicels and calyces) pubescent with mostly gland-tipped hairs; calyx united ca. ½ its length; corolla lobes notched ¼–⅔ of their length, lobe narrowed abruptly below notch; most widely distributed expression but primarily northern; associated with variety of substrates .... **subsp. *bifida*** | cleft phlox

2b. Inflorescence-herbage pilose; calyx united ca. ⅝ its length; corolla lobes notched ⅛–½ their length, narrowed gradually below notch; primarily southern portion of species range; associated with limestone or dolomite ....
.... **subsp. *stellaria*** (A. Gray) Wherry, Castanea 16: 99. 1951 | Kentucky cleft phlox

Geography

Central United States. The nominate expression of cleft phlox (subsp. *bifida*) ranges from the Ozark Plateaus (NW Arkansas and S Missouri) north through the Central Lowlands of the Interior Plains as far as EC Iowa and the lower Great Lakes region (N Illinois, NW Indiana, SW Michigan, and SE Wisconsin), with outlying occurrence in the upper West Gulf Coast Plain in NC Texas. It occurs most abundantly in Illinois. The occurrence of cleft phlox in Texas is based on a single collection made in Dallas County in 1958, described by Shinners (1961) as *Phlox bifida* var. *induta*. The identity was later interpreted as Oklahoma phlox (*P. oklahomensis*) (Cypher 1993; Wherry 1965b), but morphological attributes and ecological associations favor cleft phlox (Springer 1983).

Subspecies *arkansana* is endemic to NW Arkansas, occurring primarily on the Springfield Plateau of the Ozarks. It is known historically from eight occurrences in the counties of Benton, Carroll, Madison, Stone, and Washington. Cypher reported only four extant populations in 1993, and believed it warranted listing as an endangered species.

The historical distribution of subsp. *stellaria* is concentrated in the southern part of the overall range of the cleft phlox species complex, primarily in the Interior Low Plateaus (C Kentucky and C Tennessee). This subspecies occurs most abundantly in Kentucky, and is possibly extirpated from Illinois. In his original description of *Phlox stellaria*, Gray (1870) noted the collection locality as "Cliffs of Kentucky River (probably above Lexington), in the fissures of the most precipitous rocks," adding, "[t]he station should be rediscovered." This cryptic challenge compelled Wherry (1929a) to search for and rediscover this "long-lost *Phlox*" near Camp Nelson in Jessamine County, Kentucky, in 1923, an adventure so satisfying that it helped launch him on his nearly fifty-year study of the genus *Phlox* and the family Polemoniaceae

Environment
Plains, lake plains, dissected plateaus and broad mountain domes, on upper slopes of rolling terrain and divides, morainal features, bluffs, cliffs, rock ledges, ridges, and along streams, 395 ft. (120 m) in S Illinois to 1445 ft. (440 m) in NW Arkansas; *MAP* 33 in. (83 cm) in EC Iowa to 50 in. (127 cm) in C Tennessee; subsp. *bifida* occurs on relatively steep bluffs and hills formed in loess deposits and on glacial moraines, and in sandy habitat (typically more common in partially stabilized areas than in more densely vegetated sites); all subspecies occur in stony, shallow soils with bedrock at or near the surface, not typically on exposed rock surfaces but in shallow soil at the edge of bedrock exposures; *parent material* various, with subsp. *bifida* typically associated with sandstone, chert, and granite (rhyolite), subspp. *arkansana* and *stellaria* with limestone or dolomite.

Cleft phlox, as the species or its subspecies, is a plant of conservation concern throughout much of its range. Its rarity in certain areas is due to destruction and/or alteration of its habitat, and association with plant communities such as savannas, prairies, and glades that were historically influenced by periodic burning. Control of fire has resulted in encroachment of shrubs and other woody plants to the detriment of the herbaceous components of these plant communities.

Associations
Oak-dominated forest/woodland/savanna, grassland, riverwash, and rock outcrop communities. Where it is associated with forest and woodland, cleft phlox occurs on upper slopes and other topographic situations that result in a relatively open canopy structure. White oak (*Quercus alba*) and black oak (*Q. velutina*) are the

typical dominants, with shagbark hickory (*Carya ovata*), and bitternut hickory (*C. cordiformis*) also important. In the drier forests of the Ozarks, black hickory (*C. texana*), eastern red-cedar (*Juniperus virginiana*), and shortleaf pine (*Pinus echinata*) may be associated. Cleft phlox occurs with other spring blooming wildflowers including Dutchman's breeches (*Dicentra cucullaria*), cutleaf toothworth (*Cardamine concatenata*), and bloodroot (*Sanguinaria canadensis*). In the Ozarks, cleft phlox also occurs where woodlands grade into glades and limestone prairies dominated by herbaceous species characteristic of tallgrass prairie. Subspecies *arkansana* occurs in dry oak-hickory forest and woodland that develop over limestone or dolomite bedrock. White oak typically dominates, with black hickory and shortleaf pine also associated.

At the northern end of its range, cleft phlox occurs in plant communities that develop on windblown sand deposits adjacent to river systems and the shore of Lake Michigan, and on gravelly hills of glacial origin. In Iowa, N Illinois, and N Indiana, cleft phlox occurs in oak savanna and oak barrens characterized by white oak and black oak, with bur oak (*Quercus macrocarpa*) post oak (*Q. stellata,*), and blackjack oak (*Q. marilandica*) associated. Black oak dominates on sandier sites, as at the Indiana Dunes National Lakeshore along the southern shore of Lake Michigan. Grasses and forbs typical of tallgrass prairie intermix with the scattered oak trees in these savannas, with little bluestem (*Schizachyrium scoparium*) being the most common grass. A shrubby component of American hazelnut (*Corylus americana)* and prairie willow (*Salix humilis*) may also be present. These wooded communities grade into prairie or barrens dominated by little bluestem, along with an arenaceous flora that includes sand lovegrass (*Eragrostis trichodes*), sand reedgrass (*Calamovilfa longifolia*), and numerous forbs.

In the Mississippi and Illinois River valleys of C Illinois, cleft phlox occurs in association with hill prairies, a distinctive expression of tallgrass prairie that develops on exposed, well-drained crests and upper slopes of loess bluffs and glacial moraines. Historically, these relatively small patch communities occurred as openings in an otherwise forested landscape. Big bluestem (*Andropogon gerardii*) and little bluestem are the dominant grasses. In the Ozarks, cleft phlox occurs infrequently in riverwash habitat.

Cleft phlox (subsp. *stellaria*) is associated with a variety of herbaceous plant systems that occur as small patch communities on shallow, rocky soils within a larger matrix of forest or woodland. In C Tennessee near Nashville, it occurs in association with cedar glades that form over surfacing limestone bedrock. These unique communities are dominated by poverty dropseed (*Sporobolus vaginiflorus*), a summer annual grass, with stunted forms of post oak, blackjack oak, and eastern red-cedar at the margins. Cedar glades exhibit zonation of vegetation, with cleft phlox occurring at the edge of the driest, most exposed parts of the limestone-underlain glades, sometimes at the base of cedar trees. The cedar glades of Middle Tennessee are host to a number of rare, endemic forbs, some of which occur with cleft phlox including Gattinger's prairie-clover (*Dalea gattingeri*), southern scurf-pea (*Pediomelum subacaule*), and Tennessee coneflower (*Echinacea tennesseensis*), an endangered species. Early Tennessee botanist Augustin Gattinger (1901) wrote of the colorful presence of this plant (as *Phlox stellaria*) on the cedar glades, noting it "deserves a bed in every garden." Brand (1907) based his *P. stellaria* var. *cedaria* (of the cedars) on a specimen Gattinger collected in 1879 from the cedar barrens near La Vergne, Tennessee. In the Jessamine Gorge of the Palisades region of C Kentucky, cleft phlox (subsp. *stellaria*) occurs in cliff communities with several rare plants including Canby's mountain-lover (*Pachistima canbyi*). In S Indiana, it occurs in riverwash habitat dominated by prairie grasses, where it is an associate of Short's goldenrod (*Solidago shortii*), an endangered species (Homoya and Abrell 2005).

Cultivation

Thin leaves, wiry stems, and deeply cleft, starry flowers combine to give cleft phlox an airiness unmatched by any other member of the genus. Cleft phlox is most often used as a rock garden subject, but also is attractive festooning from a rock wall. In the wild, cleft phlox is associated with the herbaceous layer of open, relatively dry woodlands and savannas, indicating untapped potential as a groundcover for dry shade situations. Grow in full sun to partial shade in dry, light, well-drained soil. Cleft phlox self-sows readily, a positive attribute in many garden settings, but making it a potential problem in rock garden settings where space is limited. Cultivars are available. Horticultural hybrids are documented (see Appendix B).

Phlox buckleyi
Greenbrier phlox

There is a hardscrabble, hard-times feel to the shaley woods of Pocahontas County, West Virginia, where oak and pine show the pinched, hillbilly character of trees that would grow stronger and straighter somewhere else. This is the shale barrens country of the middle Appalachians, and the slopes are dotted with open, exposed places where rare and interesting wildflowers grow instead of trees. It is also the home of Greenbrier phlox, an early summer bloomer with sword-shaped leaves and showy clusters of bright pink flowers.

While endemic to the shale barrens region of Virginia and West Virginia, Greenbrier phlox occurs in dry woodland rather than on the open barrens. This rare species has the added distinction of being one of the legendary "lost plants" of Appalachia. Samuel Buckley collected his namesake in 1838 near White Sulphur Springs, West Virginia, but it would be eight decades before it would surface again—a surprising oversight, since it grows near the famed Greenbrier Resort, long a favorite haunt of apparently distracted staff from the New York Botanical Garden.

❧☙

Phlox buckleyi Wherry, J. Wash. Acad. Sci. 20: 26. 1930 [after Samuel Botsford Buckley (1809–1894), collector of the type] Rosulate perennial herb forming colonies from slender modified stems (rhizomes) that spread laterally just below the surface of the ground and are tipped by short rosettes of sword-shaped persistent leaves, these scattered rosettes elongating in spring of the following year into erect flowering shoots 15–45 cm tall with 3–7 leaf nodes below the inflorescence; *leaves* thickish, long-acuminate, the lower linear or narrow-elliptic, glabrate, the upper lanceolate, ciliate and glandular-pilose, max. length 50–100(–125) mm, max. width 5–10(–12) mm; *inflorescence* 6- to 25-flowered, compact but in age becoming lax, its herbage copiously glandular-pubescent, max. pedicel length 3–9 mm; *calyx* 7–13 mm long, united ca. ⅝ its length, lobes linear-subulate with apex sharp-cuspidate to short-aristate, membranes flat to subplicate; *corolla* tube 17–23 mm long, glandular-pubescent, lobes narrowly to broadly obovate (average dimensions 10 × 8 mm) with apex obtuse-truncate and entire or erose-emarginate, hue bright purple to pink, the eye often paled and purple-striate; *stamens* nearly as long as the corolla tube, with a single anther sometimes exserted; *style* 14–20 mm long, united to tip which is free for 1 mm; *ovary/capsule* often with 2 ovules/seeds per locule; *flowering season* early summer, May–June; tetraploid, $2n = 28$ (Smith and Levin 1967).

❧☙

Geography

Appalachian Highlands, endemic to the shale barrens region of the Valley and Ridge physiographic province in Virginia and West Virginia. Greenbrier phlox occupies the smallest range of any of the eastern phloxes, and is known from fourteen counties in western Virginia and two counties (Greenbrier and Pocahontas) in SE West Virginia. The common name commemorates the discovery of this species in Greenbrier County, and its occurrence in the beautiful watershed of the Greenbrier River.

Environment

Mountains at lower elevations, on moderately steep, unstable slopes, and at roadsides, 1000–2400 ft. (305–732 m); *MAP* 33–42 in. (84–107 cm); *habitat* dry; *soils* shallow, with little organic matter accumulation, and with surface covered by thin mantle of weather-resistant shaley rock fragments, pH acidic; *parent material* shale or shaley sandstone, primarily Devonian Braillier Formation.

While Greenbrier phlox occurs in the shale barrens region, it is associated with wooded slopes adjacent and transitional to more exposed barrens "where the forest litter is well mixed with shale flakes" (Wherry 1935c). Wherry (1930) described the "normal habitat" of Greenbrier phlox as "a thinly wooded slope toward the base of a hill of Devonian shale, the soil being usually a humus-rich gravel of subacid reaction." Wherry (1953) noted Greenbrier phlox had been successfully cultivated in rock gardens in Pennsylvania with no shale material, indicating that shale is not a requirement but instead provides habitat relatively free from the competition of more dense vegetation. Greenbrier phlox appears well-adapted to the unique environment presented by weathering shale formations, with thick, narrow leaves and trailing (soboliferous) stems that can accommodate the shifting nature of shale habitat. Its rhizomatous habit enables it to form sometimes sizable clonal colonies. One such colony on Mack Mountain in Pulaski County, Virginia, "extended for about a half-mile, with a depth of at least two hundred yards, perhaps more" (Uttall 1971).

Associations

Open woodland dominated by Virginia pine (*Pinus virginiana*) and chestnut oak (*Quercus prinus*), with scrub oak (*Q. ilicifolia*) also included. Although limited to the mid-Appalachian shale barrens region, Greenbrier phlox is seldom an associate of the shale barrens endemics, most of which occur in the open barrens (see discussion of shale barrens in *Phlox subulata* species account). The herbaceous layer of these woodlands is typically rather sparse, with associates including wild crane's-bill (*Geranium maculatum*), bowman's-root (*Porteranthus trifoliatus*), bird's-foot violet (*Viola pedata*), ericaceous plants such as teaberry (*Gaultheria procumbens*), black huckleberry (*Gaylussacia baccata*), and American wintergreen (*Pyrola americana*), along with mosses and reindeer lichen.

Cultivation

Greenbrier phlox has been grown to a limited extent as a rock garden plant. Given its natural habitat, it also should prove tolerant of challenging dry shade situations in the landscape. With its colonizing habit and attractive, evergreen foliage,

Greenbrier phlox would make an excellent groundcover for the eastern United States. Grow in partial shade in dry, well-drained soil, although soils with higher clay content would probably be tolerated.

Phlox caespitosa
Bitterroot phlox
PLATES 12–14

Brings-Down-the-Sun was a careful watcher of birds and stars and seasons. Medicine man and chief of his people, the Blackfeet, he possessed deep knowledge of the Bitterroot Mountains in what is today northwest Montana. By his earthy calendar, early summer was the "Moon of Flowers"—the time when the mountain parks burst into bloom. Lupine, geranium, balsamroot, and scarlet skyrocket are at their prime during these days, as are shrublets of Bitterroot phlox, cool daubs of white in a rowdy palette of color.

The distribution of this ankle-high phlox is centered in the Bitterroot Mountains of Montana and Idaho, where it graces glacial valley grasslands and ponderosa pine savannas. While Bitterroot phlox associates with a wide spectrum of bold and showy wildflowers, it seems particularly inclined toward the quiet company of mariposa lilies, which lift exquisite chalices brimming with mariposa-mirth.

Phlox caespitosa Nuttall, J. Acad. Nat. Sci. Philadelphia 7: 41, plate 6, figure 1. 1834 [in tufts or dense clumps, of the plant growth habit; "stem . . . caespitose"] Subshrub, sparingly and diffusely branched, forming open tufts 5–15 cm tall from a woody caudex, the numerous erect-ascending annual flowering branches 2.5–7.5 cm long, with internodes typically apparent; *leaves* pale green, linear-oblong, thickish, cuspidate, ciliate with rather coarse hairs, surficially sparse-pilose to glabrous, max. length 7.5–12.5 mm, width 1–2 mm; *inflorescence* 1- to 3-flowered, its herbage glandular-pubescent, max. pedicel length 1–6 mm; *calyx* 6–9(–11) mm long, united ⅜–⅝ its length, lobes subulate with rather prominent costa and with apex cuspidate, membranes flat; *corolla* tube 8–12 mm long, lobes obovate (average dimensions 6 × 4 mm), hue lavender to white; *style* 5–8 mm long, united to tip which is free for ca. 1 mm; *flowering season* spring and early summer, April–June.

Taxonomic notes
The identity of *Phlox caespitosa* has been confused with that of *P. condensata*, *P.*

douglasii, and *P. pulvinata*, all treated here as separate species. See Locklear (2009) for taxonomic circumscription of *P. caespitosa* in relation to these species.

Geography

Rocky Mountain System, chiefly in the Northern Rocky Mountains (SE British Columbia, N Idaho, and W Montana) where it occurs most abundantly on the eastern slope of the Bitterroot Mountains (W Montana) and in the intermountain valleys of the Rocky Mountain Trench (NW Montana), with outlying occurrence south in the Pryor Mountains of the Middle Rocky Mountains (SC Montana). The common name commemorates the Bitterroot Mountains region where the type was collected.

Environment

Hilly piedmont, mountains at low to middle elevations, and morainal features associated with river valleys, 2500–4900 ft. (7–1494 m), up to 6800 ft. (2040 m) in isolated occurrences in Bitterroot Range; *MAP* 12–25 in. (31–64 cm); *habitat* dry; *soils* stony, coarse-textured.

Associations

Conifer woodland and grassland. Bitterroot phlox is strongly associated with ponderosa pine (*Pinus ponderosa*) woodland and savanna. These open stands of ponderosa pine occur on the driest forested sites in the Northern Rocky Mountains and have an herbaceous layer dominated by a single species of xerophytic bunchgrass, primarily bluebunch wheatgrass (*Pseudoroegneria spicata*). Along with Bitterroot phlox, typical forbs include arrow-leaf balsam-root (*Balsamorhiza sagittata*), Douglas' brodiaea (*Brodiaea douglasii*), elegant mariposa lily (*Calochortus elegans*), sticky geranium (*Geranium viscosissimum*), prairie smoke (*Geum triflorum*), scarlet skyrocket (*Ipomopsis aggregata*), Pursh's silky lupine (*Lupinus sericeus*), and meadow deathcamas (*Zigadenus venenosus*). Ponderosa pine savanna characterizes the presumed type locality of Bitterroot phlox near Thompson Falls, Montana. Bitterroot phlox occurs in association with limber pine (*Pinus flexilis*) woodland in the Pryor Mountains of SC Montana.

Bitterroot phlox is common in grasslands of the Tobacco Plains in NW Montana, a narrow glacial valley in Lincoln County. It also occurs in grasslands in the valley of the North Fork of the Flathead River on the west side of Waterton-Glacier national parks. Bluebunch wheatgrass, Idaho fescue (*Festuca idahoensis*), and rough fescue (*F. campestris*) are the dominant graminoids. Typical forbs include hairy false goldenaster (*Heterotheca villosa*) and yellow wild buckwheat (*Eriogonum flavum*), plus other species found in the herbaceous layer of ponderosa pine wood-land and savanna. Scattered populations of Bitterroot phlox occur in grasslands on the slopes of the eastern foothills of the Bitterroot Mountains, and at higher elevations in grassy balds occurring in areas of coniferous forest dominated by Engelmann spruce (*Picea engelmannii*) and Rocky Mountain subalpine fir (*Abies bifolia*)

Cultivation

True Bitterroot phlox has seen little if any traffic in horticulture, despite the pervasiveness of the name *Phlox caespitosa* in the literature of rock gardening. Because of the nomenclatural baggage described above, most of what has circulated under this epithet is either *P. condensata* or *P. pulvinata*—two high elevation species that have long been pursued by rock gardeners. Taxonomic confusion aside, Bitterroot phlox is a showy species with potential as a drought-tolerant landscape plant for the Mountain West. Grow in full sun to light shade in dry, well-drained soil.

Phlox carolina
Carolina phlox

The soldier was recovering from an almost-fatal wound when he resolved to quit the hospital and the Civil War. He set out for Cold Mountain in the North Carolina highlands, bound for home and sweetheart. The way back was perilous, and horrific battlefield memories haunted him. At his lowest moments he would pull "the Bartram" from his pack. Reading the enraptured word-pictures of the landscape and plants of the Blue Ridge Mountains, the young man could see the ridges and coves and watercourses of his native land, and he gained both solace and strength.

Charles Frazier's *Cold Mountain* is fiction; William Bartram's *Travels* is not. Yet both were inspired by the mystical splendor of the Blue Ridge Mountains. Describing his botanical ramble through the region in the spring of 1775, Bartram wrings the dictionary dry singing the praises of the land and its "mountain vegetable beauties." The rich flora of the Blue Ridge includes ten different species of phlox, but none makes a bolder show than Carolina phlox in summer, when its flowering transforms certain mountain meadows into "Pink Beds."

Phlox carolina Linnaeus, Sp. Pl., ed. 2, 1: 216. 1762 [of the British colony of Carolina] Upstanding perennial herb from a thick short rhizome which from irregularly spaced nodes sends up a few flowering shoots (20–)40–100(–200) cm tall with (6–) 15–25 leaf nodes below the inflorescence, sometimes accompanied by erect non-flowering shoots; *leaves* bright green, thickish, usually linear below and widening up the stem (except subsp. *angusta*) to oblong-lanceolate or broadly elliptic, glabrous to somewhat pubescent, max. length 50–100(–180) mm and width 10–45 mm (3–10 mm for subsp. *angusta*); *inflorescence* complex-cymose to subcylindical panicle, (15-) 30- to 60- (100-) flowered, its herbage glabrous to somewhat pubescent, max. pedicel length (2–)3–6(–10) mm; *calyx* subcylindrical, (6–)7–9(–11) mm long, united ½–¾ its length, lobes narrow-triangular or broadly subulate with moderate to strongly pronounced midrib and with apex sharp-cuspidate to subaristate with an awn 0.5–1 mm long, membranes narrow, shallow, and thinnish, tending to become plicate; *corolla* tube (15–)16–24(–26) mm long, lobes obovate (dimensions varying from 7 × 5 mm to 12 × 10 mm), hue purple to pink, infrequently lilac, rarely white, the eye sometimes paled and bearing a purple stripe at the base of each corolla lobe; *stamens* with 1 or 2 anthers sometimes exserted; *style* 16–24 mm long (10–18 mm long for subsp. *angusta*), united to tip which is free for 1 mm, with stigma sometimes exserted; *flowering season* early June and July, into autumn.

$\infty\infty$

Taxonomic notes

Carolina phlox is allied with the eastern upstanding long-styled phloxes. Wherry clarified relationships within this group through extensive field and herbarium studies in the 1920s and 1930s, by which he observed wild colonies and herbarium specimens of eastern long-styled phloxes that fell outside the suite of characters that defined the relatively well-delineated species of *Phlox glaberrima*, *P. maculata*, and *P. ovata*. Noting these "aberrant" individuals resembled one another and also corresponded to the type specimen of Linnaeus's (1762) *P. carolina*, Wherry (1932a, 1945) reinstated the name *P. carolina*.

Within this group, Carolina phlox and *Phlox glaberrima* have proven the most difficult to distinguish from each other and some authorities (Ferguson 1998; Gleason and Cronquist 1991) have treated them as a single natural entity or a single polymorphic species. Wherry (1932a) separated the two by characteristics of the calyx (see key to eastern upstanding long-styled phloxes in chapter 3). In addition, there appears to be phenological and ecological differentiation, with Carolina phlox coming into bloom later and flowering longer than *P. glaberrima* and with Carolina phlox typically associated with acidic soils and *P. glaberrima* with calcareous soils. Certainly the prairie-associated expression of *P. glaberrima* (subsp. *interior*) is distinct morphologically, geographically, and ecologically. Carolina phlox and *P. glaberrima* are treated here as separate species.

Wherry (1955) recognized *Phlox carolina* subsp. *alta*, subsp. *angusta*, and subsp.

turritella, which he separated from the nominate expression (subsp. *carolina*) by differences in stature, leaf shape, the number of flowers in and geometry of the inflorescence, and habitat tolerances. Subspecies *angusta* is the most strongly differentiated from the nominate expression, and characterizes *P. carolina* in the western portion of the species range.

Key to the Subspecies of *Phlox carolina*

1a. Upper internodes elongate, exceeding length of leaves; leaves mostly linear throughout with upper ones often markedly reduced, width 3–10 mm . . . .
. . . . **subsp. *angusta*** Wherry, Gen. Phlox 108. 1955 and in Baileya 4: 98. 1956 |
narrowleaf Carolina phlox

1b. Upper internodes short, exceeded by length of leaves; leaves linear below and widening up the stem to oblong-lanceolate or broadly elliptic, width 10–45 mm
. . . . **subsp. *carolina*** | Carolina phlox

Geography

Southeastern United States in the southern Appalachian Highlands and adjacent Atlantic Coastal Plain, west along the Gulf Coastal Plain to E Texas, extending north locally onto the Interior Plains and Interior Highlands.

Environment

Coastal plains, hilly piedmont, and mountains on gently rolling terrain and in mountain valleys, river valleys, floodplain terraces, and mountain bogs; *habitat* mesic to wet-mesic, often seasonally moist. Wherry (1932a) noted "extraordinary variability in habit and foliage" of Carolina phlox, with gradations toward *Phlox glaberrima* and *P. ovata*. Such variability is greatest in the southern Appalachian Highlands, where the ranges of these species overlap and where long-term, human-mediated habitat disturbances may have disrupted and/or blurred more precise partitioning of habitat, bringing these species into closer contact with resultant gene flow and possible hybridization and introgression.

Associations

Of all the eastern upstanding long-styled phloxes, the ecology of Carolina phlox is the most difficult to delineate with precision. This has been due in part to the difficulty of distinguishing Carolina phlox from other taller phloxes within its range, particularly *Phlox glaberrima* and *P. ovata*, warranting suspicion of published reports of *P. carolina* in ecological and floristic literature. In addition, Carolina phlox typically occurs in habitat subject to natural or human-mediated disturbance, with the result that habitat descriptions are often vague and imprecise, such as "open woods," "meadows," "clearings," and "roadsides."

Carolina phlox appears to be a successional species, typically associated with intermediate stages of recovery from disturbance (Wherry 1932a; Pittillo 2000b). Occurrences are often reported from the margins of forests, woodlands, and shrub thickets, where these interface with more open, meadow habitat. In the Blue Ridge Mountains of W North Carolina, Carolina phlox is reported from the outer portions of an upland bog and fen community where it borders on climax oak-hickory forest (Tucker 1972). Periodic or chronic natural disturbance may be necessary to maintain the openness of this wetland habitat.

The most unique association reported for Carolina phlox involves the "Pink Beds" of the Blue Ridge Mountains of North Carolina (Transylvania County). So named for large populations of Carolina phlox ("the midsummer pink phlox that blooms here"), the Pink Beds are associated with Southern Appalachian bogs, a mosaic of shrub thickets and open meadow communities (Pittillo 2000a). Carolina phlox does not grow directly in the bogs but in damp meadows at the outer edges. North Carolina botanist Dan Pittillo (pers. comm.) notes the Pink Beds were cleared of timber by early homesteaders, with Carolina phlox appearing in the meadows about fifteen to twenty years following the cessation of grazing by domestic livestock. July is the peak of flowering for Carolina phlox in this habitat.

Cultivation

Carolina phlox is something of an enigma, in the wild and in the garden. In both settings its distinctiveness as a species appears to have been more apparent in the past than it is today. What does seem clear is that its genes were in the bloodlines of some of the earliest phlox selections and hybrids, and are probably circulating yet today in many of our early summer (early June into July) flowering border phlox cultivars. Cultivars of unknown parentage but suspected of being selections of *Phlox carolina* or of its hybrids with *P. maculata* should be identified with the *Phlox* Suffruticosa Group (see Appendix B). Carolina phlox can be used in herbaceous perennial gardens and borders and in mixed borders. Grow in full sun to partial shade in deep, loamy, evenly moist, well-drained soil. Carolina phlox is somewhat susceptible to foliar diseases (see *P. paniculata* species account for management practices).

Phlox caryophylla
Pagosa phlox

Like the flags of conquering empires, a succession of names has flown over the Rio Chama valley of northern New Mexico. The languages differ profoundly—Tewa, Navajo, Spanish, probably others long forgotten—but not the meaning, always conveying the same distinctive character of the landscape. *Tierra Amarilla* eventually gained the day and persists on modern maps, communicating what other names also sought to express—*yellow earth*.

The exposed and weathered shales of the Mancos Formation give this country its color, and influence its flora. A handful of plants are specialties of the harsh, clayey soils derived from the Mancos, including Pagosa phlox. This shrubby species is limited to the Rio Chama valley and the vicinity of Pagosa Springs, Colorado, mostly in open pinyon-juniper woodland. Its botanical name reflects the spicy, carnation-like scent of its flowers, rendered elegantly by Edgar Wherry in the language of science—*Flores odore caryophylli fragrantes*.

Phlox caryophylla Wherry, Notul. Nat. Acad. Nat. Sci. Philadelphia 146: 4, figure 3. 1944 [with the fragrance of carnation, *Dianthus caryophyllus* (Caryophyllaceae) "Its attractive, carnation-like fragrance . . . suggests the epithet here used for the species"] Upstanding suffrutescent perennial with erect-ascending flowering shoots 10–20 cm tall with ca. 5 leaf nodes below the inflorescence, pubescence wholly glandless; *leaves* narrow-elliptic to oblong or linear, acute to acuminate, the upper sparsely ciliate and pilose, max. length 30–50 mm and width 2–4 mm; *inflorescence* 3- to 12-flowered, its herbage pubescent with glandless hairs, max. pedicel length 12–25 mm; *calyx* 11–15 mm long, united ½–⅝ its length, lobes linear-subulate with apex cuspidate, membranes flat; *corolla* tube 15–17 mm long, lobes obovate (average dimensions 8.5 × 6 mm) with apex obtuse, entire, erose, or often emarginate, hue purple or pink, with little eye development; *stamens* with some anthers exserted; *style* 8–15 mm long, united to tip which is free for 1–1.5 mm; *flowering season* late spring, mid-May to June, with buds formed in May and flowering peaking in early to mid-June.

Geography
Colorado Plateau, limited to a small area at the interface of the eastern edge of the plateau (the Navajo section) with the western edge of the San Juan Mountains of the Southern Rocky Mountains in SW Colorado and NC New Mexico. Field

studies in the 1980s reported 23 known populations of Pagosa phlox occurring in two distinct areas, one within about 15 air miles (24 km) of Pagosa Springs (Archuleta County), Colorado, the other a band 7 miles (11 km) wide and 22 miles (35 km) long between the towns of Chama and Nutrias in New Mexico (Rio Arriba County). Colorado occurrences are centered in the middle valley of the Rio San Juan, while New Mexico occurrences are dispersed along drainages of the Rio Chama.

Environment

Mountains at middle elevations, on benches, flats, and slopes, 6500–7800 ft. (2000–2400 m); *MAP* 22 in. (56 cm); *habitat* dry to dry-mesic; *soils* fine-textured clay loam, moderately deep; *parent material* Mancos and Lewis shale Cretaceous marine formations. The distribution of Pagosa phlox corresponds to that of surface exposures of the Mancos shale formation in the region. Several other plant species are local endemics of Mancos shale.

Associations

Primarily pinyon-juniper woodland codominated by two-needle pinyon pine (*Pinus edulis*) and one-seed juniper (*Juniperus monosperma*). Pagosa phlox occurs to a lesser extent in ponderosa pine (*P. ponderosa*) woodland, with Gambel oak (*Quercus gambelii*) associated, and in meadowlike openings comprised of mountain big sagebrush (*Artemisia tridentata* subsp. *vaseyana*) in association with graminoids including Indian mountain-ricegrass (*Achnatherum hymenoides*), blue grama (*Bouteloua gracilis*), needle-and-thread (*Hesperostipa comata*), and muttongrass (*Poa fendleriana*). Pagosa phlox is reported to "thrive in communities at mild disclimax" and appears to tolerate some disturbance, as along roadsides. See Knight and O'Kane (1986) and O'Kane (1988) for valuable information on the ecology and conservation of Pagosa phlox.

Cultivation

Pagosa phlox has not been reported in cultivation. It is considered a species of conservation concern in Colorado and New Mexico because of its relative rarity and collecting seed or plants from wild populations is discouraged except for authorized conservation purposes.

Phlox cluteana
Navajo phlox

To the Navajo eye, all of *Dine Bikeyah* pulses with personality and power. Yet it is in the mountains of their desert homeland where they discern deepest being. The most revered of these is Navajo Mountain on the Arizona-Utah border. Solitary, massive, rounded of form, it is regarded as the head of a gigantic female earth figure—*Pollen Mountain*. Her male counterpart in the expansive Navajo geo-imagination is the Chuska Range, a lanky, 90-mile-long chain of mountains striding to the southeast along the Arizona–New Mexico border.

Sundered by leagues of desert scrub, Navajo Mountain and the Chuskas must settle for botanical tokens of union—crowns of evergreen forest and colonies of Navajo phlox. Known only from these and a scattering of other montane places across northern Arizona, Navajo phlox thrives in the cool shade and loamy duff of the forest floor. I was fortunate to see this rare plant in the Lukachukai Mountains of the Chuska Range, from which I also first glimpsed ethereal Shiprock, icon of the American Southwest, medicine pouch of the Chuska-man.

Phlox cluteana A. Nelson, Amer. Bot. 28: 24. 1922 [after Willard Nelson Clute (1869–1950), collector of the type] Rosulate perennial herb forming colonies from slender modified stems (rhizomes) that spread laterally just below the surface of the ground and are tipped by rosettes of short persistent leaves, these scattered rosettes elongating in spring of the following year into erect flowering shoots 15–20 cm tall with ca. 5 leaf nodes below the inflorescence; *leaves* narrow-elliptic to oblong, acutish, the upper sparse-ciliate and pilose, max. length 35–50 mm and width 3–5 mm; *inflorescence* 6- to 12-flowered, rather compact, its herbage glandular-pubescent, max. pedicel length 8–18 mm; *calyx* 8–9 mm long, united ½–⅝ its length, lobes subulate with apex cuspidate, membranes flat; *corolla* tube 15–18 mm long, glabrous, lobes obovate (average dimensions 10 × 7 mm) with apex obtuse and entire or emarginate, hue purple with barely developed pale eye or striae; *stamens* borne mid- to upper corolla tube, with some anthers slightly exserted; *style* 12–18 mm long, united to tip which is free for 1 mm, with stigma placed among the anthers at or slightly exserted beyond corolla orifice; *flowering season* early summer, June–July.

Geography
Colorado Plateau. Navajo phlox is most abundant in the Navajo section (NE Arizona, NW New Mexico, and SE Utah) of the Colorado Plateau, where it was first

collected on Navajo Mountain, an isolated, dome-shaped laccolith on the border of Utah and Arizona. It is also reported from the vicinity in Long Canyon in SE Utah and from Keet Seel, Long, and Waterlily canyons in NE Arizona. The southeastern limit of its distribution is the Chuska Range, a northwest-to-southeast trending chain in NE Arizona and NW New Mexico comprised of the Carrizo, Lukachukai, and Chuska mountains. The range of Navajo phlox extends west to the Grand Canyon section of the Colorado Plateau, with occurrences on the Shivwits Plateau and the Black Rock Mountains on the north side of the Grand Canyon in NW Arizona. The common name commemorates the discovery of this species on Navajo Mountain, and its occurrence in the homeland of the Navajo Nation.

Environment

Mountains at middle to upper elevations, on moderate to steep slopes, often with a northern aspect; 4800–9000 ft. (1500–2700 m), up to 10,000 ft. (3000 m) on Navajo Mountain; *MAP* 9 in. (23 cm); *habitat* dry to dry-mesic; *soils* sandy to stony, shallow; *parent material* sandstone of the Dakota, Morrison, Entrada, and Chuska formations in the eastern part of its range, basalt in NW Arizona.

Associations

Primarily conifer forest and woodland. Navajo phlox can be locally abundant and sometimes occurs in large, conspicuous populations, with stands covering several square meters in area. In the eastern part of its range on Navajo Mountain and in the Chuska Mountains, Navajo phlox occurs in montane forest dominated by stands of ponderosa pine (*Pinus ponderosa*) and Douglas-fir (*Pseudotsuga menziesii*), with quaking aspen (*Populus tremuloides*) and white fir (*Abies concolor*) in association at higher elevations. These forested mountains are surrounded by desert grassland and Great Basin desert scrub vegetation. Navajo phlox occurs in light to heavy shade beneath the trees, often in a dense duff of pine needles. Associated herbs include New Mexico groundsel (*Packera neomexicana*) and lanceleaf stonecrop (*Sedum lanceolatum*). Navajo phlox descends to lower elevations on these mountains into pinyon-juniper woodland dominated by two-needle pinyon pine (*Pinus edulis*) and one-seed juniper (*Juniperus monosperma*), and in montane scrub dominated by Gambel oak (*Quercus gambelii*) and Utah serviceberry (*Amelanchier utahensis*). Navajo phlox occurs in grassy openings in ponderosa pine woodland in the western part of its range, also descending into pinyon-juniper woodland codominated by single-leaf pinyon pine (*P. monophylla*) and Utah juniper (*J. osteosperma*) and into sagebrush-steppe dominated by big sagebrush (*Artemisia triden-*

tata) in association with grama grasses (*Bouteloua* spp.) and James' galleta (*Pleura-phis jamesii*). On Black Rock Mountain in NW Arizona, Navajo phlox occurs in basalt-derived soils in association with Black Rock Townsend-daisy (*Townsendia smithii*), a rare local endemic. See Clute (1919) and Phillips et al. (1982) for valuable information on the ecology and conservation of Navajo phlox.

Cultivation

Navajo phlox has not been reported in cultivation. It is considered a species of conservation concern in New Mexico and Utah because of its relative rarity and collecting seed or plants from wild populations is discouraged except for authorized conservation purposes.

Phlox colubrina
Hells Canyon phlox
PLATE 15

"Such wild confusion as to benumb the imagination," wrote Ira Gabrielson of the Hells Canyon region of Idaho and Oregon. This fierce gorge, hewn by the Snake River and its tributaries through thousands of feet of basalt, is deeper than any other on the North American continent. The canyon's tumultuous topography and varied microclimates support a unique flora with at least thirty endemic species. Hells Canyon phlox is one of them, stippling the gray-green, bunchgrass-knit ramparts with dashes of pink in spring.

While limited to a relatively small geographic area, Hells Canyon phlox is far from rare in the region. In fact, it is common enough throughout the canyon that it is the principal indicator species of several unique, local plant communities. With stems, leaves, and flower stalks so thin and wiry as to be invisible, the festooning Hells Canyon phlox all but disappears into the grassy matrix, its presence perceived only as sprays of flowers, floating down slope on air.

❧❧

Phlox colubrina Wherry & Constance, Am. Midl. Nat. 19: 433. 1938 [of the canyon of the Snake River; "from one of the Latin terms for snake, in reference to its region of occurrence"] Upstanding suffrutescent perennial 20–40 cm tall with spreading or decumbent stems giving rise to slender ascending flowering shoots 10–25 cm long with ca. 5 leaf nodes below the inflorescence, the habit more lax and sprawling when growing in among shrubs, with stems up to 50 cm long; *leaves* linear to narrowly lanceolate, thickish, long-acuminate, glabrous, max. length 40–80 mm and width

1–2.5 mm; *inflorescence* 2- to 12-flowered, its herbage glabrous except for pilosity inside the calyx lobes, max. pedicel length 18–50(–70) mm; *calyx* 8–12 mm long, united ½–⅝ its length, lobes subulate with moderately pronounced midrib and with apex cuspidate, membranes carinate; *corolla* tube 9–15 mm long, lobes variable but mostly elliptic-oblanceolate and much longer than wide (average dimensions 15 × 6 mm) with apex acutish to obtusish and apiculate, hue pink to white, sometimes with a pale eye and 1 or 2 deep-hued striae; *stamens* borne on mid corolla tube, with anthers included; *style* 1.5–4 mm long, united ca. ½ its length, with stigma deeply included and placed below the anthers; *flowering season* spring, April–June.

Geography

Columbia Plateau, endemic to the Blue Mountain section where it is further limited to the vicinity of Hells Canyon (WC Idaho, NE Oregon, and SE Washington). Hells Canyon phlox is a common forb in canyon grasslands and is the most common *Phlox* of the Hells Canyon region. The Hells Canyon region includes Hells Canyon of the Snake River and the canyons of associated tributaries, most notably the lower portions of the Salmon, Imnaha, and Grande Rhonde rivers. Hells Canyon, also referred to as the Snake River Canyon and the "Grand Canyon" of the Snake River, is the deepest gorge on the North American continent and one of the deepest on earth. Hells Canyon phlox also occurs in the mountains of the Seven Devils complex, an uplift between the canyons of the Salmon and Snake rivers in Idaho County, Idaho.

Environment

Canyons of deeply dissected plateaus, also mountains, on steep, talus-covered slopes, 800–4000 ft. (1200–5900 m); *MAP* 13–18 in. (33–45 cm); *habitat* primarily dry, sometimes xeric; *soils* stony, with bedrock at or near the surface; *parent material* basalt, rarely limestone. The landscape of the Hells Canyon region is influenced by the characteristics of Columbia River Basalt (primarily Imnaha and Yakima flows), the dominant surface rock of the region. Weathering of the basalt follows a vertical cleavage pattern that produces a series of vertical cliffs, each with a steep and often long talus slope at its base. Hells Canyon phlox occurs on slopes ranging from 7–100 percent, with 43–70 percent typical. The complex topography, large and steep relief, and varied edaphic conditions result in a complex pattern of microclimates that supports a diverse flora with at least thirty endemic plants. Hells Canyon phlox occurs in association with an isolated outcrop of limestone (Limekiln Formation) at the extreme northern end of Hells Canyon, with the local endemic Asotin milkvetch (*Astragalus asotinensis*) (Bjork and Fishbein 2006).

Associations

Primarily grassland. Hells Canyon phlox occurs in an Idaho fescue (*Festuca idahoensis*) / prairie Junegrass (*Koeleria macrantha*) community type, an Idaho fescue / bluebunch wheatgrass (*Pseudoroegneria spicata*) community type, and a bluebunch wheatgrass / curly bluegrass (*Poa secunda*) community type. These communities have been referred to collectively as "Canyon Grasslands," a relatively small but distinctive vegetation region of the Pacific Northwest (Tisdale 1986). Hells Canyon phlox (under the name "Snake River phlox") is a principal indicator species of a number of plant associations within the canyon, including an Idaho fescue / bluebunch wheatgrass / Snake River phlox association and a bluebunch wheatgrass / curly bluegrass / Snake River phlox association (Johnson and Simon 1987).

On dry basaltic ledges, Hells Canyon phlox occurs in shrub-dominated communities characterized by open stands of Cusick's serviceberry (*Amelanchier alnifolia* var. *cusickii*), curl-leaf mountain-mahogany (*Cercocarpus ledifolius*), Nevada greasebush (*Glossopetalon spinescens* var. *aridum*), smooth sumac (*Rhus glabra*), and desert gooseberry (*Ribes velutinum*). Hells Canyon phlox has a tendency to grow within shrubs such as sumac, where it assumes a more lax habit in the protection and partial shading provided by the branches. Shrubby forms of netleaf hackberry (*Celtis laevigata* var. *reticulata*) are also common in these communities, and prickly-pear (*Opuntia polyacantha*) is often associated. Hells Canyon phlox occurs in open woodland dominated by netleaf hackberry and, to a lesser extent, ponderosa pine (*Pinus ponderosa*). These shrubland and woodland communities have a well-developed herbaceous layer comprised of grasses and forbs typical of the grassland communities found in the canyon.

The forbs arrow-leaf balsam-root (*Balsamorhiza sagittata*), shaggy fleabane (*Erigeron pumilus*), and Pursh's silky lupine (*Lupinus sericeus*) are often associated with Hells Canyon phlox, as are a number of regional endemics including cross-haired rockcress (*Arabis crucisetosa*), Cusick's milkvetch (*Astragalus cusickii*), Snake Canyon milkvetch (*Astragalus vallaris*), Davis' fleabane (*Erigeron engelmannii* var. *davisii*), Hazel's prickly-phlox (*Leptodactylon pungens* subsp. *hazeliae*), Snake Canyon desert-parsley (*Lomatium serpentinum*), Snake Canyon nemophila (*Nemophila kirtleyi*), whorled penstemon (*Penstemon triphyllus*), Snake Canyon squaw current (*Ribes cereum* var. *colubrinum*), Barton's blackberry (*Rubus bartonianus*), large-flowered tonella (*Tonella floribunda*), and the endangered MacFarlane's four-o'clock (*Mirabilis macfarlanei*). A number of these endemics are plants of conservation concern.

Cultivation

Hells Canyon phlox has not been reported in cultivation, although seed has been available from commercial sources. Its unique festooning habit would make it a good plant for rockeries and rock walls where its large pink flowers could be shown to full effect. Grow in full sun to light shade in dry, well-drained soil.

Phlox condensata
bristlecone phlox
PLATES 16–20

The line between life and death cuts through the *kampfzone*. This lofty biological divide, the "zone of struggle" to European ecologists, is where tree-life staggers, drops to its knees, and gives up the fight. The twisted timber of bristlecone pine marks this invisible verge on high mountains ranges in the western United States. Hammered on the anvil of the alpine, the contorted frame of a bristlecone pine speaks of monumental yet incremental wrestlings. The cost of living here is plain to see but hard to fathom, and you avert your eyes from the travail.

The rarified world of this high altitude pine is also home to bristlecone phlox, a cushion-forming plant that is its frequent companion in the Southern Rockies and on a scattering of ranges across the Great Basin. Often associated with limestone or dolomite rock, bristlecone phlox grows in the most xeric of alpine environments. Its most hostile habitat is found in California's White Mountains, where it grows at the feet of 4000-year-old bristlecone pines, the oldest living things on earth, and on dolomite barrens of magnificent desolation.

∽∂∾

Phlox condensata (A. Gray) E. E. Nelson. BASIONYM: *P. caespitosa* Nuttall var. *condensata* A. Gray, Proc. Amer. Acad. Arts 8: 254. 1870 [condensed, of the growth habit; "pulvinato-caespitosa"] Caespitose perennial herb forming dense cushions 2–7.5 cm tall from a short-branched caudex, the numerous flowering shoots 1–3 cm long; *leaves* linear-subulate (subsp. *condensata*) to narrow-elliptic and concave with margin markedly thickened (subsp. *covillei*), ciliate with coarse hairs, surficially pilose to rarely glabrate, max. length 5–10 mm and width 0.75–2 mm; *inflorescence* 1- or 2-flowered, its herbage pubescent, max. pedicel length 0.5–3 mm; *calyx* 4–7.5 mm long, united ca. ⅜–⅝ its length, lobes subulate with prominent midrib and with apex cuspidate, membranes flat; *corolla* tube 6–10 mm long, lobes obovate (dimensions ranging from 3 × 1.5 mm to 5 × 3 mm), hue white (subsp. *condensata*) or lavender to white (subsp. *covillei*); *stamens* borne on upper corolla tube, with anthers included; *style* 1.5–3 mm long, united to tip which is free for 0.5 mm, with stigma placed below the anthers; *flowering season* alpine spring and summer, July–August.

∽∾

Taxonomic notes

Elias Nelson (1899) described *Phlox covillei* from type material collected in the White Mountains of E California. Wherry (1955) recognized this at the species rank, but Cronquist (1959) incorporated it into *P. condensata*. Wherry (1969) disagreed, distinguishing the two by the morphological attributes detailed in the key below, but most subsequent workers have followed Cronquist (1959, 1984). Given morphological differences and complete geographic separation, Locklear (2009) recognized them as subspecies under the first published name.

Key to the Subspecies of *Phlox condensata*

1a. Plant comprised of discrete divergent shoots; pubescence rather sparse, leaves linear-subulate, ciliate; inflorescence-herbage glandular-pubescent; Southern Rocky Mountains . . . . **subsp. *condensata*** | bristlecone phlox
1b. Plant comprised of interlacing shoots; pubescence rather copious; leaves narrow-elliptic, copiously coarse-ciliate, the margin markedly thickened; inflorescence-herbage densely pubescent with coarse hairs, only some gland-tipped; Sierra Nevada and mountains of the Great Basin . . . . **subsp. *covillei*** (E. E. Nelson) Locklear, J. Bot. Res. Inst. Texas 3: 646. 2009 | California bristlecone phlox

Geography

The distribution of the bristlecone phlox species complex is bi-centric, with the eastern subsp. *condensata* occurring in the Southern Rocky Mountains and the western subsp. *covillei* occurring in the Sierra Nevada and on isolated mountain ranges in the Great Basin. This distribution approximates and parallels that of the bristlecone pine species complex (hence the common name), with Colorado bristlecone pine (*Pinus aristata*) in the Southern Rocky Mountains and Intermountain bristlecone pine (*P. longaeva*) in the Great Basin.

Subspecies *condensata* occurs in many mountain ranges in Colorado—Front Range, Mosquito Range, Sawatch Range, Elk Mountains, Cochetopa Range, San Juan/La Plata Mountains, and Sangre de Cristos Mountains. The northern limit of its distribution occurs in the area between James Peak in Middle Park region and Longs Peak at the southern end of Rocky Mountain National Park, which also corresponds to the northern limit of bristlecone pine in Colorado. At the southern limits of its range in N New Mexico, it occurs in the Sangre de Cristos Mountains on Pecos Baldy, Truchas Peak, and Wheeler Peak.

Subspecies *covillei* occurs on a number of mountain ranges in the Great Basin, including the White and Inyo Mountains of E California and the Monitor Mountains, Ruby Mountains, Snake Range, Toiyabe Mountains, Toquima Range, and Charleston Mountains (Spring Mountains) of Nevada. An isolated station is reported in association with hot springs near Bridgeport, Mono County, California. It also occurs in the Sierra Nevada of California, mostly toward the southern end of the mountain range on the drier east slope and along the Sierra Crest. The northernmost occurrence (originally described as *Phlox dejecta*) is on the summit of Mount Rose in W Nevada, a high spur of the Sierra Nevada influenced by the dry atmospheric conditions of the Great Basin. The southernmost occurrence is in the San Bernardino Mountains of S California.

Environment

The general environment of bristlecone phlox is that of mountains at high elevations, on exposed slopes and ridges subject to desiccating winds and intense solar radiation, associated with rock ledges and pavements, fell-fields, talus, and scree slopes. Soils are minimally developed, stony, with little organic matter, shallow with bedrock at or near the surface, deeper and better developed on moderate slopes and more protected sites. Parent material is typically calcareous, including limestone, dolomite, travertine, marble, or other carbonate rock formations. Subspecies *condensata* occurs at elevations of 11,000–14,000 ft. (3400–4300 m) in the Rocky Mountains; subsp. *covillei* occurs at elevations from 6100–9200 ft. (1800–2800 m) in the Charleston Mountains (Nevada), 8400–13,500 ft. (2600–4100 m) in the White Mountains (California), to 8500–13,000 ft. (2600–3900 m) in the Sierra Nevada. Precipitation mostly occurs as snow in winter, with high winds limiting amount and duration of accumulation.

Bristlecone phlox (both subspecies) is associated with the most xeric alpine environments in North America. Of the eight plant associations described by Kiener (1967) from the alpine zone of Longs Peak in the Southern Rocky Mountains, subsp. *condensata* was limited to the three most xeric. In the White Mountains, subsp. *covillei* is restricted to dolomite barrens nearly devoid of vegetation due to the combined effects of dolomite substrate, thin, dry air, and intense solar radiation, which is greater in intensity for this latitude than anywhere else on earth. A study of alpine plant physiology and distribution in relation to rock types in the Sierra Nevada found subsp. *covillei* to be most abundant on outcrops of marble, which had higher calcium and magnesium concentrations and correspondingly higher pH than other rock types (Wenk and Dawson 2007).

Associations

Primarily subalpine conifer woodland and dry alpine communities. Subalpine woodland occurs at the upper limits of tree growth and is comprised of open stands of conifers, the trees typically stunted in size and deformed by wind into krummholz. Alpine plant associations occur above treeline and are comprised of dwarf shrubs, graminoids, and perennial herbs. Bristlecone phlox is typically one of the most abundant herbs present in these communities and in some associations is the dominant forb and is recognized as a principal indicator species.

The eastern expression of bristlecone phlox (subsp. *condensata*) occurs in subalpine woodland dominated by Colorado bristlecone pine (*Pinus aristata*), with Engelmann spruce (*Picea engelmannii*) and limber pine (*Pinus flexilis*) often associated. Above treeline, it occurs in alpine turf and fell-fields communities. Turf communities are dominated by dwarf, fibrous-rooted graminoids, with kobresia (*Kobresia myosuroides*), rock sedge (*Carex rupestris*), and timberline bluegrass (*Poa glauca* subsp. *rupicola*) being characteristic species. Fell-field communities have less graminoid cover, and are comprised primarily of cushion-forming plants. In all of these communities, bristlecone phlox is associated with a characteristic assemblage of low-growing forbs that includes arctic forget-me-not (*Eritrichium nanum*), old-man-of-the-mountain (*Tetraneuris grandiflora*), alpine stitchwort (*Minuartia obtusiloba*), alpine oreoxis (*Oreoxis alpina*), Gray's point-vetch (*Oxytropis podocarpa*), Rocky Mountain nailwort (*Paronychia pulvinata*), moss campion (*Silene acaulis*), whip-root clover (*Trifolium dasyphyllum*), and dwarf clover (*T. nanum*).

The western expression of bristlecone phlox (subsp. *covillei*) is relatively constant and even abundant in the sparse herbaceous layer of subalpine woodland in the White Mountains and other high elevation ranges of the Great Basin where Intermountain bristlecone pine (*Pinus longaeva*) and/or limber pine are the dominant species. In the San Bernardino Mountains of California, bristlecone phlox occurs in subalpine woodland comprised of Sierran lodgepole pine (*P. contorta* var. *murrayana*) and limber pine. In the Charleston Mountains of Nevada, it occurs in openings in high elevation white fir (*Abies concolor*) forests. It sometimes descends into lower elevation woodlands, particularly on carbonate substrates. It occurs in pinyon-juniper woodland in the White Mountains and in the Charleston Mountains, comprised of single-leaf pinyon pine (*P. monophylla*) and Utah juniper (*Juniperus osteosperma*). At its lowest elevation occurrences in the Great Basin, it is associated with curl-leaf mountain-mahogany (*Cercocarpus ledifolius*) and mountain big sagebrush (*Artemisia tridentata* subsp. *vaseyana*) shrublands.

Bristlecone phlox (subsp. *covillei*) is an important component of alpine plant communities in the High Sierra Nevada, particularly on dry, rocky, exposed sites on the east slope and along the Sierra crest. Here it occurs in some of the most xeric of Sierran alpine communities, most notably bunchgrass-cushion plant communities where it is a codominant along with the grass bottlebrush squirrel-tail (*Elymus elymoides*). Typical associates include dwarf alpine Indian paintbrush (*Castilleja nana*), pygmy fleabane (*Erigeron pygmaeus*), Sierran beardtongue (*Penstemon heterodoxus*), Sierra podistera (*Podistera nevadensis*), shining goldenweed (*Pyrrocoma apargioides*), and narrow false oats (*Trisetum spicatum*) with various other species depending on whether the soil parent material is granite and metamorphics or marble.

Above treeline in the White Mountains, and on certain mountain ranges in the Great Basin, bristlecone phlox (subsp. *covillei*) is associated with subalpine and alpine barrens all but devoid of vegetation. Characteristic herbaceous associates include Indian mountain-ricegrass (*Achnatherum hymenoides*) and timberline bluegrass, along with forbs including spiny milkvetch (*Astragalus kentrophyta*), dwarf alpine Indian paintbrush, shining goldenweed, White Mountain wild buckwheat (*Eriogonum gracilipes*), and prairie flax (*Linum lewisii*). Certain of the isolated, high elevation mountain ranges of the Great Basin possess suites of alpine endemics of which bristlecone phlox is often an associate, particularly ranges with significant exposures of limestone or dolomite.

Cultivation

Bristlecone phlox has been cultivated to a limited extent as a rock garden plant. Its diminutive size and the exacting care required to mimic its alpine cultural requirements would limit its appeal to only the most serious of collectors. Grow in full sun in dry, well-drained soil.

Phlox cuspidata
Navasota phlox

The Navasota River rolls through prime scissortail real estate. This undulating East Texas landscape presents an orchardlike, prairie-laced-with-trees mosaic that scissortail flycatchers find irresistible, luring these birds back each spring from their wintering grounds in South America. The trees, mostly post oak, are spreading and broad-crowned and possess the friendly open character of a life lived with plenty of elbowroom. It is pastoral, welcoming country, made even more gracious by these beautiful birds with their long streaming tails.

The savanna habitat fancied by scissortails also favors Navasota phlox, an annual that is starting to bloom as these birds return in March. Similar in appearance to the much better-known Drummond's phlox, Navasota phlox occurs in heavier soils. While the heart of its distribution is in the post oak woodlands of East Texas, Navasota phlox ranges northward through Blackland prairies and Cross Timbers woodlands into Oklahoma, and southward to the Gulf Coast, where it sometimes occurs on the mysterious mima mounds of remnant coastal prairies.

❧❦

Phlox cuspidata Scheele, Linnaea 23: 139. 1850 [having a cusp or stiff sharp point on the tip of the leaf; "Folia . . . acuminato-cuspidata"] Annual herb with erect stem, simple to sparingly branched upward, occasionally branched at base, plant varying in stature and extent of branching in relation to environmental conditions, 5–25 cm tall, moderately pubescent with the upper hairs gland-tipped; *leaves* opposite at 4–6 nodes, alternate above, the lower oblanceolate and mucronate, the upper linear and cuspidate, max. length 15–35 mm and width 3–6 mm; *inflorescence* a compound cyme of several 3- or 4-flowered cymules, its herbage glandular-pubescent, max. pedicel length 3–6 mm; *calyx* 7–10 mm long, united ⅜–⁷⁄₁₆ its length, lobes linear-subulate the apex aristate with an awn 1–1.5 mm long, membranes flat; *corolla* tube 8–13 mm long, glandular-pubescent, lobes oblanceolate (dimensions varying from 6 × 3 mm to 11 × 8 mm) with apex obtusish to apiculate, hue purple to lilac, often with paled eye, with weak red or purple striae; *style* 1.5–2.5 mm long, united ¼–½ its length, included; *flowering season* spring, March–May.

❧❦

Taxonomic notes

Whitehouse (1945) described *Phlox cuspidata* var. *grandiflora* and var. *humilis*, which she separated from the nominate expression (var. *typica*) by differences in stature and size of floral parts. Erbe and Turner (1962) considered var. *cuspidata* (syn. var. *typica*) and var. *humilis* "well-marked forms," noting only one or the other is found in a given occurrence. Wherry recognized no variants of *P. cuspidata* in his 1955 monograph, but provided a key to Whitehouse's three varieties in his treatment for the *Flora of Texas* (1966), noting they freely intergrade. Turner (1998a) considered these "but forms of a single variable species, the variability compounded by the occasional hybrid and/or backcrosses with *Phlox drummondii*." While the taxon *humilis* appears to be rather strongly delineated, as presently understood the morphological variation represented by these varieties does not show clear correlation with geographic/ecological segregation.

Geography

West Gulf Coastal Plain, chiefly in EC Texas from near the Gulf Coast north to the Red River Valley (S Oklahoma and N Texas), extending into SW Louisiana. Northern populations were classified by Whitehouse (1945) as var. *humilis*. The common name reflects the prevalence of this species in the watershed of the Navasota River in EC Texas.

Environment

Coastal plains on upper slopes of gently rolling to hilly terrain, from near sea level to 750 ft. (229 m); *MAP* 30–55 in. (76–140 cm); *habitat* mesic; *soils* sandy clay loam to clay loam. Navasota phlox is most often associated with *Alfisols*, soils with very slowly permeable, clayey subsoil, with a loam or clay-loam surface horizon. While not as dry as the sandy soils that support *P. drummondii*, Alfisols have lower clay content than soils that support typical Blackland tallgrass prairie in Texas. Alfisol soils often feature small semi-circular mounds called mima or pimple mounds that add some topographic relief to the generally subdued landscape. Erbe and Turner (1962) described soils supporting var. *humilis* in coastal areas as "generally gray, poorly drained, slightly acid and of low organic content," while Whitehouse (1945) portrayed the habitat of this variety as "black waxy prairies."

Associations

Primarily woodland and savanna dominated by post oak (*Quercus stellata*). Whitehouse (1945) described Navasota phlox as occurring "along borders of post oak woods." These open woodlands occur in mosaic with adjacent tallgrass prairie communities, with the herbaceous layer comprised of prairie grasses and forbs. The post oak woodlands of EC Texas differ floristically from that of the Cross Timbers region of NC Texas, occurring on soils with somewhat higher clay content, often underlain by a claypan subsoil, while the Cross Timbers develop on coarse, sandy soils. Navasota phlox can be locally abundant and often occurs in large, conspicuous populations that consist of thousands of individuals.

Near the Texas Gulf Coast Navasota phlox occurs in Upper Coastal prairie, an expression of tallgrass prairie dominated by seaside bluestem (*Schizachyrium scoparium* subsp. *littoralis*), brown-seed paspalum (*Paspalum plicatulum*), and yellow Indiangrass (*Sorghastrum nutans*) with a significant component of annual forbs. Navasota phlox is sometimes associated with mima mounds within these communities, and occurs with Texas prairie dawn (*Hymenoxys texana*), an imperiled annual species, in such habitat near Houston (Butler 1979). The type locality of Navasota phlox—"Pine Island" in Waller County, Texas—is located in what historically was Upper Coastal prairie, the name referring to an isolated stand of pines growing within the expanse of prairie.

North of the Gulf Coastal Plain, Navasota phlox occurs in Blackland prairie that develops on upland habitat with Alfisol soils and comprised of species similar to that of Upper Coastal prairie. Navasota phlox is also reported from the Red River Valley along the Texas-Oklahoma border, where a band of sandy soils supports unique vegetation, including a rare grassland community dominated by silveus' dropseed (*Sporobolus silveanus*) and Mead's sedge (*Carex meadii*). This grassland, with a composition not comparable to any other tallgrass prairie community in North America, occurs over Alfisols and features mima mounds similar to those found on the Gulf coastal plain.

Cultivation

Wherry (1935b) dismissed Navasota phlox as "a plant of no particular horticultural interest." With smallish flowers of undistinguished pink, Navasota phlox can't hold a candle to the flashy, red-flowered forms of *Phlox drummondii*, its immensely popular cousin. However, given the preference of Navasota phlox for heavier soils, a breeding program involving crosses between the two annuals could result in horticultural hybrids adapted to a wider range of soil types.

Phlox diffusa
Cascade phlox
PLATES 21–23

Any summit of any substance in the Pacific Northwest will be adorned with Cascade phlox. From the Bitterroots of Montana to the maritime ranges of British Columbia, down the Cascades and through the Sierras, this is the most common mountain phlox west of the Continental Divide. While it sometimes ventures into the alpine, the mat-forming Cascade phlox flourishes in subalpine meadows and at timberline. In these sky-swept heights, poet Sandra McPherson bears witness, "It lies flat on its back and looks at stars. . . . A throw pillow bolted to granite."

It is hard to overstate the eminence of Cascade phlox throughout its range, where it often characterizes and sometimes carpets the subalpine: it is the most common phlox of California's Sierra Nevada; Mount Rainier, Mount Shasta, and the other great slumbering volcanoes each have their displays; it is so abundant in the Olympics of Washington it shows up in mountain goat droppings. The success of Cascade phlox seems due in part to its ability to colonize disturbed places, from marmot mounds to the post-eruption rubble of Mount Saint Helens.

ᗢᗡᗢ

Phlox diffusa Bentham, Pl. Hartw. 325. 1849 [spreading, of the branches; "ramosissima, diffusa"] Subshrub, freely and diffusely branched from a woody base, ranging in habit from the nominate expression (subsp. *diffusa*) which forms open tufts 5–15 cm tall with internodes typically apparent, to more strongly condensed cushion-forming expressions 2.5–7.5 cm tall (subsp. *longistylis*); *leaves* subulate, sparsely ciliate, short-pilose with glandless hairs to glabrate or glabrous, max. length 8–15 mm (to 20 mm on vigorous nonflowering shoots) and width 1–2.5 mm; *inflorescence* 1- to 3-flowered, its herbage glabrate to pubescent with copious to sparse glandless hairs, max. pedicel length 0.5–7.5(–12) mm; *calyx* 5.5–11 mm long, united 7⁄16–5⁄8 its length, lobes linear-subulate with moderately strong midrib and with apex cuspidate, membranes flat; *corolla* tube 9–14 mm long, lobes obovate to elliptic (dimensions varying from 6 × 4 mm to 7 × 5 mm) or rarely orbicular and somewhat larger, hue purple (occasionally deep purple in local populations), lilac, pink, lavender, or white; *stamens* borne on upper corolla tube, with anthers mostly included; *style* 3–7 mm long (6–11 mm in subsp. *longistylis*), united to tip which is free for ca. 1 mm, with stigma included and placed below the anthers; *flowering season* spring and summer, May–August.

ᗢᗡᗢ

Taxonomic notes

Wherry (1955) recognized *Phlox diffusa* subsp. *diffusa* plus subsp. *longistylis*, subsp. *scleranthifolia*, and subsp. *subcarinata*. Subspecies *diffusa* and subsp. *longistylis* are the most strongly differentiated. The nominate subsp. *diffusa* is the most widespread expression, with a distribution centered in the Sierra Nevada of California, extending north into the Klamath Mountains region of N California and S Oregon. Subspecies *longistylis* is of smaller stature, has longer styles (7–10 mm long versus 3–6 mm for other expressions), and has a distribution to the northwest centered in the Northern and Middle Cascade Mountains of Washington. Cronquist (1959) called subsp. *longistylis* "well-marked," noting it "might with almost equal propriety be treated at the specific level."

Wherry's other subspecies are more problematic. Subspecies *subcarinata* is distinguished by a more compact growth habit and a calyx similar to the carinate (raised and keel-like) calyx-membranes of *Phlox austromontana*. Wherry considered it transitional between *P. diffusa* and *P. austromontana*, but Cronquist (1959) reduced it to synonymy under the latter. Wherry considered subsp. *scleranthifolia* a narrow-leaved, lax-growing expression of *P. diffusa*, but Cronquist considered it a "subglabrate form" of *P. hoodii*. Wherry (1962, 1965b, 1969) disagreed with Cronquist's treatment of these two subspecies, and offered reasoning for allying them with *P. diffusa*. As presently understood, the morphological variation represented by these subspecies does not show clear correlation with geographic/ecological segregation.

Geography

Chiefly Pacific Mountain System, centered along the Cascade-Sierra axis from British Columbia south through California. Cascade phlox occurs in all of the mountain ranges along the Pacific Coast, from the Maritime Range on Vancouver Island (British Columbia), south through the Olympic Mountains (NW Washington), the Oregon Coast Ranges, and the North Coast Ranges of California. It ranges farther inland in the Northern, Middle, and Southern Cascade Mountains (British Columbia south into California), the Klamath Mountains (NW California and SW Oregon), the Sierra Nevada (California), and the Transverse Ranges of S California. Outside of this core range, Cascade phlox occurs in the Northern Rocky Mountains (S British Columbia and N Idaho), ranging as far east as the western slope of the Bitterroot Mountains in W Montana, and on scattered mountain ranges in the northern Great Basin.

Cascade phlox occurs on all of the principal Cascade volcanoes, including (from north to south) Mount Garibaldi in British Columbia; Mount Baker, Glacier Peak, Mount Rainier, Mount Saint Helens, and Mount Adams in Washington; Mount Hood, Mount Jefferson, and Mount Mazama (Crater Lake) in Oregon; and Mount Shasta and Lassen Peak in California.

Outside this well-established range of Cascade phlox are reports of more distant occurrences that require further investigation to confirm taxonomic identity. At least some reports of Cascade phlox from the northern Great Basin in Oregon are *Phlox douglasii*, a similar-appearing species of more xeric habitat. Wherry's (1955) mapping of highly disjunct occurrences of *P. diffusa* subsp. *scleranthifolia* in the Black Hills of SW South Dakota and Pine Ridge of NW Nebraska may represent a local expression or hybrid of Hood's phlox (*P. hoodii*). Cascade phlox has been reported from the Kaibab Plateau of N Arizona (Wherry 1955), but this most likely represents *P. austromontana* (Wilken and Porter 2005).

Environment

Mountains at middle to high elevations, on slopes, ridges, cliffs, rock ledges, fell-fields, talus, and scree slopes, ranging from near sea level on the Pacific coast of Oregon up to 11,800 ft. (3600 m) in the Sierra Nevada of California; *MAP* varied throughout species range, precipitation at high elevations mostly as snow in winter, with high winds limiting amount and duration of accumulation; *habitat* primarily dry to dry-mesic, sometimes xeric; *soils* medium- to coarse-textured, stony, shallow with bedrock at or near the surface, sometimes subject to soil frost activity; *parent material* granite, basalt, serpentine, unstable pyroclastic (volcanic) materials.

Cascade phlox is tolerant of various kinds of natural habitat disturbance. In the North Cascades and Olympic Mountains, it is associated with vegetation stripes, long parallel strips of vegetation comprised of cushion plants that develop in response to downslope creep of the unstable rock mantle. In the relatively dry northeast sector of the Olympic Mountains, Cascade phlox occurs in association with patterned ground features termed "turf-banked terraces," where it is one of the dominant species on two of the three distinctive plant communities that occur on the terraces (Hansen-Bristow and Price 1985). In the Northern Rocky Mountains in Idaho, Cascade phlox occurs in avalanche chutes, high altitude areas where snow accumulation is great and the topography steep enough that snow-slides and avalanches are common, limiting the growth of trees and shrubs and creating an open area populated by cushion plants (Moseley 1990). In a dramatic display of tenacity, Cascade phlox was one of the first vascular plants to re-establish on Mount Saint Helens and nearby mountains following the eruptions of this volcano in 1980 (del Moral 1983; Lupp 1999).

Associations

Conifer forest and woodland, subalpine meadow and parkland, and alpine communities. Cascade phlox is typically one of the most abundant herbs present in these communities and in some associations is the dominant forb and is recognized as a principal indicator species.

Throughout most of its distribution, Cascade phlox is associated with conifer-dominated forest and woodland, where it ranges from mid-elevation forest up through subalpine woodland. Within forested systems, Cascade phlox is associated with small patch openings, rocky slopes, ridges, and other dry microhabitats vegetated by shrubs and herbaceous plants. In the North Cascades, Cascade phlox is associated with subalpine woodland dominated by subalpine larch (*Larix lyallii*), subalpine fir (*Abies lasiocarpa*), or whitebark pine (*Pinus albicaulis*). In the Klamath Mountains, Cascade phlox ranges from the mid-elevation California red fir (*Abies magnifica*) zone through the higher mountain hemlock (*Tsuga mertensiana*) and western white pine (*Pinus monticola*) zones, up to the whitebark pine zone on

ridge-crests and summits. On high elevation serpentine outcrops in these mountains, Cascade phlox also occurs in woodland dominated by foxtail pine (*P. balfouriana* subsp. *balfouriana*), a California endemic of limited distribution. Cascade
phlox is associated with several forest and woodland systems in the Sierra Nevada,
ranging from lower montane forests dominated by ponderosa pine (*P. ponderosa*) or
Jeffrey pine (*P. jeffreyi*) through mid-elevations forest of Sierra lodgepole pine (*P.
contorta* var. *murrayana*) up to subalpine woodland dominated by mountain hemlock, western white pine, and whitebark pine. It is also associated with Sierra juniper (*Juniperus occidentalis* subsp. *australis*) on exposed granite outcrops and with
foxtail pine (*P. balfouriana* subsp. *austrina*) in the High Sierra. Cascade phlox is
most common on the west slope (Pacific side) of the Sierra Nevada, while the
similar-appearing desert phlox (*Phlox austromontana*) ascends to relatively high
elevations on the more xeric, rain-shadowed east slope. In the Northern Rocky
Mountains, Cascade phlox occurs from the lodgepole pine (*Pinus contorta* var.
latifolia) forest zone up into subalpine woodland comprised of subalpine larch or
whitebark pine.

In the Klamath Mountains, Cascade phlox occurs in montane chaparral communities associated with exposures of serpentine rock and dominated by huckleberry oak (*Quercus vacciniifolia*), green-leaf manzanita (*Arctostaphylos patula*), and
small-leaved cream-bush (*Holodiscus microphyllus*).

Cascade phlox frequently occurs in subalpine meadows and grasslands dominated by dwarf, fibrous-rooted graminoids, particularly sedges. Where moisture
gradients or mosaics are present in these systems, Cascade phlox is associated with
drier portion of these communities. It is a very common component of subalpine
meadows in the Olympic Mountains, and is sometimes the dominant species present. In the western Cascade Range of Oregon, Cascade phlox occurs in montane
meadows dominated by Idaho fescue (*Festuca idahoensis*) and Pennsylvania sedge
(*Carex pensylvanica*). It also is important in subalpine meadows on Mount Rainier
and in the Sierra Nevada. Cascade phlox is a major species in a unique subalpine
fell-field community on Mount Pinos, part of the Transverse Ranges of S California (Gibson et al. 2008).

Cascade phlox occurs in a variety of dry to dry-mesic alpine systems throughout its range, including dwarf-shrubland, turf communities, herb-fields, and fell-
fields. These communities are sparsely vegetated, mostly by cushion plants, with
herb-fields having a higher percentage of plant cover (greater than 50 percent)
than fell-fields. In the upper subalpine and lower alpine zones of the North Cascades, Cascade phlox is associated with meadowlike communities dominated by
dwarf shrubs of the heath family (Ericaceae), primarily western bell-heather (*Cas-

siope mertensiana), pink mountain-heath (*Phyllodoce empetriformis*), and Rainier blueberry (*Vaccinium deliciosum*). Cascade phlox is also a component of dry graminoid turf communities dominated by northwestern showy sedge (*Carex spectabilis*) or mountain hare sedge (*C. phaeocephala*).

Cascade phlox occurs to a limited extent in the alpine zone of the Sierra Nevada, particularly in less xeric habitat in the northern and central High Sierras. Near treeline in the Sierra Nevada, Cascade phlox occurs in xeric, dwarf-shrubland dominated by Rothrock's artemisia (*Artemisia rothrockii*). Cascade phlox is also reported in isolated occurrences of alpine habitat in the Warner Mountains of NE California and the San Gabriel Mountains of SW California.

Cascade phlox occurs in sparsely vegetated communities associated with rock outcroppings and unstable pyroclastic material (cinders, pumice, tephra). Chappell (2006) identified two associations from mid-elevation bluffs and balds in the southern Cascades of Washington and the Olympics in which Cascade phlox is the dominant species. In Yosemite National Park in the Sierra Nevada, Cascade phlox is most common on rocky outcrops and open ridges within the mixed conifer and montane forest zones.

Cultivation

Cascade phlox has proven to be one of the more adaptable of the western cushion-forming species. The species itself plus some exciting color selections have become more available as cultivars in recent years through specialty nurseries in the Pacific Northwest. Grow in full sun in dry, well-drained soil.

Phlox dispersa
High Sierra phlox
PLATES 24, 25

If John Muir was right, and California's Sierra Nevada is composed, not of rock, but of light, then High Sierra phlox is rooted in photons. Known only from the neighborhood of Mount Whitney, the highest peak in the Lower Forty-Eight, this rare phlox grows in raw gravels essentially unaltered but for size from the Sierra's granitic core. Fine and loose, with almost no organic matter, the gravelly sands that accumulate in the alpine basins and canyons of the High Sierra Nevada present a very dry substrate that supports a very meager flora.

Unlike other alpine phloxes that form dense, sometimes-sizable cushions, High Sierra phlox grows as small, scattered tufts connected by slender stolons that run through the minuscule rubble. This unique, flexible, dispersed growth habit is an elegant solution to an extreme environment. The beauty of it dawned on me when I noticed several tufts of High Sierra phlox growing about the perimeter of a rock and discovered they were but one plant, connected by hidden stolons and finding comfort in the thin cache of moisture clinging to the rock's surface.

Phlox dispersa Sharsmith, Aliso 4: 128. 1958 [scattered, of the diffuse caudex giving rise to subterranean stolons that terminate in "small scattered caespitose tufts"] Caespitose perennial herb arising from a diffuse caudex, the branches of caudex numerous, developed as very slender rhizomes 8–18 cm long, uniformly 1–2 mm thick, terminating in densely caespitose, small leafy tufts 1.5–3 cm tall and 3–8 cm wide; *leaves* narrowly lanceolate to linear, 4–11 (averaging ca. 6) mm long, 1.0–1.5 mm wide, subappressed to somewhat spreading, firm, pungent, dark or dull green, densely glandular-puberulent, the margins scarcely thickened, mostly nonciliate, rarely ciliate with a few simple hairs; *inflorescence* 1-flowered; *calyx* 7–8.5 mm long, lobes firm, midrib apparent, pungent, densely glandular-puberulent, membranes flat; *corolla* tube 9–12 mm long with puberulent zone within just above base, lobes obovate, 6.0–6.5 mm long, subentire to entire, hue white, with throat sometimes purplish; *style* (including style branches) 1.5–3.0 mm long; *flowering season* alpine summer, late June to August.

Geography

Pacific Mountain System, endemic to the Sierra Nevada of California where it is limited to the Southern High Sierras (Fresno, Inyo, and Tulare counties) along the crest and eastern slope in the general vicinity of Mount Whitney, the second highest peak in the United States after Denali of Alaska. High Sierra phlox has been collected on the Boreal Plateau, Cirque Peak, Mount Guyot, Mount Langley, Mount Whitney, and Olancha Peak, among others, with most known populations occurring in Kings Canyon National Park and Sequoia National Park. The distribution of High Sierra phlox approximates that of the Sierra Nevada subspecies of foxtail pine (*Pinus balfouriana* subsp. *austrina*). Of all Sierran conifers, foxtail pine is most strongly associated with the xeric environment in which High Sierra phlox occurs.

Environment

Mountains at high elevations, on high plateaus and in glaciated valleys, 11,000–13,900 ft. (3400–4200 m); *MAP* may be less than 1 in. (5 cm) due to the rain shadow effect caused by western peaks of the Sierra Nevada, precipitation mostly as snow in winter with high winds and intense solar radiation limiting amount and duration of accumulation; *habitat* xeric; *soils* not developed, restricted to fine loose gravel, scree, and coarse sands that accumulate on gentle to moderate slopes and drain rapidly, presenting a very dry rooting medium, churned by needle ice in summer, possibly limiting tree growth and more dense herbaceous vegetation; *parent material* granite.

The unique growth habit of High Sierra phlox is apparently suited to this specialized substrate. Sharsmith (1958) noted:

> On level sites in gravel fields a given individual may occupy a roughly circular area of two feet, a fact not at all evident in the undisturbed plant. Only careful digging discloses the small scattered caespitose tufts as interconnected by elongate and very slender subterranean stolons.

The small tufts of High Sierra phlox sometimes occur at intervals around the perimeter of a rock, possibly taking advantage of the accumulation and retention of moisture and organic matter at the edge of the rock.

Associations

Upper edge of subalpine woodland comprised primarily of open stands of foxtail pine or Sierran lodgepole pine (*Pinus cortorta* var. *murrayana*). High Sierra phlox also occurs in very sparsely-vegetated subalpine meadows comprised of tufted, cushion-forming, and rosette-forming herbaceous plants. These unique meadows, found only in the extreme southern High Sierra, occur in areas of granitic sand and are situated between woodland and the more dense herbaceous vegetation typical of subalpine meadows (Benedict 1983; Benedict and Major 1982). These areas are often surrounded by mature, almost pure stands of foxtail pine. The herbaceous layer of the woodlands is very sparse, and above treeline the density of herbaceous plants in the meadows is considerably less than one plant per square meter. Herbaceous associates of High Sierra phlox are few, and include Mt. Hood pussy-paws (*Cistanthe umbellata*), hoary wild buckwheat (*Eriogonum incanum*), Lemmon's whitlow-grass (*Draba lemmonii*), field ivesia (*Ivesia campestris*), alpine bitterroot (*Lewisia pygmaea*), Clement's mountain-parsley (*Oreonana clementis*), creeping beardtongue (*Penstemon davidsonii*), Watson's spikemoss (*Selaginella watsonii*), and narrow false oats (*Trisetum spicatum*).

Cultivation

High Sierra phlox has not been reported in cultivation. It is considered a species of conservation concern in California because of its relative rarity and collecting seed or plants from wild populations is discouraged except for authorized conservation purposes.

Phlox divaricata
timber phlox

It is April in the Virginia woods, and a young woman is afoot along the banks of Tinker Creek. Springtime wildflowers are stirring, and it registers with her that "phlox is at its peak," but she gives it only passing notice. This day her lively, strange curiosity is arrested by the sight of a freshly unfurled leaf atop a tuliptree sapling. Contemplating it conjures up apprehension of the primal might seething beneath the kindly demeanor of trees, and leads her to a grand, green epiphany— "I suspect that the real moral thinkers end up, wherever they may start, in botany."

Annie Dillard should not be faulted for looking past timber phlox, one of the most familiar of all American woodland wildflowers. Timber phlox will be encountered on almost any springtime ramble in almost any woods in the eastern half of the United States, from the rich cove forests of Appalachia to barely-wooded prairie draws in Kansas. Depending on the light, the bluish color of its flowers, cool and refreshing, can also lend to its anonymity, making individuals and even colonies hard to discern from patches and poolings of shade on the forest floor.

Phlox divaricata Linnaeus, Sp. Pl. 1: 152. 1753 [strongly divergent, of the branches of the inflorescence; "Rami *alterni, basi divaricati*"] Upstanding perennial herb with both decumbent nonflowering stems that root at the nodes and erect-ascending flowering shoots 18–35(–45) cm tall with 4–7 well-spaced leaf nodes below the inflorescence; *leaves* on nonflowering shoots partly persistent and broadly elliptic, on flowering shoots deciduous and ovate to lanceolate, obtusish, acutish, or the upper acuminate, max. length 23–50(–55) and width 8–25 mm, lower leaves glabrate, upper ones ciliate and sparsely glandular-pubescent to pilose; *inflorescence* 9- to 24-(30-) flowered, at first fairly compact but in age becoming lax, its herbage glandular-pubescent, max. pedicel length 4–18 mm; *calyx* 6–11 mm long, united ⅜–½(–⅝) its length, lobes linear-subulate with apex sharp-cuspidate, membranes flat to subplicate; *corolla* tube 11–17(–20) mm long, glabrous, lobes oblanceolate to obovate (dimensions varying from 13 × 8 mm to 19 × 13 mm), the apex in subsp. *divaricata* usually notched 0.5–2.5(–4) mm deep (but in a few clones in many colonies and numerous clones in exceptional ones obtuse and entire, erose or obliquely apiculate), the tip in subsp. *laphamii* entire to apiculate, hue often described as blue but actually light violet, ranging to lavender to near-white, with deeper hues (violet to purple) in the western subsp. *laphamii*, the eye often paler and may bear faint violet or purple striae, in subsp. *laphamii* at times coalescing into a purple ring; *stamens* with anthers included; *style* 1.5–3 mm long, united ⅜–⅝ its length, with stigma included; *flowering season* spring, beginning in March in the south and April in the north, continuing through May into early June.

Key to the Subspecies of *Phlox divaricata*

1a. Corolla lobes often notched, though in occasional variants entire; average height 28 cm; eastern portion of species range .... **subsp. *divaricata*** | timber phlox
1b. Corolla lobes entire, obtuse, or apiculate, only anomalously notched; average height 35 cm; western portion of species range ....
..... **subsp. *laphamii*** (Alph. Wood) Wherry, Baileya 4: 97. 1956 | midland timber phlox

Geography

Eastern and central United States and extending into southeastern Canada. The nominate expression of timber phlox (subsp. *divaricata*) occurs in the eastern portion of the species range, from the Appalachian Highlands and Interior Low Plateaus through the north-central Interior Plains to the lower Great Lakes region (N Indiana and S Michigan). Subspecies *laphamii* ranges from the Great Lakes region west to the edge of the Great Plains, and south through the Interior Highlands and onto the Gulf Coastal Plain.

Environment

Coastal plains and hills, dissected plains and plateaus, hilly piedmont, and mountains at low to middle elevations, on the lower slopes of rolling terrain, ridges, hills, and ravines, and river valleys, floodplain terraces, stream banks, riverwash, 100–3000 ft. (30–900 m); *habitat* mesic; *soils* alluvial or loamy, humus-rich, moderately deep to sometimes shallow with bedrock at or near the surface. While floodplain habitat occupied by timber phlox is moist fall through spring, it can be very dry during summer. The eastern subsp. *divaricata* is typically associated with humus-rich, circumneutral to subacid soils, while the western subsp. *laphamii* tends to occur in association with calcareous soils, forming in loess, glacial till, or residuum weathered from chert, limestone, or dolomite.

Timber phlox is tolerant of more disturbance than most woodland herbs, including recurrent submergence and deposition of alluvium in floodplain habitat and the flooding regime of gravel bar habitat in the Ozarks and the Potomac River Gorge. The distribution of vegetation of floodplain habitat tends to be stratified according to small differences in elevation, with timber phlox occuring most often on the higher terraces of a floodplain.

Associations

Primarily hardwood forest and woodland, where it is characteristic of climax associations. Timber phlox can be locally abundant and sometimes occurs in large, conspicuous populations, particularly subsp. *laphamii* which is generally more frequent within the core of its range than is subsp. *divaricata*. The occurrence of timber phlox within a particular ecological community reflects differences in elevation, soil texture, light, moisture, and other factors, and is often expressed as an association with a particular aggregation, "union," or "guild" of forest herbs.

The distribution of the eastern expression of timber phlox (subsp. *divaricata*) is centered in the mixed mesophytic forest of the Appalachian Plateau and adjacent regions where sugar maple (*Acer saccharum*), buckeye (*Aesculus* spp.), American beech (*Fagus grandifolia*), tuliptree (*Liriodendron tulipifera*), oaks, and American basswood (*Tilia americana*) are the most important tree species. It also ranges northward into the beech-maple (*F. grandifolia–A. saccharum*) forest of the Great Lakes region.

While typically associated with alluvial soils, timber phlox occurs to some extent in rocky habitat. In southern Ontario, Canada, it occurs in association alvars, plant communities that develop on exposures of horizontal limestone/dolomite bedrock, in a grassy woodland dominated by black maple (*Acer nigrum*) and chinquapin oak (*Quercus muhlenbergii*). Along the gorge of the Potomac River near Washington, D.C., timber phlox occurs in riverwash habitat in herbaceous communities dominated by grasses and forbs typical of the tallgrass prairie.

Subspecies *laphamii* occurs in the relatively drier forest and woodland associations at the western and southern limits of the eastern deciduous forest, most notably the oak-hickory forest of the Ozark Plateau and Midwest, the maple-basswood (*A. saccharum–T. americana*) forest of the upper Midwest, and the oak-pine-hickory forest stretching from the Piedmont Plateau of Georgia onto the upper Coastal Plain from Alabama to E Texas. More so than the eastern subspecies, subsp. *laphamii* is strongly associated with floodplain habitat and at its westernmost occurrences in the tallgrass prairie region is associated with narrow gallery forests of river valleys.

In N Florida and SW Georgia, timber phlox (subsp. *laphamii*) occurs in unique mixed hardwood forests associated with calcareous soils on slopes, ravines, and other dissected terrain, notably the Apalachicola Bluffs and Ravines. This specialized habitat is cooler and more mesic than the surrounding sandy pinelands, and harbors remnants of a more temperate flora. Known locally as "hammocks" or "hardwood hammocks," these forests are dominated by oaks, hickories, American beech, and large-flower magnolia (*Magnolia grandiflora*), along with shrubs and herbs characteristic of Appalachian forests. Timber phlox is also associated with calcareous bluffs in Mississippi.

Throughout its range, timber phlox is a consistent member of the rich spring herb communities that characterize the eastern deciduous forest of North America. Typical associates include small white leek (*Allium tricoccum*), jack-in-the-pulpit (*Arisaema triphyllum*), Canada wild ginger (*Asarum canadensis*), rattlesnake fern (*Botrychium virginianum*), cutleaf toothwort (*Cardamine concatenata*), dwarf larkspur (*Delphinium tricorne*), dutchman's breeches (*Dicentra cucullaria*), white trout-lily (*Erythronium albidum*), wild crane's-bill (*Geranium maculatum*), round-lobe hepatica (*Hepatica nobilis*), John's-cabbage (*Hydrophyllum virginianum*), Virginia bluebells (*Mertensia virginica*), mayapple (*Podophyllum peltatum*), bloodroot (*Sanguinaria canadensis*), numerous trillium species (*Trillium* spp.), and large-flower bellwort (*Uvularia grandiflora*). Timber phlox (subsp. *laphamii*) is one of the last members of this herbaceous flora to persist at the western limits of the eastern deciduous forest, occurring in gallery forests within the context of tallgrass prairie on the eastern edges of Nebraska, South Dakota, and North Dakota.

Cultivation

It speaks volumes that horticulturists in eighteenth century Europe just left timber phlox alone. While they started refining the other phloxes coming in from the American colonies as soon as they got their hands on them—making selections and breeding new hybrids—they found little reason to fiddle with timber phlox. What could be done to improve upon the earthy, cerulean mood it brought to the springtime garden? Today, timber phlox is an essential element of the shade and woodland garden on both sides of the Atlantic, valued for the bluish hues of its fragrant flowers and its early season of bloom. English plantsman B. H. B. Symons-Jeune magically captured the sentiment for timber phlox when he wrote, "as to my mind the tone and make-up suggest coolness, the coolness of bluebells in a beech wood."

The phlox cultivar 'Chattahoochee' is a popular garden subject in North America and Europe, valued for its striking flower color, combining deep lavender corolla-lobes with a corolla-eye of intense red-purple. The original material of 'Chattahoochee' was collected from the wild in NW Florida in the 1940s and appears to be a strain of *Phlox divaricata* subsp. *laphamii*, but at least two different entities have circulated in horticulture under this name, neither of which represents the original. These are treated here as a cultivar group under the name the *Phlox* Chattahoochee Group (see discussion in Appendix B).

Timber phlox can be used as a groundcover, or as a subject for herbaceous perennial gardens and borders, mixed borders, and woodland gardens. Grow in shade to partial shade in evenly moist, well drained, humus-rich soil. The impact (color and fragrance) of timber phlox is most effective when plants are grouped in masses. Colonies can form by the spread and rooting of decumbent stems and by abundant seed production and dissemination, an admirable attribute when a groundcover is desired but problematic in a rock garden. Many cultivars are available. A horticultural hybrid with *Phlox paniculata*, rendered *P.* x *arendsii*, is in wide circulation (see discussion under *P. paniculata*).

Phlox dolichantha
Bear Valley phlox

Like minions of the Mojave Desert, Joshua trees look to be laying siege to California's San Bernardino Mountains. Bristling with leafy bayonets and striking agitated poses, you get the feeling these tree-yuccas are intent on overthrow as you pass through the desert on your way up into the forested mountains. The Mojave does make an island of these mountains, and thirty endemic plants have arisen in the imposed isolation. Bear Valley phlox is one of them, known only from the pine and oak forests of the San Bernardinos.

With thin, stretched-out stems and the longest flowers in the genus, Bear Valley phlox is a wisp of a plant. It reaches almost a foot in height, but its slight build makes it all but disappear in the shade cast by Jeffrey pines, incense cedars, and oaks. These woodlands develop on heavy clay soils that are unique to these mountains and that open into barren, gravelly flats in a number of places. Most of the San Bernardino endemics occur on these "pebble plains," except for the retiring Bear Valley phlox, which hangs back in the shadow and shelter of the forest.

❧ ❧

Phlox dolichantha A. Gray, Proc. Amer. Acad. Arts 22: 310. 1887 [long-flowered; "corollae . . . tubo sesquipollicari calyce triplo longiore"] Upstanding perennial herb with a few slender sparingly branched erect-ascending flowering shoots 10–30 m) tall with ca. 7 leaf nodes below the inflorescence; *leaves* elliptic to lanceolate, acute to acuminate, the upper sparse ciliate, max. length 30–60 mm and width 3–8 mm; *inflorescence* (1-) 3- to 12-flowered, its herbage glandular-pubescent, max. pedicel length 10–25 mm; *calyx* 10–16 mm long, united ½–⅝ its length, lobes subulate with apex sharp-cuspidate, membranes plicate; *corolla* tube 35–45 mm long, lobes narrowly obovate (average dimensions 9 by 6 mm) with apex entire or erose, hue bright pink; *stamens* with anthers somewhat exserted; *style* 30–40 mm long, united to tip which is free for 1.5–2 mm; *flowering season* late spring and early summer, late May into early July.

❦❧

Geography

Pacific Mountain System, endemic to the San Bernardino Mountains of S California, most abundantly in the Big Bear and Holcomb valleys in the northeastern sector of the mountains. The San Bernardino Mountains has one of the highest rates of floristic endemism in continental North America for an area of comparable size, with thirty plants strictly endemic to the range. Twenty of these endemics, including Bear Valley phlox, occur in the Big Bear Valley area.

Environment

Mountains, primarily in mid-elevation draws and slopes, typically with a north aspect, 6600–9600 ft. (2000–2900 m); *MAP* 15–30 in. (38–76 cm) *habitat* dry to dry-mesic; *soils* clay to clay-loam; *parent material* conglomerate sediments.

The Big Bear Valley area of the San Bernardino Mountains is noted for the presence of open, gravelly flats and benches referred to as "pebble plains." These dense clay soils are armored by a pavement of Saragosa quartzite pebbles, pushed to the surface by frost heaving. While devoid of the surrounding forest vegetation, these flats have a distinctive flora of low-growing cushion- and mat-forming herbaceous species, a number of which are endemic to the pebble plains. Bear Valley phlox does not occur on the most exposed and open parts of the pebble plains, but it is found in open forest adjacent to this unique habitat.

Associations

Primarily forest and woodland dominated by Jeffrey pine (*Pinus jeffreyi*), with incense cedar (*Calocedrus decurrens*) often associated. Bear Valley phlox also occurs in drier habitat where Jeffrey pine is accompanied by California black oak (*Quercus kelloggii*), single-leaf pinyon pine (*P. monophylla*), and Sierran juniper (*Juniperus occidentalis* var. *australis*). California black oak becomes dominant in some lower elevation occurrences of Bear Valley phlox. Understory shrubs of curl-leaf mountain-mahogany (*Cercocarpus ledifolius*), Mexican manzanita (*Arctostaphylos pungens*), and mountain whitethorn (*Ceanothus cordulatus*) are often present. The highest elevation occurrences of Bear Valley phlox are found on Sugarloaf Mountain (9600 ft. [2926 m]), in association with forest of Sierran lodgepole pine (*P. contorta* var. *murrayana*) and white fir (*Abies concolor*). Bear Valley phlox occurs with several San Bernardino Mountain endemics, including Parish's rockcress (*Arabis parishii*), Bear Valley sandwort (*Arenaria ursina*), Heckard's Indian paintbrush (*Castilleja montigena*), silver-haired ivesia (*Ivesia argyrocoma*), and San Bernardino butterweed (*Packera bernardina*). It also occurs with the bizarre and showy saprophyte, snow-plant (*Sarcodes sanguinea*) in Jeffrey pine forests in the San Bernardino Mountains. See Krantz (1983, 1990, 1994) for valuable information on the ecology and conservation of Bear Valley phlox.

Cultivation

Bear Valley phlox has not been reported in cultivation. It is considered a species of conservation concern in California because of its relative rarity and collecting seed or plants from wild populations is discouraged except for authorized conservation purposes.

Phlox douglasii
Columbia phlox
PLATES 26–28

Sir William Jackson Hooker knew a good "scientific traveler" when he saw one, and David Douglas proved to be his finest discovery. Hewn from Scottish stock, as a youth Douglas displayed a passion for natural history. His keen interest in plants secured him a series of gardener positions which in 1820 brought him into contact with the eminent English botanist. Three years later, Douglas was on a ship sailing for North America, dispatched by the Royal Horticultural Society of London on the first of several plant collecting expeditions.

The year 1826 found Douglas "botanising" the Columbia Plateau and Blue Mountains region of the Pacific Northwest. Among the scores of new species he discovered was a low, shrubby, needle-leaved beauty which Hooker would later name *Phlox douglasii*. Douglas first encountered Columbia phlox in eastern Washington, where it occurs in bunchgrass steppe and ponderosa pine savanna, a landscape the hardy but homesick Scot called "by far the most beautiful that I have seen . . . extensive plains, with groups of pine-trees, like an English lawn."

Phlox douglasii Hooker, Fl. Bor.-Amer. (Hooker) 2: 73. 1838 [after David Douglas (1799–1834), collector of the type] Subshrub, sparingly and diffusely branched, forming open tufts 10–20 cm tall from a woody base, the numerous erect-ascending annual flowering branches 5–15 cm long with ca. 4 nodes apparent, the habit more strongly condensed in xeric situations with flowering shoots 2–4 cm long; *leaves* subulate, somewhat acerose, thinnish and dark green, sparsely ciliate and surficially pilose to glabrate with fine gland-tipped hairs, max. length 7–12 mm and width 0.75–1.5 mm; *inflorescence* 1- to 3-flowered, its herbage copiously glandular-pubescent, max. pedicel length 0.5–6 mm; *calyx* 7.5–11 mm long, united ½ to ca. ¾ its length, lobes linear-subulate with rather prominent midrib and with apex cuspidate, membranes flat; *corolla* tube 10–14 mm long, glabrous or exceptionally pubescent, lobes obovate (average dimensions 7.5 × 5 mm), hue lavender, pink or white; *style* 4–8 mm long, united to tip which is free for ca. 1 mm; *flowering season* spring to early summer, April–May, into June.

Taxonomic notes

Wherry (1941c, 1955) recognized *Phlox douglasii* as a distinct species, but Cronquist (1959) placed the name into synonymy under *P. caespitosa*. Wherry (1962, 1965a) strongly disagreed, but most subsequent workers have followed Cronquist. Examination of the type material of both *P. douglasii* and *P. caespitosa*, coupled with field study, affirms Wherry's interpretation (Locklear 2009).

Wherry (1955) recognized *Phlox douglasii* subsp. *rigida*, which he separated from the nominate expression (subsp. *douglasii*) by differences in stature and size of floral parts. Subspecies *rigida* appears to be an environmentally reduced expression occurring in more xeric habitat. A tall (10–15 cm) form of Columbia phlox common in the Spokane River valley of E Washington was described as *P. piperi* by Elias Nelson in 1899, but Wherry (1955) considered this a "[l]uxuriant" expression of *P. douglasii*. As presently understood, the morphological variation represented by these subspecies does not show clear correlation with geographic/ecological segregation.

Geography

Chiefly Columbia Plateau, from the Walla Walla Plateau and Blue Mountain sections (W Idaho, E Oregon, and E Washington) south to the Harney section (SC Oregon), extending north into the Okanagan Mountains of the Northern Rocky Mountains (NC Washington) and south into the northern Great Basin (NE California, S Oregon, and NW Nevada).

Environment

Dissected plateaus, hilly piedmont and mountains at low to middle elevations, on upper slopes of rolling terrain and divides, escarpments, bluffs, scablands, and ridges, 1000–7000 ft. (300–2100 m); *MAP* 8–14 in. (20–36 cm); *habitat* xeric to dry; *soils* medium-textured, often stony, shallow with bedrock at or near the surface; *parent material* primarily basalt, but also glacial outwash, pumice and other volcanic substrates, and outwash derived from the volcanic deposits.

While Columbia phlox is typically associated with xeric, rocky habitat with coarse-textured soils, it also occurs in soils with a moderately to strongly structured B horizon that is very close to the soil surface and high in clay content. These properties impede water drainage and result in conditions of poor aeration in the rooting zone during the winter and spring months due to the development of a perched water table above the dense clay horizon. Columbia phlox is commonly associated with dwarf sagebrush (*Artemisia arbuscula*) in such habitat. The "*A. arbuscula* scab flats" of the Modoc Plateau of NE California, where Columbia phlox is a common species, are described as "virtually lakes in the spring when snow melts" due to the clayey soils (Young et al. 1977).

Associations

Conifer woodland, shrubland, grassland, and rock outcrop communities. Columbia phlox is associated with ponderosa pine (*Pinus ponderosa*) woodland and savanna throughout most of the northern portion of its range on the Columbia Plateau. These open stands of ponderosa pine have an herbaceous layer typically dominated by a single species of xerophytic bunchgrass, either bluebunch wheatgrass (*Pseudoroegneria spicata*), Idaho fescue (*Festuca idahoensis*), or needle-and-thread (*Heperostipa comata*), with a rich diversity of forbs. Along with Columbia phlox, typical forbs include arrow-leaf balsam-root (*Balsamorhiza sagittata*), yellow mission-bells (*Fritillaria pudica*), bulbous woodland-star (*Lithophragma glabrum*), Pursh's silky lupine (*Lupinus sericeus*), sagebrush buttercup (*Ranunculus glaberrimus*), and Douglas' blue-eyed-grass (*Olsynium douglasii* var. *inflatum*). These communities often "front" more densely forested areas of higher elevations, as in the Okanogan Highlands of N Washington and Idaho and the Blue Mountains of NE Oregon and adjacent Washington.

In C Oregon and extending into extreme NE California, Columbia phlox occurs in western juniper (*Juniperus occidentalis* var. *occidentalis*) woodland. These open stands of western juniper, the most xeric forested community in the Pacific Northwest, occur with a shrub understory of Wyoming big sagebrush (*Artemisia tridentata* subsp. *wyomingensis*), dwarf sagebrush, or antelope bitterbrush (*Purshia tridentata*), along with an herbaceous layer dominated by bluebunch wheatgrass or Idaho fescue. Columbia phlox is often a prominent forb in the herbaceous layer of juniper woodlands, notably in the Ochoco Mountains of C Oregon and the Warner Mountains of NE California.

Columbia phlox occurs in several types of shrubland and shrub-steppe on the Columbia Plateau and northern Great Basin. It is strongly associated with dwarf sagebrush in stony, shallow-soil habitat. These communities occur in open barrens or scabland habitat, sometimes in a complex mosaic of openings within woodland, shrubland, and grassland communities. In the Hart Mountain National Antelope Refuge in SE Oregon, Columbia phlox occurs in association with extensive stands of dwarf sagebrush on upland flats. It also occurs in shrubland and shrub-steppe dominated by antelope bitterbrush and, less commonly, big sagebrush.

Columbia phlox is a component of grassland communities throughout its range on the Columbia Plateau. It is particularly common in the bunchgrass communities of the watershed of the Spokane River in E Washington and adjacent Idaho dominated by Idaho fescue or bluebunch wheatgrass. David Douglas appeared to be writing of Columbia phlox in his journal while in the vicinity of present-day Spokane, Washington, on 10 May 1826, when he noted a "small beautiful species of *Phlox* which I found some time since on the Columbia gave the whole open places a fine effect."

Columbia phlox is often a dominant element in the flora of xeric rock outcrop habitat in the Pacific Northwest. It is very common along the rim-rock edges of the basalt plateaus that form the spectacular canyons of the Deschutes and Crooked Rivers in NC Oregon. It also is one of the showiest forbs growing on the comparatively recent volcanic substrates in Lava Beds National Monument on the Modoc Plateau in NE California. Columbia phlox is a dominant plant in the sparsely vegetated badland habitat derived from eroded volcanic ash and tuff deposits of the John Day and Clarno formations in C Oregon. In NE Washington, it occurs on rocky mounds and rock stripes in patterned ground habitat.

Cultivation

As Ira Gabrielson observed, "*P. douglasii* is a name covering . . . a multitude of botanical sins." Despite the classification of many beloved cushion phlox cultivars under the label "Douglasii Hybrids," true Columbia phlox has seen only limited horticultural circulation in North America, and there is no evidence that David Douglas himself ever brought seed of it back to England (see discussion of *Phlox* Scotia Alpines Group, Appendix B). This confusion aside, Columbia phlox is a showy species with potential not only as a subject for the rock garden but also as a drought-tolerant landscape plant for drier portions of the Pacific Northwest. Grow in full sun to light shade in dry, well-drained soil.

Phlox drummondii
Drummond's phlox
PLATE 29

The view from atop Enchanted Rock reveals a peculiar side of Texas. Bald masses of pink rock rise, here and there, above oak-heavy woodlands, a scene reminding writer Rick Bass of the capitol domes of an ancient civilization. As a boy, Bass reveled in his visits to Hill Country's "humped chest of granite," drinking in its wildness before his vacationing family drug him back to the suburbs of Houston. This rough-cut region is a world apart from the domestic mildness of the coastal plain, but the two landscapes have at least one thing in common—Drummond's phlox.

Blooming in concert with bluebonnet and paintbrush, Drummond's phlox helps put the splash in Texas springtime. This showy annual occurs in great drifting colonies in the sandy prairies and post oak savannas of East Texas, its true empire. Its variants are also found on Gulf Coast barrier islands, Panhandle dunelands, and in the granitic sands of Bass's boyhood *Walden*—sands which yielded a familiar *crunch* when first his little daughter walked with him here, a childhood echo so potent it made him feel faint, "so light that I thought I might take flight."

Phlox drummondii Hooker, Curtis's Bot. Mag. 62: plate 3441. 1835 [after Thomas Drummond (1780–1835), collector of the type] Annual herb with erect stem, simple to sparingly branched upward, sometimes branched at base with basal branches spreading-ascending to the height of the central stem or decumbent and sprawling (as in subsp. *glabriflora*), plant varying in stature and extent of branching in relation to environmental conditions, the eastern subsp. *drummondii* 10–50 cm tall, the more western subspecies typically under 30 cm tall; *leaves* opposite at 3–5 nodes, alternate above, passing into bracts below the inflorescence, lower leaves oblanceolate, taper-petiolate, upper oblong or essentially so, sessile to subclasping, acute or short-acuminate, subaristate, max. length (20–)30–75 mm and width ranging from 3–9 mm (subsp. *tharpii*) to 10–20 mm (subsp. *drummondii*), leaves in depauperate individuals much reduced; *inflorescence* a helicoid compound cyme of aggregated 3- or 6-flowered cymules, max. pedicel length 5–10(–20) mm; *calyx* 8–12 mm long, united ⅜–½ its length, lobes linear-subulate with apex aristate with an awn 1–1.5 mm long, membranes flat; *corolla* tube (10–)12–17(–20) mm long (18–25 mm long in subsp. *johnstonii*), mostly glandular-pubescent (but glabrous in subsp. *glabriflora*), lobes broadly obovate (dimensions varying markedly but averaging 12 × 9 mm) with apex obtuse, often apiculate, hue purple to lavender to lilac to pink, to intense red to reddish purple in subsp. *drummondii*, eye bearing deep-hued striae coalescing into a conspicuous ring or star (subspecies *drummondii* and *tharpii*), or eye paled (subspecies *glabriflora*, *johnstonii*, and *mcallisteri*); *stamens* with anthers included; *style* 1.5–4 mm long, united ca. ½ its length, with stigma included and placed below the anthers; *flowering season* spring and early summer, late February through May, some subspecies blooming longer in years of abundant moisture.

Taxonomic notes

In her seminal paper on the taxonomy of the annual phloxes, Whitehouse (1945) recognized three varieties of *Phlox drummondii* plus five closely related species—*P. glabriflora*, *P. goldsmithii*, *P. littoralis*, *P. mcallesteri*, and *P. tharpii*. Wherry (1955) took up most of Whitehouse's epithets, but reduced them to subspecies of either *P. drummondii* or *P. glabriflora*. In their paper of the biosystematics of the annual phloxes, Erbe and Turner (1962) reduced each of Whitehouse's species to subspecies or varieties under *P. drummondii*. Wherry (1961) subsequently described another annual phlox, *P. johnstonii*.

After three more decades of field and herbarium study, Turner (1998b) published an updated treatment of the *Phlox drummondii* complex that recognized five naturally occurring varieties (vars. *drummondii*, *johnstonii*, *littoralis*, *mcallesteri*, and *tharpii*). These are recognized here at the rank of subspecies, with subsp. *glabriflora* incorporating var. *littoralis*.

Key to the Subspecies of *Phlox drummondii*

1a. Populations mostly local, highly variable, especially in flower color, corolla hue varying from white to pink, to lavender and crimson; introduced, but persistent, cultivars . . . . **var. *peregrina*** Shinners, Field & Lab.19: 127. 1951
1b. Populations relatively uniform, especially in flower color (exceptional hybrid individuals of *P. drummondii* × *P. cuspidata* excluded) 2

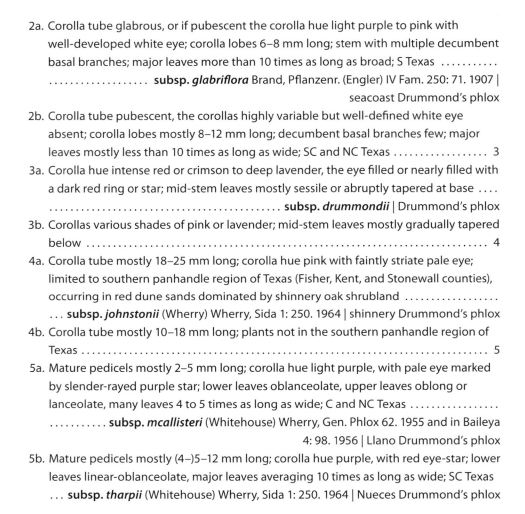

2a. Corolla tube glabrous, or if pubescent the corolla hue light purple to pink with well-developed white eye; corolla lobes 6–8 mm long; stem with multiple decumbent basal branches; major leaves more than 10 times as long as broad; S Texas . . . . **subsp. *glabriflora*** Brand, Pflanzenr. (Engler) IV Fam. 250: 71. 1907 | seacoast Drummond's phlox

2b. Corolla tube pubescent, the corollas highly variable but well-defined white eye absent; corolla lobes mostly 8–12 mm long; decumbent basal branches few; major leaves mostly less than 10 times as long as wide; SC and NC Texas 3

3a. Corolla hue intense red or crimson to deep lavender, the eye filled or nearly filled with a dark red ring or star; mid-stem leaves mostly sessile or abruptly tapered at base . . . . **subsp. *drummondii*** | Drummond's phlox

3b. Corollas various shades of pink or lavender; mid-stem leaves mostly gradually tapered below 4

4a. Corolla tube mostly 18–25 mm long; corolla hue pink with faintly striate pale eye; limited to southern panhandle region of Texas (Fisher, Kent, and Stonewall counties), occurring in red dune sands dominated by shinnery oak shrubland . . . . **subsp. *johnstonii*** (Wherry) Wherry, Sida 1: 250. 1964 | shinnery Drummond's phlox

4b. Corolla tube mostly 10–18 mm long; plants not in the southern panhandle region of Texas 5

5a. Mature pedicels mostly 2–5 mm long; corolla hue light purple, with pale eye marked by slender-rayed purple star; lower leaves oblanceolate, upper leaves oblong or lanceolate, many leaves 4 to 5 times as long as wide; C and NC Texas . . . . **subsp. *mcallisteri*** (Whitehouse) Wherry, Gen. Phlox 62. 1955 and in Baileya 4: 98. 1956 | Llano Drummond's phlox

5b. Mature pedicels mostly (4–)5–12 mm long; corolla hue purple, with red eye-star; lower leaves linear-oblanceolate, major leaves averaging 10 times as long as wide; SC Texas . . . **subsp. *tharpii*** (Whitehouse) Wherry, Sida 1: 250. 1964 | Nueces Drummond's phlox

Geography

The distribution of the Drummond's phlox species complex is concentrated on the West Gulf Coastal Plain. It is nearly endemic to Texas, occurring with greatest abundance in SC Texas, extending into the southern Interior Plains (Central Lowland physiographic province) in NC Texas and southwest into northeastern Mexico. Drummond's phlox and its cultivars have escaped cultivation in parts of North America and on other continents, in some cases becoming a weed and in some areas with mild climates becoming a permanent part of the flora.

Each subspecies has a fairly well-defined geographical range, yet, with the exception of subsp. *johnstonii*, have zones of overlap with at least one other subspecies. The subspecies may have been more isolated in the past, now brought into contact by human-caused disturbances such as forest and brush clearing, overgrazing, and road building. Subspecies *johnstonii* has the smallest and most remote range of the subspecies, being limited to three counties (Fisher, Kent, and Stonewall) in the watersheds of the Salt Fork and Double Mountain Fork of the Brazos River in the Panhandle of NW Texas, about 100 miles (161 km) distant from the main distribution of the complex. Subspecies *johnstonii* is a plant of conservation concern in Texas.

Environment

The general environment of Drummond's phlox is that of upper slopes of gently rolling to hilly terrain, and on coastal dunes, beach sands, and inland sand sheets and dune fields, from sea level to 1000 ft. (300 m); *MAP* 15–40 in. (39–102 cm), driest at the western edge of the distribution of the species complex, where subspp. *johnstonii* and *tharpii* occur; *soils* range from pure sand to sandy loam; *parent material* granite, sandstone.

Whitehouse (1945) concluded the distribution of each "species" (syn. subspecies) of Drummond's phlox "is apparently limited by the particular formation on which it occurs," adding, "variations in this group of phlox have largely resulted from changes in the soil." Comparative analysis of soils from nine different occurrences of Drummond's phlox showed significant differences in textural composition, percent organic matter, total available nitrogen, phosphorus, potassium, calcium, and magnesium content, but these factors appear to act in concert with a complex of environmental variables in affecting morphological differentiation between populations of Drummond's phlox (Schwaegerle and Bazzaz 1987).

The nominate expression of Drummond's phlox (subsp. *drummondii*) occurs in sandylands in soils derived from sandstone formations including the Carrizo and Wilcox formations. The distribution of the Carrizo Sandstone Formation in particular appears to define the distribution of this subspecies. The Carrizo Sandstone and the "Carrizo Sands" derived from it form a narrow (6–12 miles [10–20 km] wide) diagonal band extending roughly from Carrizo Springs northeast about 450 miles (724 km) to Texarkana (McBryde 1933).

Subspecies *glabriflora* is associated with dunes and beach sands along the southern part of the Texas Gulf Coast, and on the partially stabilized, semi-halophytic foredunes of barrier islands including Mustang and Padre Islands. This subspecies also occurs on the South Texas Sand Sheet (Coastal Sand Plain) in Kenedy and Brooks counties, a wind-deposited plain where extremely pure, fine sand at the surface is susceptible to blowing into dunes. Dunes are typically stabilized to partially stabilized by vegetation.

Subspecies *johnstonii* occurs in deep, shifting sand dunes on the Red Bed Plains of the southern Texas Panhandle. This habitat is distinctive from that of all other subspecies for its combination of very low rainfall and short growing season.

Subspecies *mcallisteri* occurs along streams in gravel wash and shallow sandy soils (Trinity Formation) derived from granite outcrops unique to the Llano Uplift region of Texas. This subspecies also occurs to the north in the Lampasas Cut Plain, Comanche Plateau, and Palo Pinto regions of NC Texas along streams in coarse, sandy soils derived from various sandstone formations.

Subspecies *tharpii* occurs in deep, loose sandy soils, primarily derived from the Carrizo Sandstone Formation. This subspecies also occurs on the South Texas Plains at the southwestern extremity of the Carrizo sands, in the driest part of the 450-mile (724 km) extent of this formation.

Associations

Woodland, savanna, and grassland, in areas where perennial vegetation cover is somewhat open due to disturbance. Drummond's phlox is absent from the more densely vegetated tallgrass prairies of Texas where perennial grasses and forbs dominate the heavy clay soils. Although Drummond's phlox is an annual, it does not behave as a colonizing or weedy species, but is instead a major component of distinct, late successional annual plant community endemic to sandy soils in C and SC Texas.

The nominate expression of Drummond's phlox (subsp. *drummondii*) is a characteristic component of xeric sandyland communities that occur within the context of the post oak savanna region of Texas. The regional vegetation is dominated by scattered individuals of post oak (*Quercus stellata*) in a matrix of tallgrass prairie dominated by little bluestem (*Schizachyrium scoparium*).

Subspecies *glabriflora* occurs in beach sands and semi-halophytic coastal dune communities dominated by seaoats (*Uniola paniculata*), gulfdune paspalum (*Paspalum monostachyum*), and other grasses, along with the sprawling vines railroad morning-glory (*Ipomoea pes-caprae*) and fiddle-leaf morning-glory (*I. stolonifera*). This subspecies occurs with a wealth of other spring-blooming annual forbs. It also is a significant component of upland prairie and savanna communities of the "Wild Horse Desert" of the South Texas Sand Sheet, dominated by seacoast bluestem (*Schizachyrium scoparium* subsp. *littoralis*) with scattered groves (*mottes*) of plateau live oak (*Quercus fusiformis*) and honey mesquite (*Prosopis glandulosa*) intermixed. Spring-blooming annual forbs are an important part of the flora of these communities.

Subspecies *johnstonii* is restricted to sand shinnery oak shrubland, a unique plant community dominated by shinnery oak (*Quercus havardii*), a dwarf, clonal shrub. Other important plants include sand sagebrush (*Artemisia filifolia*), grasses little bluestem and sand dropseed (*Sporobolus cryptandrus*), and rich complement of arenaceous forbs.

Subspecies *mcallisteri* occurs in open oak woodland dominated by plateau live oak or post oak, with blackjack oak (*Quercus marilandica*) and black hickory (*Carya texana*) in association. This subspecies is also associated with Cross Timbers vegetation in northern part of its range, a mosaic of forest, woodland, savanna, and prairie vegetation, characterized by scattered post oak and blackjack oak trees in a matrix of tallgrass prairie.

Subspecies *tharpii* occurs at the dry, southwestern limits of the species complex, in post oak woodland that grades into plateau live oak woodland and eventually into honey mesquite savanna. Little bluestem, slender pinweed (*Lechea tenuifolia*), and palmate copperleaf (*Acalypha radians*) are typical herbaceous species, with annual plants characterizing disturbed, early-successional sites. This subspecies appears to have caught the attention of Theodore Roosevelt while on a peccary hunting trip in southern Texas in 1892. Traversing the country between the Frio and Nueces rivers southwest of San Antonio, Roosevelt (1893) "kept catching glimpses through the mesquite trees of lilac stretches which [he] had first thought must be ponds of water. On coming nearer they proved to be acres on acres thickly covered with beautiful lilac-colored flowers." Whitehouse (1936) surmised the soon-to-be president of the United States (1901–1909) was observing subsp. *tharpii*, as his travels would have placed him in the heart of the range of this purple-hued subspecies at the height of its flowering season.

Natural history notes

The ecology of Drummond's phlox is closely tied to that of pocket gophers (genus *Geomys*). These burrowing mammals alter habitat physically by excavating and turning over tremendous volumes of soil and by below- and above-ground feeding activities, causing significant disturbance of perennial vegetation cover and influencing soil nutrient composition. The relationship between pocket gopher activity and the bunchgrass–annual forb community of the Texas coastal prairie vegetation (of which Drummond's phlox is an important component) is so strong that it may be coevolved (Williams et al. 1986).

In a survey of several hundred naturally occurring populations of Drummond's phlox throughout its range, pocket gopher activity was noted at each site (Leverich 1983). Annual plants including Drummond's phlox dominate the vegetation of these sites, accounting for 70 percent of the total flora at one site (Schaal and Leverich 1982). The composition of these communities is apparently regulated by the massive disturbance caused by the activity of pocket gophers, which may be essential for the persistence of Drummond's phlox at some sites. A study of environmental effects on growth and fruit production in Drummond's phlox found gopher disturbances enhanced growth and reproduction in natural stands, apparently by reducing competition from neighboring *Phlox* plants and surrounding vegetation (Schwaegerle and Levin 1990).

Cultivation

Drummond's phlox has a long and remarkable history in horticulture (see chapter 2), and is one of the most widely grown annual bedding plants in the world. Drummond's phlox can be used in flower beds and borders or in containers. Grow in full sun in moist, light, sandy soil. Sow seed of this winter annual directly in the garden in the fall to germinate and produce flowering plants the following spring, or in a greenhouse in early spring to produce transplants for containers or gardens after danger of frost has passed. Many seed-propagated strains are available.

Phlox floridana
Florida phlox

They're still fishing for answers in Apalachicola. How do you explain the rich biological diversity concentrated around the Apalachicola River in northwest Florida? One line of thinking suggests the region was part of an ancient highland— "Orange Island"—where plants and animals found refuge during alternating periods of glacial advance and rising seas. While plant geographers find such unwitnessed, primeval scenarios plausible, everyday Floridian E. E. Callaway had a theory of his own: Apalachicola was once the Garden of Eden.

Whatever was its past, the Apalachicola country of the Florida panhandle and adjacent Alabama and Georgia is today home to a wealth of endemic species, including Florida phlox. A big, strapping cousin of downy phlox, Florida phlox occurs most often in sandhill habitat dominated by open stands of longleaf pine. These sunny savannas, *high pine* to the locals, have an understory of scrub oak and a jubilant, flowery brawl of herbs, into which Florida phlox throws punches of pink and purple throughout much of the growing season.

☙❧

Phlox floridana Bentham, Prodr. (DC.) 9: 304. 1845 [of the region of Florida] Upstanding perennial herb with a few simple erect-ascending flowering shoots 25–50(–80) cm tall, glabrous below the inflorescence, with (8–)10–20(–25) leaf nodes below the inflorescence; *leaves* linear to oblong or lanceolate, acute to short-acuminate, max. length 40–80(–90) mm and width 3–6(–7) mm, the largest at mid-stem and the upper ones successively reduced in size and usually appressed to the stem, all but uppermost glabrous, *inflorescence* (6-) 12- to 24-flowered, compact, conspicuously exceeding the expanse of the upper leaves, its herbage glandular-pubescent, max. pedicel length (3–)4–8(–10) mm; *calyx* 7–11 mm long, united ⅜–½ its length, lobes subulate with apex cuspidate to subaristate, membranes often plicate; *corolla* tube 15–20(–22) mm long, glabrous, lobes obovate (average dimensions 11 × 8 mm) with apex obtuse, hue purple to pink, the eye often pale and bearing at the base of each corolla blade 1, 2, or rarely 3 deep-hued striae; *stamens* with anthers included; *style* 1.5–3 mm long, united ¼–½ its length, with stigma included; *flowering season* summer, primarily May and June, but ranging from March to November; tetraploid, 2*n* = 28 (Levin 1966; Smith and Levin 1967).

Geography
East Gulf Coastal Plain, concentrated in the Apalachicola River Basin (SE Alabama, NW Florida, and SW Georgia), extending south into northern peninsular Florida. Florida phlox also occurs in the Marianna Lowlands and the Tallahassee Red Hills regions of N Florida.

Environment
Coastal plains and hills on upper slopes of rolling to dissected terrain, from near sea level to 350 ft. (100 m); *habitat* dry to dry-mesic; *soils* range from deep, dry sands of sandhills to more mesic clay loams; *parent material* sandy limestone formations in the Marianna Red Lands and Tallahassee Red Hills of N Florida. Wherry (1931b) described the typical habitat of Florida phlox as "chiefly in open oak-pine woods on dry sand-clay of subacid reaction." Florida phlox occurs in natural communities that were maintained historically by low intensity surface fires, and the characteristic species of these communities are adapted to periodic fire.

Associations
Primarily longleaf pine (*Pinus palustris*) woodland. This savanna-like plant community, known colloquially as "high pine," occurs on rolling, deep sands and is comprised of open stands of longleaf pine with an understory of scrub oaks (typically *Quercus laevis*) and an herbaceous layer dominated by wiregrass (*Aristida stricta*) in association with a rich diversity of forbs. On more level, mesic sites,

Florida phlox occurs in association with pine flatwoods, woodlands characterized by a relatively open overstory of pines (typically longleaf pine), an extensive low shrub stratum, and a variable and often sparse herbaceous layer. Florida phlox does not occur in the more xeric sand pine scrub communities of Florida.

Florida phlox also occurs in mixed hardwood forest and woodland on upland sites having soils with somewhat higher clay content. These communities have been described as "mesic hardwood" in Alabama, "hammocks" and "pine-oak-hickory woods" in Florida, and "oak woods" and "oak barrens" in Georgia. Florida phlox occurs in the drier expressions of these communities at their ecotone with longleaf pine woodlands. Evergreen oaks, most notably sand laurel oak (*Quercus hemisphaerica*), live oak (*Q. virginiana*), and laurel-leaf oak (*Q. laurifolia*), dominate these open, dry-mesic to subxeric forests in N Florida. Hickories including sweet pignut hickory (*Carya glabra*) and mockernut hickory (*C. tomentosa*) are also important.

Cultivation

Florida phlox is a graceful, upstanding species with showy clusters of flowers and a long season of bloom. With a natural range limited to the Gulf Coast region, it ought to be a target of southern horticulturists, particularly those seeking plants for drier landscape situations. Mary Gibson Henry discovered a colony of a dwarf, pink-flowered variant of Florida phlox growing near the coast in 1939 which proved entirely cold hardy in her Pennsylvania garden, enduring below zero temperatures (Wherry 1943b). Grow in full sun to partial shade in dry to slightly moist, light, well-drained soil.

Phlox glaberrima
smooth phlox
PLATES 30, 31

A gentle but prolonged summer rain had been falling on central Alabama, and the Little Cahaba River was running brown and edging out of its banks. I had missed the bloom of the celebrated Cahaba lily, grassy plants that lift spidery white flowers above this river's rocky shoals, but I did notice patches of pink bobbing at the water's edge. It was a colony of smooth phlox, the plants up to their elbows in the rise of the Little Cahaba and looking, for all the world, to be enjoying the ride.

Smooth phlox ranges throughout the eastern United States, gracing damp habitat with knee-high clusters of flowers. Like my encounter in Alabama, I have had other sodden but sublime brushes with this plant. One was a wet prairie in

Missouri, wind-churned and flower-bright, the air crystalline with the fluting of upland sandpipers. Another was a floodplain forest in southern Indiana, near the muggy merging of the Wabash with the Ohio, the understory drenched in watery green light and hushed but for warbler-song echoing out of a nearby cypress swamp.

∽∾

Phlox glaberrima Linnaeus, Sp. Pl. 1: 152. 1753 [smooth, of the foliage; "foliis lineari-lanceloatis glabris"] Upstanding perennial herb from a thick short rhizome which from irregularly spaced nodes sends up a few to multiple flowering shoots (30–)50–150 cm tall with 7–15 leaf nodes below the inflorescence, sometimes accompanied by erect nonflowering shoots; *leaves* linear below, widening upward to broad-linear or narrow-lanceolate (to ovate in the Appalachian Highlands), the largest ca. 3 nodes below the inflorescence, glabrous to rarely sparse-pilose, max. length 50–150 mm and width 5–10(–20) mm; *inflorescence* a panicle of small cymes, 15- to ca. 125-flowered, max. pedicel length 3–12(–25) mm; *calyx* subcampanulate, 6–12 mm long, united ½–⅝ its length, lobes broad-subulate to narrowly triangular with prominent midrib and with apex sharp-cuspidate to aristate, membranes firm, broad, flat or somewhat carinate; *corolla* tube (15–)18–24(–27) mm long, lobes obovate to orbicular (dimensions ranging from 7 × 5 mm to 14 × 12 mm), hue light purple to pink or sometimes white, with little eye-development but sometimes bearing a purple stripe at the base of each corolla blade; *stamens* with one anther often exserted; *style* (10–)15–23(–26) mm long, tending to be shortest in subsp. *interior*, united to tip which is free for 0.5–1 mm; *flowering season* late spring and early summer, June–July.

∽∾

Taxonomic notes
See *Phlox carolina* species account for justification of recognition of *P. glaberrima* as a distinct species.

Wherry (1955) recognized *Phlox glaberrima* subsp. *interior* and subsp. *triflora*, which he distinguished from the nominate expression (subsp. *glaberrima*). Delineation of subspecies is difficult in the eastern and central portions of the range of smooth phlox, but subsp. *interior* is distinguished by its western distribution and its strong ecological association with tallgrass prairie. Subspecies *glaberrima* and subsp. *triflora* are combined here under subsp. *glaberrima* as representing the eastern, nominate expression of *P. glaberrima*.

Key to the Subspecies of *Phlox glaberrima*

1a. Nonflowering shoots well-developed to occasional; leaves linear on lower stem, widened up the stem to narrow-lanceolate or ovate; calyx 7–12 mm long; eastern distribution, primarily forested communities.......**subsp. *glaberrima*** | smooth phlox
1b. Nonflowering shoots rare; leaves linear on lower stem, becoming broad-linear to lanceolate upward; calyx 6–7.5 mm long; western distribution, primarily grass- or sedge-dominated communities ....
......**subsp. *interior*** (Wherry) Wherry, Gen. Phlox 111. 1955 and in Baileya 4: 98. 1956 | Wabash smooth phlox

Geography
Eastern and central United States. The nominate expression of smooth phlox (subsp. *glaberrima*) is concentrated in the southern Appalachian Highlands and Interior Low Plateaus, extending locally onto the Atlantic Coastal Plain. Subspecies *interior* extends the range of the smooth phlox species complex from the Interior Low Plateaus north through the Interior Plains to the lower Great Lakes region (N Illinois, NW Indiana, and SE Wisconsin) and west to the Interior Highlands (C Arkansas, S Missouri, and SE Oklahoma).

Environment
Coastal plains and hills, lake plains, dissected plateaus, hilly piedmont, and mountains at low to middle elevations, associated with fens, groundwater seeps, springs, sinkholes, river valleys, floodplain terraces, stream edge, river shoals, and river-wash, 100–3000 ft. (30–900 m); *habitat* mesic to wet; *soils* alluvial or loamy, moderately deep to sometimes shallow with bedrock at or near the surface; *parent material* typically calcareous, limestone or dolomite.

Associations
Forest, grassland, stream edge, and wetland communities. The eastern expression of smooth phlox (subsp. *glaberrima*) occurs primarily in association with floodplain and riparian forest systems. While the species composition of the regional upland forest systems (such as oak-hickory and beech-maple) differs across the range of smooth phlox, the associated floodplain and riparian forests of the eastern United States are often comprised of similar species, including silver maple (*Acer saccharinum*), river birch (*Betula nigra*), sycamore (*Platanus occidentalis*), eastern cottonwood (*Populus deltoides*), black willow (*Salix nigra*), and American elm

(*Ulmus americana*). Smooth phlox also occurs at the lower edges of mesic upland hardwood forests where these intersect floodplain and riparian forest. Sugar maple (*A. saccharum*), tuliptree (*Liriodendron tulipifera*), and American beech (*Fagus grandifolia*) are often components of these forests. Smooth phlox occurs with the federally-endangered Alabama leatherflower (*Clematis socialis*) in wooded floodplain habitat in NE Alabama, and is so closely associated that it is considered an indicator of Alabama leatherflower habitat (Boyd and Hilton 1994).

Smooth phlox occurs in wet-mesic plant communities associated with swamps, sinkhole ponds, fens, and seeps. It typically occurs at the margins of such habitat, where the wetland transitions to more mesic plant communities. Meadow phlox (*Phlox maculata*) may also be present in these communities, but in the wetter portion. Smooth phlox occurs on calcareous fens and seeps on the Highland Rim in Tennessee, where it is an associate of the federally-endangered Tennessee yellow-eyed-grass (*Xyris tennesseensis*) (Sommers 1994). Smooth phlox also occurs in wet-mesic prairie barrens dominated by big bluestem (*Andropogon gerardii*) on the Eastern Highland Rim of Tennessee.

Smooth phlox is associated with a variety of stream edge communities. On the Cumberland Plateau in Kentucky and Tennessee, it occurs on boulder and gravel bars that develop on medium-high gradient streams. Scouring floods in the spring and drought stress on the well-drained, course-textured substrate in summer reduce tree cover, resulting in shrubby and/or prairielike herbaceous vegetation dominated by grasses like big bluestem and little bluestem (*Schizachyrium scoparium*). Smooth phlox is closely associated with the federally endangered Cumberland false rosemary (*Conradina verticillata*) in this type of habitat on the Cumberland Plateau in Tennessee and Kentucky, and is considered an indicator plant for Cumberland rosemary habitat (Shea and Roulston 1996). In C Alabama, smooth phlox occurs with the rare Cahaba lily (*Hymenocallis coronaria*) on rocky shoals and in stream edge habitat. In S Indiana, smooth phlox occurs in gravel wash prairie that harbors the federally endangered Short's goldenrod (*Solidago shortii*) (Homoya and Abrell 2005). Subspecies *interior* occurs in floodplain and stream edge communities, notably in the Ozarks of S Missouri where it occurs on gravel bars in river-wash communities dominated by Ozark witch-hazel (*Hamamelis vernalis*), pale dogwood (*Cornus amomum* subsp. *obliqua*), and other shrubs.

The western expression of smooth phlox (subsp. *interior*) is primarily associated with tallgrass prairie formations. It is an important forb in wet-mesic tallgrass prairies and sedge meadows in N Illinois and adjacent Wisconsin and Indiana, and is an associate of several rare species found in these prairies including eastern prairie white-fringed orchid (*Platanthera leucophaea*). In the same region, it occurs in tallgrass prairie in association with surfacing dolomite bedrock, where it is an associate of leafy prairie-clover (*Dalea foliosa*), a federally endangered species. In SW Missouri, it occurs in wet prairie dominated by prairie cordgrass (*Spartina pectinata*), where it is sometimes associated with mima mounds. Historically, smooth phlox was among the most abundant forbs in the tallgrass prairies of the Chicago region. Clute (1911a, 1911b) described smooth phlox as "the most conspicuous plant in the landscape" of the Chicago lake plains during its flowering season, adding it "spreads away in brilliant masses over meadow and prairie as far as the eye can reach."

Cultivation

Though it has the distinction of being the first species in the genus to be cultivated (see chapter 2), early introductions of smooth phlox from the Atlantic seaboard never generated much enthusiasm as a garden plant. New selections should be made from the showier expressions of the species that occur in the southern Appalachian Highlands and in the tallgrass prairies of the Midwest. The early summer (early June into July) flowering season of smooth phlox can fill a void in the floral progression of the garden, and its propensity for relatively moist situations would make smooth phlox forgiving of typical landscape situations. Smooth phlox can be used in herbaceous perennial gardens and borders, mixed borders, rain gardens, and regional native plant horticulture. Grow in full sun to partial shade in evenly moist, well-drained soil, although smooth phlox is tolerant of seasonally wet conditions and soils with higher clay content. Smooth phlox is attractive to butterflies. Cultivars are available.

Phlox gladiformis
Red Canyon phlox
PLATES 32, 33

Driving into Utah's Red Canyon for the first time is like crash landing onto another planet. You are suddenly surrounded by a dreamland geology of pinnacles, knobs, hoodoos, and battlements, all carved out of a pinkish orange limestone named the Claron Formation. A comfortable mantle of conifers dominates north-facing sides of the canyon, but opposite slopes look to have been landscaped by Dr. Seuss, with contorted ponderosa pines and igloo-shaped manzanita shrubs sprinkled across a desolate carrot-colored backdrop.

In botany there is an unwritten axiom that the stranger the place the more peculiar the plants, and Red Canyon provides classic validation. Nearly a dozen species are endemic to the canyon and its vicinity including Red Canyon phlox, which forms dense, trailing mats on the barren, unstable slopes of the Claron Formation. The close link between Red Canyon phlox and the Claron marks it as an *edaphic specialist*—a plant whose entire world is limited to a particular rock type.

Phlox gladiformis (M. E. Jones) E. E. Nelson, Rev. W. N. Amer. Phloxes 21. 1899. BASIONYM: *P. longifolia* Nuttall var. *gladiformis* M. E. Jones. Proc. Calif. Acad. Sci. II 5: 711. 1895 [sword-shaped, of the leaves] Caespitose perennial herb forming mats 6–15 cm tall from a branching caudex, the numerous decumbent to erect-ascending flowering shoots 3–9 cm long; *leaves* linear-lanceolate, firm and sharp-pointed, ciliate and pilose with gland-tipped hairs, max. length 12–25 mm and width 1.5–3 mm; *inflorescence* 1- to 3- (6-) flowered, copiously glandular-pubescent, max. pedicel length 0.5–5 mm; *calyx* 6.5–9.5 mm long, united ⅜–⅝ its length, lobes subulate with apex sharp-cuspidate, membranes plicate-carinate; *corolla* tube 8–15 mm, glandular-pubescent, lobes obovate (average dimensions 7 × 5 mm), hue pale lilac to lavender and to white with a bluish sheen; *stamens* borne on mid- to upper corolla tube, with anthers mostly included; *style* 2.5–5.5 mm long, united to tip which is free for 0.5–1 mm, stigma included and placed below the anthers; *flowering season* late spring, mid-May through June, into July.

Geography
Chiefly on the Colorado Plateau, concentrated in the Utah High Plateaus section where it occurs on the Markagunt and Paunsaugunt plateaus (SW Utah), with outlying occurrences on the eastern edge of the Great Basin in the Mormon and Pahroc mountains (SE Nevada).

Environment
Dissected plateaus and mountains at low to middle elevations, on slopes of eroding substrates, 3500–8300 ft. (1050–2500 m); *MAP* 16 in. (41 cm), most coming as winter snow fall; *habitat* xeric; *soils* minimally developed, influenced by characteristics of the parent material (described below); *parent material* limestone.

Red Canyon phlox is an edaphic specialist, largely restricted to exposures of the Claron Formation (known formerly as the Wasatch Formation), a calcareous limestone that weathers into unstable, highly eroded slopes (20–30 degrees). Little or no soil development occurs in this habitat, other than fine clay particles overlain by gravel and scree derived from the parent material, resulting in a xeric environment where infiltration of precipitation is low and runoff is high. On steeper slopes, trailing mats of Red Canyon phlox tend to line up along a distinct micro-topographical erosion pattern of mounds and rills caused when gravel-sized rock chips and soils particles are washed downslope during summer thunderstorms in June through August. At least ten other rare perennial herbs are restricted to the Claron Formation in Red Canyon and the immediate vicinity (Hreha 1993, 1995; Mohlenbrock 1988). Like most of the plants endemic to the Claron Formation, Red Canyon phlox reproduces primarily by vegetative means rather than from seed.

Associations
Primarily open ponderosa pine (*Pinus ponderosa*) woodland, but ranges from pinyon-juniper woodland comprised of two-needle pinyon pine (*P. edulis*) and Rocky Mountain juniper (*Juniperus scopulorum*) through the ponderosa pine zone up into subalpine woodland of Intermountain bristlecone pine (*P. longaeva*). Scattered green-leaf manzanita (*Arctostaphylos patula*) shrubs may also be present, along with creeping Oregon-grape (*Mahonia repens*). These woodlands have a very open stand structure similar to that occurring on serpentine outcrops in California and Oregon. Vegetative cover is less than 10 percent on the harshest sites. Ripley (1944) observed Red Canyon phlox "growing in fissures of rounded boulders [of a calcareous rock formation] at the base of the Pahroc Range" in Nevada, accompanied by rock purpusia (*Ivesia arizonica* var. *saxosa*) and stonecrop scorpion-weed (*Phacelia saxicola*).

Cultivation
Red Canyon phlox has not been reported in cultivation. It has potential as a rock garden plant. Grow in full sun in dry, well-drained soil, although soils with higher clay content would probably be tolerated.

Phlox griseola
pinyon phlox
PLATES 34–36

The soapy sweet fragrance of lilac hangs in the air, but the setting is hardly a cottage garden. Here in the arid Pine Valley Mountains of southwest Utah stunted pinyon pine and juniper trees scratch out a living in rudimentary almost-soils derived from limestone and volcanic rock. The heavy perfume is being cranked out by an expansive colony of pinyon phlox, flower-smothered frisbies of pink and white strewn across the rusty-brown slopes.

Westward into Nevada, on sterile clay knolls even more resistant to plant life, the already low-growing pinyon phlox becomes dwarfed to the extreme, forming dense, rock-hard hubcaps of stem and leaf. Transplanted British botanical explorer Rupert Barneby marveled at Nevada's desert cushion plants, which reminded him of the *vegetable sheep* of the New Zealand alpine. Such striking plants inspire picturesque simile: rock gardeners, per European tradition, refer to them as *buns*; ranchers, less familiar with pastries, liken them to *cowpies*.

Phlox griseola Wherry, Notul. Nat. Acad. Nat. Sci. Philadelphia 113: 10. 1942 [grayish, of the stems and leaves; "Planta . . . griseola"] Caespitose perennial herb 3–6 cm tall from a short-branched caudex, ranging in habit from the nominate expression (subsp. *griseola*) which forms loose cushions with numerous erect-ascending flowering shoots 2–4 cm long to more strongly condensed forms in more xeric situations, the most extreme (subsp. *tumulosa*) becoming tumulose and forming dense hard mounds with close-packed flowering shoots 1–2 cm long; *leaves* linear-subulate to oblong, thick, with prominent pale margin and cuspidate tip, copiously coarse-ciliate, surficially minutely pilose, ranging from 3–6 mm long and 0.75–1.25 mm wide for subsp. *tumulosa* to 6–12 mm long and 1–2 mm wide for subsp. *griseola*; *inflorescence* 1-flowered, its herbage densely pubescent with coarse glandless hairs, max. pedicel length 0.5–3 mm; *calyx* length ranging from 4.5–6.5 mm for subsp. *tumulosa* to 8–10 mm for subsp. *griseola*, united ½–⅝ its length, lobes broad-subulate to narrow-triangular with thick margin and weak midrib and with apex cuspidate, membranes flat; *corolla* tube length ranging from 8–10 mm for subsp. *tumulosa* to 11–13 mm for subsp. *griseola*, lobes obovate (dimensions ranging from 4.5 × 3 mm for subsp. *tumulosa* to 6 × 4 mm for subsp. *griseola*), hue light purple, pink, or white; *stamens* borne on mid- to upper corolla tube, with anthers mostly included, one or two slightly exserted; *style* 2–4 mm long, united to tip which is free for ca. 1 mm, with stigma included and placed below the anthers; *flowering season* spring, April–May.

Key to the Subspecies of *Phlox griseola*

1a. Plants forming open to compact cushions, shoots 3–6 cm tall; leaves 5–12 mm long, 1–2 mm wide; calyx 8–10 mm long; corolla tube 11–13 mm long . . . . **subsp.** *griseola* | pinyon phlox

1b. Plants forming tight cushions and mounds as much as 50 cm across, stems closely packed, shoots 1–2 cm tall; leaves appressed, 1.5–5 mm long, 0.5–1.25 mm wide; calyx 4.5–6.5 mm long; corolla tube 8–10 mm long . . . . **subsp.** *tumulosa* (Wherry) Wherry, Gen. Phlox 140. 1955 and in Baileya 4: 98. 1956 | dwarf pinyon phlox

Geography

Great Basin, limited to the southeastern sector (NW Arizona, E Nevada, and W Utah). The nominate expression of pinyon phlox (subsp. *griseola*) occurs in the eastern portion of this overall range, most abundantly in the foothills of the Pine Valley Mountains (SW Utah). Subspecies *tumulosa* occupies the western portion of the overall range and is essentially limited to the Calcareous Mountains region of the eastern Great Basin, occurring in association with valleys and foothills of the Egan, Fortification, Monitor, and Snake ranges of EC Nevada, and the Mineral and Wah Wah ranges of W Utah.

Environment

Intermountain basins, hilly piedmont, and mountains at low to middle elevations, on upper slopes of rolling to dissected terrain, bluffs, and ridges, 3900–6500 ft. (1200–2000 m) for subsp. *griseola*, 4900–6600 ft. (1,500–2,000 m) for subsp. *tumulosa*; MAP less than 10 in. (25 cm); *habitat* xeric; *soils* minimally developed, stony, with little organic matter, shallow with bedrock at or near the surface.

The nominate expression of pinyon phlox (subsp. *griseola*) is associated with the Claron, Quichapa, and Muddy Creek formations, parent materials that yield relatively barren substrates in the Pine Valley Mountains of SW Utah. The Claron Formation is a calcareous limestone that weathers into reddish, gravelly soils that in other places in Utah support a large number of endemic plant species, including Red Canyon phlox (*Phlox gladiformis*). Soils derived from the volcanic Quichapa Formation are iron-rich, coarse, and strewn with fragments of red basalt and hematite. Subspecies *tumulosa* occurs in EC Nevada on barren knolls, benches, and mounds derived from limestone or tuffaceous sedimentary deposits associated with hot springs. Soils are gravelly, sandy or silty-clay loam, calcareous/alkaline, sometimes chalky, and white from encrustation with salt and gypsum.

Associations

Conifer woodland, chaparral, shrubland, and barrens communities. Pinyon phlox can be locally abundant and sometimes occurs in large, conspicuous populations. Pinyon phlox is primarily associated with conifer woodland dominated by open stands of singleleaf pinyon (*Pinus monophylla*) in association with Utah juniper (*Juniperus osteosperma*) or, on drier sites, by Utah juniper alone. It also is associated with shrubland dominated by basin big sagebrush (*Artemisia tridentata* subsp. *tridentata*), with black sagebrush (*A. nova*), shadescale (*Atriplex confertifolia*), and rabbitbrush (*Chrysothamnus* spp.) also associated. The common name reflects strong association with pinyon-juniper woodland, particularly in the Pine Valley

Mountains where the nominate expression of pinyon phlox is most abundant. The Pine Valley Mountains are notable for the unusually large variety of pine species present, including two species of pinyon pine—*P. edulis* and *P. monophylla*.

At higher elevations in the Pine Valley Mountains, pinyon phlox (subsp. *griseola*) occurs in chaparral comprised of birch-leaf mountain-mahogany (*Cercocarpus montanus*), Gambel oak (*Quercus gambelii*), Utah serviceberry (*Amelanchier utahensis*), and skunkbush (*Rhus trilobata*). The Pine Valley Mountains region is noteworthy for the disjunct occurrence of true evergreen chaparral, floristically similar to that of C Arizona, and containing many of the dominant genera of the California chaparral. Ground-cover milkvetch (*Astragalus humistratus*) is often associated with and even grows up through mats of subsp. *griseola* in the Pine Valley Mountains (Warrick 1987).

Subspecies *tumulosa* occurs in conifer woodland and shrubland as described above, and in sparsely vegetated barrens comprised of low-growing, mound-forming plants, typically dominated by pygmy sagebrush (*Artemisia pygmaea*). Other associates include mound daisy (*Erigeron compactus*), Shockley wild buckwheat (*Eriogonum shockleyi*), southwestern peppergrass (*Lepidium nanum*), matted prickly-phlox (*Leptodactylon caespitosum*), and rock-loving point-vetch (*Oxytropis oreophila*). Barneby (1947) called these plant communities "the most remarkable association of aretioid or vegetable-sheep life forms known to me in western America." This subspecies of pinyon phlox occurs with several rare regional endemics in these barrens including Welsh's cat's-eye (*Cryptantha welshii*), Sunnyside green-gentian (*Frasera gypsicola*), and Tiehm's stickleaf (*Mentzelia tiehmii*).

Cultivation

The dwarf expression of pinyon phlox (subsp. *tumulosa*) is the sort of novelty that drives rock gardeners mad with desire. Famed alpinist Dwight Ripley found it "so absurdly bunlike that by comparison most other dwarf Phloxes look about as compact as sweet-peas." Given its association with harsh, cold desert habitat, this phlox would be as hard to cultivate and keep in character as the most challenging of alpine species. But the more open-growing expression of pinyon phlox (subsp. *griseola*) is the most beautiful of all the cushion phloxes and should prove more amenable to horticulture. The highly fragrant flowers come in striking shades of pink and rose, and are wonderfully set off against the grayed foliage, although they are often produced in such profusion as to completely cover the plant. Grow in full sun in dry, well-drained soil, although soils with higher clay content would probably be tolerated.

Phlox hendersonii
Henderson's phlox

Summer is warming the Wenatchee alpine, and pikas are not the only creatures scrambling to make hay. Henderson's phlox shares the fell-fields with these feisty, elfish mammals, where the opportunity to grow and procreate passes quickly. This cushion-forming phlox is known from only a handful of high places in Washington and Oregon, including Mount Hood and Mount Adams, two of the Northwest's looming volcanoes, plus a scattering of high altitude stations in the North Cascades and Olympic Mountains.

The rarity of Henderson's phlox reflects the scarcity of its habitat. Compared to the Rockies with their broad, expansive tracts of tundra, there is little quarter for alpine between timberline and ice on the volcanoes and high mountains of the Pacific Northwest. Steep, unstable slopes, permanent snowfields, and glaciers conspire to curb the extent of alpine habitat on these peaks. Yet Henderson's phlox prospers in its narrow niche, finding foothold where exposure and slope lessen the accumulation of cloaking, ocean-stoked snows.

Phlox hendersonii (E. E. Nelson) Cronquist, Vasc. Pl. Pacific NorthW. 130. 1959. BASIONYM: *P. condensata* (A. Gray) E. E. Nelson var. *hendersoni* E. E. Nelson, Revis. W. N. Am. Phlox. 14. 1899 [after Louis Forniquet Henderson (1853–1942), collector of the type] Caespitose perennial herb forming compact cushions 2–6 cm tall from a short-branched caudex, the numerous close-packed erect flowering shoots 1–3 cm long; *leaves* crowded and appressed, subulate, thickish-margined, sparsely ciliate and surficially pilose with gland-tipped hairs, max. length 6–12 mm and width 0.75–1.5 mm; *inflorescence* 1-flowered, its herbage copiously glandular-pubescent, max. pedicel length 1 mm; *calyx* 5–8 mm long, united ½–⅝ its length, lobes linear-subulate with apex cuspidate, membranes obscure; *corolla* tube 8–10 mm long, lobes obovate (average dimensions 5 × 3 mm), hue white; *stamens* borne on mid- to upper corolla tube, with anthers mostly included; *style* 1.5–3 mm long, united to tip which is free for ca. 1 mm, with stigma included and placed below the anthers; *flowering season* alpine summer, July–August.

Geography
Pacific Mountain System, chiefly in the Northern Cascade Mountains in NC Washington (from Mount Stuart in the Wenatchee Mountains north along the Sawtooth Ridge to Star Peak) and in the Olympic Mountains of NW Washing-

ton. Henderson's phlox also occurs on two isolated volcanic cones in the Middle Cascade Mountains—Mount Adams (S Washington) and Mount Hood (N Oregon), the latter being the only occurrence outside of Washington.

Environment

Mountains at high elevations on exposed slopes and ridges subject to desiccating winds, associated with cliffs, rock ledges and outcrops, fell-fields, and talus, 6800 ft. (2100 m) on Mount Hood, 7600 ft. (2300 m) on Mount Adams, 7200–8500 ft. (2200–2600 m) in the North Cascades, and 6500 ft. (2000 m) in the Olympic Mountains; *MAP* mostly as snow in winter, with steep slopes and exposure to high winds limiting amount and duration of accumulation; *habitat* dry to xeric; *soils* minimally developed, stony, with little organic matter, shallow with bedrock at or near the surface; *parent material* granite, quartz diorite, basalt, pyroclastic (volcanic) materials.

Henderson's phlox is associated with the driest alpine environments in the Pacific Northwest. In addition to local environmental factors such as slope, exposure, and high winds that create xeric habitat, the distribution of Henderson's phlox is limited to the eastern portions of mountain systems where precipitation is relatively low due to the rain shadow effect. Cascade phlox (*Phlox diffusa*), the widespread mountain phlox of the Pacific Northwest, often occurs in proximity to Henderson's phlox, but at somewhat lower elevations and in less stressful habitat.

Associations

Subalpine conifer woodland and with dry alpine communities. Subalpine woodland occurs at the upper limits of tree growth and is comprised of open stands of conifers, the trees typically stunted in size and deformed by wind into krummholz. Alpine plant associations occur above treeline and are comprised of dwarf shrubs, graminoids, and perennial herbs. Henderson's phlox is often one of the most abundant herbs present in these communities and in some associations is the dominant forb and is recognized as a principal indicator species.

Whitebark pine (*Pinus albicaulis*) is the most common subalpine tree species within the range of Henderson's phlox, particularly in the Northern Cascade Mountains. Other timberline conifers with which Henderson's phlox is associated include subalpine fir (*Abies lasiocarpa*), subalpine larch (*Larix lyalii*), and Engelmann spruce (*Picea engelmannii*). The herbaceous layer of these woodlands is sparse and comprised of species from adjacent alpine vegetation. Above treeline in the Northern Cascade Mountains, Henderson's phlox occurs in a variety of relatively dry alpine plant associations including turf, herb-field, and fell-field communities.

Turf communities are dominated by dwarf, fibrous-rooted graminoids, with Brewer's sedge (*Carex breweri*) and Nard sedge (*C. nardina*) being typical dominants. Commonly associated forbs include alpine stitchwort (*Minuartia obtusiloba*), alpine yellow fleabane (*Erigeron aureus*), alpine lupine (*Lupinus lobbii*), and alpine smelowskia (*Smelowskia ovalis*). Herb-fields and fell-fields occur on level to gently sloping areas where coarse rock detritus has accumulated. They are sparsely vegetated, mostly by tufted and cushion-forming plants including fescue sandwort (*Arenaria capillaris*), alpine stitchwort, snow douglasia (*Douglasia nivalis*), and arctic forget-me-not (*Eritrichium nanum*). Herb-fields have a higher percentage of plant cover (greater than 50 percent) than fell-fields. Henderson's phlox is a dominant species in dry bedrock and scree habitat on Mount Stuart in the Wenatchee Mountains and in a dry graminoid community on Star Peak in the North Cascades (del Moral 1979a, 1979b [listed as *Phlox pulvinata*]).

Above treeline in the Olympic Mountains, Henderson's phlox occurs in sparsely vegetated alpine communities comprised of tufted and cushion-forming plants. These low-growing plants sometimes occur in association with cryptobiotic crusts where layers of lichens and mosses cover the soil surface (Gold et al. 2001). Associated herbaceous plants include rosy pussytoes (*Antennaria rosea*), sedges, cliff douglasia (*Douglasia laevigata*), dwarf mountain fleabane (*Erigeron compositus*), woodrushes (*Luzula* spp.), and yellow-spot saxifrage (*Saxifraga bronchialis*). On talus slopes and in the crevices of cliffs, Henderson's phlox is an associate of yellow mountain-avens (*Dryas drummondii*) and two Olympic Mountain endemics—Olympic harebell (*Campanula piperi*) and Flett's violet (*Viola flettii*).

On Mount Hood and Mount Adams, two isolated volcanic cones, Henderson's phlox occurs in the relatively narrow zone between treeline and ice in sparsely vegetated alpine communities comprised of tufted and cushion-forming plants. Associates of Henderson's phlox on Mount Adams include Mt. Hood pussy-paws (*Cistanthe umbellata*), dwarf mountain fleabane, small-flower beardtongue (*Penstemon procerus* subsp. *tolmiei*), elegant polemonium (*Polemonium elegans*), and alpine smelowskia.

Cultivation

Henderson's phlox has been cultivated to a limited extent as a rock garden plant. Like most true alpines, it would be difficult to cultivate Henderson's phlox in a manner that would maintain its attractive wild character except under the specialized conditions of a rock garden, trough garden, or alpine house. Grow in full sun in dry, well-drained soil.

Phlox hirsuta
Yreka phlox

Plants have a love-hate relationship with *serpentine*, love being the exception rather than the rule. Serpentine rocks and the skeletal soils derived from them are not only low in elements essential to plant life, they are often laced with hurtful heavy metals like nickel, chromium, and cobalt. As a result, most plants avoid serpentine like the plague, and significant exposures of this rock type impart a lunar sterility to the landscape. But some plants grow nowhere else, tolerating the toxic in exchange for less competition and more breathing room.

The Siskiyou Mountains of northern California and adjacent Oregon are rich in serpentine exposures and loaded with serpentine endemics. Yreka phlox is one of them, a short, shrubby species with hairy foliage blanketed in spring with rose-pink flowers. Known only from the vicinity of Yreka, California, it grows in open woodlands of Jeffrey pine and western juniper. While serpentine specialists are often uncommon, Yreka phlox is among the rarest of the rare—an endangered species whose present and posterity is limited to only a handful of populations.

Phlox hirsuta E. E. Nelson, Revis. W. N. Amer. Phlox. 28. 1899 [covered with coarse hairs; "coarsely hirsute throughout with long jointed hairs"] Subshrub 5–15 cm tall with spreading stems giving rise to numerous erect-ascending annual flowering branches with ca. 3 close-spaced leaf nodes below the inflorescence; *leaves* thickish, the lower elliptic-ovate and upper linear-lanceolate, somewhat acerose, ciliate and pilose with copious long hairs, max. length 15–30 mm and width 4–7 mm; *inflorescence* 3- to 15-flowered, compact, its herbage pubescent with both long and short gland-tipped hairs, max. pedicel length 2–6 mm; *calyx* 11–14 mm long, united ca. ½ its length, lobes linear with apex cuspidate, membranes slightly plicate; *corolla* tube 11–15 mm long (little if at all exserted from the calyx), lobes narrowly obovate (average dimensions 8 × 5 mm) with apex entire, hue bright rose-pink to white; *stamens* with anthers included; *style* 5–8 mm long, united to tip which is free for 1 mm, with stigma included; *flowering season* spring, late April to early June, mature plants can form hundreds of flowers and when in bloom appear covered in flowers.

Geography
Pacific Mountain System, endemic to the Klamath Mountains in NW California where it is restricted to four populations in an area within about 20 miles (32 km)

of the town of Yreka in Siskiyou County. Yreka phlox was listed as an endangered species by the U.S. Fish and Wildlife Service in 2000 due to its small number of populations, extremely limited distribution, past habitat destruction, and current vulnerability to habitat loss.

Environment
Mountains at middle elevations, on steep slopes and ridge-crests, 2800–4400 ft. (850–1300 m); *MAP* 19.6 in. (50 cm), most of it as snow; *habitat* xeric; *soils* stony, shallow, formed over massive rock outcrops; *parent material* serpentine. Yreka phlox is a serpentine endemic, only occurring in association with exposures of serpentine rock formations.

Associations
Conifer woodland comprised of open stands of Jeffrey pine (*Pinus jeffreyi*), with incense cedar (*Calodedrus decurrens*) and Douglas-fir (*Pseudotsuga menziesii*) also associated. One small occurrence of Yreka phlox is associated with western juniper (*Juniperus occidentalis* var. *occidentalis*) woodland. While of extremely limited distribution, populations of Yreka phlox are relatively large and conspicuous. Yreka phlox grows in the needle duff and shade provided by these conifers. A shrub layer is often present in these woodlands, comprised of sedge-leaf whitethorn (*Ceanothus cuneatus*), birchleaf mountain-mahogany (*Cercocarpus montanus*), and rubber rabbitbrush (*Ericameria nauseosa*). Idaho fescue (*Festuca idahoensis*) is the most commonly associated grass. Associated forbs include rough eyelash-weed (*Blepharipappus scaber*), Tolmie's mariposa lily (*Calochortus tolmiei*), common spring-gold (*Crocidium multicaule*), common woolly-sunflower (*Eriophyllum lanatum*), large-fruit desert-parsley (*Lomatium macrocarpum*), silverleaf scorpion-weed (*Phacelia hastata*), and Beckwith's violet (*Viola beckwithii*). Scott Valley phacelia (*P. greenei*), another local serpentine endemic, is found at some occurrences of Yreka phlox. See Elam (1998) for valuable information on the ecology and conservation of Yreka phlox.

Cultivation
Yreka phlox has been cultivated to a limited extent, and its beauty and heavy flowering have earned it admiration in rock-gardening circles. But because of its extreme rarity, no further collections of seed or plants should be made from wild populations except for authorized conservation purposes.

Phlox hoodii
Hood's phlox
PLATES 37, 38

Hood's phlox is immense, not in scale, but in scope. The ubiquity of this carpet-forming plant in western North America reflects, in a creaturely way, what the Puritans meant when they spoke of the immensity of God, who is "everywhere present." Wry plantsman Panayoti Kelaidis was reaching for similar expansive imagery when he wrote that the states of Wyoming, Utah, Nevada, and Idaho are actually held together by one uninterrupted mat of Hood's phlox.

Throughout the greater part of its vast distribution, Hood's phlox is an unremarkable, blue-collar constituent of mile-upon-mile of sage and grassland habitat. But on the bluffs, buttes, and badlands of the Great Plains it is the springtime glory of stark, rocky barrens. Climb Scotts Bluff, Pumpkin Buttes, or any celebrated plains landmark, tread nameless escarpments where traces of bison jump and eagle trap remain, and wind-sculpted cushions of Hood's phlox will be there, exulting in redemption from winter's rough hand with an explosion of tiny white flowers.

Phlox hoodii Richardson, Bot. App. (Richardson) 733 (734). 1823 [after Lieutenant Robert Hood (1797–1821)] Caespitose perennial herb forming carpetlike mats or compact cushions 2.5–10 cm tall from a short-branched to spreading caudex, the numerous erect-ascending flowering shoots 1–5 cm long; *leaves* linear-subulate, basally pilose to arachnoid-tomentose but toward the tip canescent to fine-pilose to glabrous, max. length 6–12 mm and width 0.5–1(–1.5) mm; *inflorescence* 1- to 3-flowered, its herbage copiously pubescent with fine glandless hairs (sometimes mixed with gland-tipped hairs) to glabrous, max. pedicel length 0.5–2(–4) mm; *calyx* varying from 4.5–7.5 mm long for subsp. *hoodii* to 6–9(–12) mm long for subsp. *canescens*, united ⁷⁄₁₆–⅝ its length, lobes subulate with apex cuspidate, membranes flat; *corolla* tube varying from 5–8 mm long for subsp. *hoodii* to 8–12 mm long for subsp. *canescens*, lobes obovate (dimensions ranging from 4.5 × 2.5 mm for subsp. *hoodii* to 6 × 4 mm for subsp. *canescens*), in some populations of subsp. *canescens* (variant treated as *P. lanata*) lobe surface minutely pubescent toward the eye (very rare in *Phlox*), hue pale lavender (occasionally deep lavender in local populations) to white; *stamens* borne on upper corolla tube, with some anthers exerted; *style* ranging from 2.5–5 mm long for subsp. *hoodii* to 3–7 mm long for subsp. *canescens*, united to tip which is free for ca. 1 mm, with stigma included and placed below the anthers; *flowering season* spring, April–June, depending on elevation and latitude.

Taxonomic notes

Wherry (1955) recognized *Phlox hoodii* subsp. *hoodii* plus subsp. *canescens*, subsp. *glabrata*, subsp. *muscoides*, and subsp. *viscidula*. Cronquist (1959) restored *P. muscoides* to its original full species status (followed here), and subsequently (1984) distilled down variation in *P. hoodii* to two "geographically significant but morphologically confluent varieties"—vars. *hoodii* and *canescens*. Wilken (1986) noted considerable morphological variation in *P. hoodii* in the Great Plains, but found "little geographical segregation," adding "it is often difficult to apply a [subspecies] name to samples from any particular population." Two subspecies of *P. hoodii* are recognized here, the eastern subsp. *hoodii* (incorporating subspp. *glabrata* and *viscidula*) and the western subsp. *canescens*. Pending further study, subsp. *canescens* incorporates *P. lanata* (Piper 1902), which Wherry (1955) recognized as a distinct species.

Key to the Subspecies of *Phlox hoodii*

1a. Flowers larger, corolla tube mostly 8–12 mm long, 2–5 mm longer than calyx, corolla lobes mostly 5–7(–8) mm long; the more western and southern expression, best developed in the Great Basin . . . . **subsp. *canescens*** (Torrey & A. Gray) Wherry, Proc. Acad. Nat. Sci. Philadelphia 90: 139. 1938 | Utah Hood's phlox

1b. Flowers smaller, corolla tube mostly 5–8 mm long, 0–3 mm longer than calyx, corolla lobes mostly 4–5 mm long; the more eastern and northern expression, best developed east of the continental divide on the northern Great Plains . . . . **subsp. *hoodii*** | Hood's phlox

Geography

Western interior of the United States and Canada, remote to Alaska and Yukon. The Hood's phlox species complex has the largest distribution of any of the western cushion-forming phloxes, ranging from the Great Plains westward through the Rocky Mountains to the eastern edge of the Cascade–Sierra axis. The nominate expression of Hood's phlox (subsp. *hoodii*) represents the eastern expression of the species, with a range concentrated on the northern Great Plains and in the Rocky Mountains east of the Continental Divide. Highly disjunct populations occur in the upper Yukon River region of C and SW Yukon Territory and in interior Alaska.

Subspecies *canescens* occupies the western portion of the overall range of the species, with its distribution concentrated in the Great Basin and extending northward onto the Columbia Plateau and eastward onto the Colorado Plateau.

Environment

Plains, intermountain basins, dissected plateaus, hilly piedmont, and mountains at low to middle elevations, on upper slopes of rolling terrain and divides, badlands, escarpments, bluffs, buttes, and ridges, 1500–9000 ft. (450–2700 m), occasionally reaches alpine zone; *MAP* 8–14 in. (20–36 cm); *habitat* xeric to dry; *soils* loamy sand, sandy loam, loam, or clay loam, often stony, with bedrock at or near the surface; *parent material* igneous, metamorphic, sedimentary.

Associations

Conifer woodland/savanna, shrubland, shrub-steppe, steppe/grassland, and rock outcrop communities. Hood's phlox often occurs in large, conspicuous populations. It is typically one of the most abundant herbs present and in some situations is the dominant forb.

Hood's phlox is associated with ponderosa pine (*Pinus ponderosa*) woodland and savanna in the foothill and lower montane zones of the Rocky Mountains, from Alberta and British Columbia south through Colorado, and in the Black Hills and other uplifted mountain "islands" east of the Rockies, as well as the Pine Ridge, Missouri Breaks, and similar escarpment landforms of the western Great Plains. On drier, more exposed sites, it occurs in woodland dominated by limber pine (*P. flexilis*), Rocky Mountain juniper (*Juniperus scopulorum*), or Utah juniper (*J. osteosperma*). In NW Montana and the southern portions of Manitoba, Saskatchewan, and Alberta, Hood's phlox occurs in grassy openings in quaking aspen (*Populus tremuloides*) parkland, with prairie Junegrass (*Koeleria macrantha*) or rough fescue (*Festuca campestris*) the dominant graminoids.

The western expression of Hood's phlox (subsp. *canescens*) occurs in a diversity of conifer-dominated woodland and savanna ecological systems, typically on dry, exposed sites at the ecotone with shrubland or grassland. On mountain ranges in the Great Basin, it occurs in pinyon-juniper woodland codominated by single-leaf pinyon pine (*Pinus monophylla*) and Utah juniper. On mesas, plateaus, and escarpments on the Colorado Plateau, it occurs in pinyon-juniper woodland codominated by two-needle pinyon pine (*P. edulis*) and Utah juniper. It is often a prominent component of the herbaceous layer of these woodlands and savannas, comprised of graminoids and low-growing forbs.

Hood's phlox occurs in almost every major upland shrubland and shrub-steppe ecological system from the western Great Plains though the Rocky Mountains. These large-scale communities are typically comprised of a dominant species of shrub, notably big sagebrush (*Artemisia tridentata*), saltbush (*Atriplex* spp.), mountain-mahogany (*Cercocarpus ledifolius* or *C. montanus*), rabbitbrush (*Chrysothamnus* spp. and *Ericameria* spp.), antelope bitterbrush (*Purshia tridentata*), or black grease-wood (*Sarcobatus vermiculatus*), often in association with a codominant species of bunchgrass. On rocky, exposed ridges and slopes in the Wyoming Basin and Rocky Mountains, Hood's phlox occurs in dwarf-shrubland dominated by black sagebrush (*Artemisia nova*). On shale-derived clay soils in the intermountain basins of Wyoming and Montana, Hood's phlox occurs in dwarf-shrubland dominated by bird's-foot sagebrush (*A. pedatifida*). While the shrub and bunchgrass species comprising these communities varies depending on slope, aspect, elevation, edaphic, and other factors, Hood's phlox is a fairly constant component of the herbaceous layer across the spectrum of these shrublands.

The western expression of Hood's phlox (subsp. *canescens*) occurs in dwarf-shrubland systems that are often small patch communities restricted to particularly harsh sites or unique edaphic situations. On the western edge of the Columbia Plateau in C Washington, it occurs in lithosol ("scabland") or patterned ground habitat associated with exposures of basalt bedrock that support unique communities dominated by one of several dwarf shrubs—scabland sagebrush (*Artemisia rigida*), snow wild buckwheat (*Eriogonum niveum*), Douglas' wild buckwheat (*E. douglasii*), or thyme-leaf wild buckwheat (*E. thymoides*). Hood's phlox is often a dominant forb in this habitat, forming large mounded plants that give these communities the appearance of tussock-covered fell-fields. In the northern Great Basin, Hood's phlox is a common associate of dwarf sagebrush (*A. arbuscula*) shrub-steppe.

Hood's phlox is associated with most of the upland grassland communities of the northern Great Plains (except those of sandy habitats), from Alberta and Saskatchewan south into NE Colorado. In the northwestern portion of its range, Hood's phlox is associated with mixedgrass prairie dominated by needle-and-thread (*Hesperostipa comata*), blue grama (*Bouteloua gracilis*), and western wheatgrass (*Pascopyrum smithii*). It is often one of the two or three most dominant forbs in the mixedgrass prairie, with prairie sagebrush (*Artemisia frigida*) usually first. Where vegetation cover is relatively thin, Hood's phlox occupies spaces between grasses and shrubs in association with dense spike-moss (*Selaginella densa*) and soil lichens (*Xanthoparmelia* spp.) and bryophytes. On the slopes of buttes, escarpments, badlands, and other erosional landforms of the Great Plains, Hood's phlox also occurs in an herbaceous community dominated by little bluestem (*Schizachyrium scoparium*), blue grama, sideoats grama (*B. curtipendula*), and thread-leaved sedge (*Carex filifolia*). At the southern limits of its range in the Great Plains, Hood's phlox is associated with shortgrass prairie dominated by blue grama and buffalograss (*Buchloe dactyloides*). Hood's phlox occasionally ascends into dry alpine/subalpine grassland and alpine turf communities in the Northern Rocky Mountains, but is more commonly associated with the montane grasslands of mid-elevations where Idaho fescue (*Festuca idahoensis*) or rough fescue are dominant. On the west slope of the Rockies in Colorado and on the Colorado Plateau, subsp. *canescens* occurs in semi-desert grassland dominated by bluebunch wheatgrass (*Pseudoroegneria spicata*) and Indian mountain-ricegrass (*Achnatherum hymenoides*).

In rocky, shallow soil habitat within these larger ecological systems, Hood's phlox occurs in sparsely vegetated barrens and openings dominated by low-growing forbs, of which it is often an abundant and conspicuous element. In the Great Plains, Hood's phlox is associated with a suite of cushion plants that includes Hooker's sandwort (*Arenaria hookeri*), three-leaf milkvetch (*Astragalus giliflorus*), tufted milkvetch (*A. spatulatus*), bluff fleabane (*Erigeron ochroleucus*), few-flower wild buckwheat (*Eriogonum pauciflorum*), alpine bladderpod (*Lesquerella alpina*), low nailwort (*Paronychia sessiliflora*), and stemless four-nerve-daisy (*Tetraneuris acaulis*).

In interior Alaska and Yukon Territory, Hood's phlox occurs in small, isolated grass-dominated associations that occur on steep, south-facing slopes and bluffs with dry, well-drained soils. Sometimes termed "subarctic steppe," these communities are dominated by prairie sagebrush, purple reedgrass (*Calamagrostis purpurascens*), and white bluegrass (*Poa glauca*), with Hood's phlox often a subdominant. Several other species disjunct from their main ranges on the northwestern Great Plains occur with Hood's phlox in these isolated stands, including thread-leaved sedge, bluff fleabane, Yukon wild buckwheat (*Eriogonum flavum* var. *aquilinum*), and Hooker's Townsend-daisy (*Townsendia hookeri*). These stands are thought by some to be remnants or modern analogues of an ice-age (Late Pleistocene) grass-dominated ecosystem that extended across Alaska and northeast Asia via the Bering Land Bridge, termed "Mammoth Steppe" because it supported extensive herds of woolly mammoth, horses, bison, and other large mammals. The stands show parallels with islandlike occurrences of steppe in northeast Asia in which *Phlox sibirica* occurs (see species account).

Cultivation

The stature and dimensions of Hood's phlox fit the horticultural definition of an "alpine," but this cushion-forming species is a plant of plains, plateaus, and basins rather than the mountains, and is easier to cultivate than a true ecological alpine. The ornamental attributes of Hood's phlox varies considerably across its vast range, with some expressions rather mundane but others much more interesting. The showiest members of the Hood's phlox complex include a dwarf, silvery form from the Great Basin sometimes treated as *Phlox lanata*, and a relatively robust expression with large lavender-blue flowers that occurs on the western edge of the Columbia Plateau. Selections from such populations could be used as drought-tolerant landscape plants as well as rock garden subjects. Grow in full sun in dry, well-drained soils, although soils with higher clay content would probably be tolerated.

Phlox idahonis
Idaho phlox

The Lochsa River gathers more than snowmelt on the flanks of the Bitterroot Range. Each captured drop crackles with mountain light, and sparks fly over the Lochsa's rapids as water and rock collide. Countless, endless explosions coalesce into flashes and flares that light up the downstream forests of Idaho's Clearwater Basin like fireworks. Dense and brooding, darkened by towering Pacific coast red cedar, hemlock, and fir, these magnificent forests become surreal in the shine of river light.

The inland maritime forests of Idaho are host to a number of interesting plants. Some, like the trees, are "oceanic elements" isolated hundreds of miles from Pacific coast forests. Several are local endemics, including Idaho phlox, which is limited to a handful of meadows near the logging village of Headquarters. Although highly localized, Idaho phlox flourishes in these lush, moose-tromped openings with their quiet meandering streams, stretching to top the waist-high canopy of sedges, forbs, and shrubs with showy heads of pink in summer.

Phlox idahonis Wherry, Notul. Nat. Acad. Nat. Sci. Philadelphia 87: 6. 1941 [of the state of Idaho] Upstanding perennial herb from a slender shallow rhizome which at its apex turns up into a solitary flowering shoot 50–100 cm tall with 15–30 leaf nodes below the inflorescence; *leaves* oblong below, becoming ovate and cordate above, the longest near mid-stem and the widest just below the inflorescence, max. length 60–80 mm and width 15–25 mm; *inflorescence* a compound cyme of 9–50 (up to ca. 100) flowers, its herbage glandular-pubescent, max. pedicel length 8–15 mm; *calyx* 9–14 mm long, united ca. ½ its length, lobes broadly subulate with apex short-aristate, membranes plicate; *corolla* tube 15–20 mm long, lobes obovate (average dimensions 9 × 7 mm), hue lilac to lavender, rarely white; *stamens* borne on upper corolla tube, with some anthers exserted; *style* 14–18 mm long, united to tip which is free for 1–1.5 mm, with stigma placed among the anthers at or near corolla orifice; *flowering season* early summer, peaking late June to early July, then sporadically into August.

Geography

Northern Rocky Mountains, endemic to the Clearwater Basin of NC Idaho where it is limited to four meta-populations (eight occurrences) in the upper tributaries of the Reeds Creek Drainage, all within a 4-mile (6 km) radius of Headquarters in Clearwater County. Idaho phlox is a plant of conservation concern in Idaho due to its extreme rarity, past habitat alteration/loss, and ongoing disturbances. Livestock grazing exerts significant pressure on Idaho phlox and its habitat, although light to moderate grazing may actually be beneficial. Ninety-eight percent of its habitat is on private land owned by a timber products company. Idaho phlox is the rarest plant in the flora of Idaho.

Environment

Mountains in small valleys traversed by low gradient, meandering streams, 2800–3200 ft. (850–960 m); *MAP* 41 in. (104 cm), with 17 in. (42 cm) during the growing season; *habitat* seasonally moist to wet, inundated early in growing season but typically dry as summer progresses; *soils* deep, sandy-loam alluvium. Idaho phlox is shade-intolerant, with stem numbers decreasing and flowering inhibited under low light intensities. It is tolerant of cool, fall fires which remove litter and kill overtopping shrubs and trees but which do not produce enough heat to kill the phlox rhizomes.

Associations

Small patch riparian shrub and meadow communities surrounded by a matrix of conifer forest dominated by western red cedar (*Thuja plicata*) and grand fir (*Abies grandis*). These shrublands are characterized by stands of speckled alder (*Alnus incana*) or alderleaf buckthorn (*Rhamnus alnifolia*) with an herbaceous layer in a mosaic with more open meadow vegetation dominated by graminoids including water sedge (*Carex aquatilis*) and blue-joint reedgrass (*Calamagrostis canadensis*), along with two introduced pasture grasses—Kentucky bluegrass (*Poa pratensis*) and meadow timothy (*Phleum pratense*). These meadows are also rich in forbs, including common yarrow (*Achillea millefolium*), Siskiyou aster (*Symphyotrichum lanceolatum* subsp. *hesperium*), large-leaved avens (*Geum macrophyllum*), cow-parsnip (*Heracleum maximum*), tall bluebells (*Mertensia paniculata*), small-flower beardtongue (*Penstemon procerus*), Canada goldenrod (*Solidago canadensis*), and California false hellebore (*Veratrum californicum*). See Moseley and Crawford (1995) and Lichthardt and Gray (2001) for valuable information on the ecology, reproductive biology, and conservation of Idaho phlox.

Cultivation

Idaho phlox has been cultivated locally in Idaho, but it has otherwise not been reported in horticulture. It is considered a species of conservation concern in Idaho because of its relative rarity and collecting seed or plants from wild populations is discouraged except for authorized conservation purposes.

Phlox kelseyi
Kelsey's phlox
PLATES 39, 40

Montana's Teton River leads a short but splendid life. From its headwaters in the precipitous Rocky Mountain Front, it tumbles out onto the plains and sets its face like flint for the Missouri, less than one hundred miles on raven wing. Its glory lies in giving birth and sustenance to Pine Butte Swamp, a rich matrix of fen, wetland, and riparian habitat that provides one of the last places on the continent where grizzly bears still trace ancestral rounds between mountains and plains, bulking up on berries and ants before they lumber back into the fastness of the montane.

The same conditions that make this prime grizzly wallow also support Kelsey's phlox, a colorful, mat-forming species with an ecology tied to wetland habitats in the Rocky Mountains. Pine Butte Swamp marks the northern limits of Kelsey's phlox, which has a fragmented distribution southward into the mountains and basins of Wyoming and Idaho. It jumps much further south in two puzzlingly remote areas of painfully severe habitat—travertine hot springs mounds in central Nevada and alkaline flats in Colorado's famous High Creek Fen.

Phlox kelseyi Britton, Bull. Torrey Bot. Club 19: 225. 1892 [after Francis Duncan Kelsey (1849–1905), collector of the type] Caespitose perennial herb forming diffuse mats 7–15 cm tall (5–10 cm tall in subsp. *salina*) from a branched, spreading caudex, the numerous prostrate to ascending flowering shoots with 3 or 4 spaced nodes; *leaves* linear-lanceolate, acuminate and cuspidate, thickish and succulent, fleshy (becoming wrinkly on drying), the margin slightly thickened, basally coarse-ciliate, surficially glabrous to moderately pubescent, max. length varying from 12–15 mm for subsp. *salina* to 12–25 mm for subsp. *kelseyi*, width 1.5–3.5 mm; *inflorescence* 1- to 5-flowered, its herbage glabrous to pubescent, the hairs gland-tipped in most colonies but glandless in some (including subsp. *salina* and type locality of subsp. *kelseyi*), max. pedicel length 3–10(–20) mm; *calyx* varying from 7–9 mm long for subsp. *salina* to 8–12 mm long for subsp. *kelseyi*, united ⅜–⅝ its length, lobes linear-subulate with weak midrib and with apex cuspidate; *corolla* tube 9–15 mm long, lobes obovate (average dimensions 7.5 × 5 mm), hue lilac to lavender to pink and to white with a bluish sheen; *stamens* borne on mid- to upper corolla tube, with some anthers exserted; *style* (4–)6–8 mm long, united to tip which is free for 1 mm, with stigma placed among the anthers near corolla orifice; *flowering season* late spring, May–July depending on altitude.

Taxonomic notes

Wherry (1955) recognized *Phlox kelseyi* subsp. *salina*, which he separated from the nominate expression (subsp. *kelseyi*) by differences in habit, stature, and size of floral parts, and by the association of subsp. *salina* with highly alkaline clay habitats. Campbell (1992) upheld recognition of these as separate subspecies in her thesis on the *P. kelseyi* complex. As presently understood, the morphological variation represented by these subspecies does not show clear correlation with geographic/ecological segregation.

Geography

Chiefly Rocky Mountain System, concentrated in the Northern Rocky Mountains (E Idaho and W Montana) extending south into the Middle Rocky Mountains (NW Wyoming) and Wyoming Basin (SE Wyoming). This distribution is discontinuous, and populations are often isolated from one another by considerable distances, with highly remote occurrences in the South Park region of the Southern Rocky Mountains (C Colorado), the Black Hills region of the Great Plains (NE Wyoming), and the Steptoe and Monitor intermountain valleys of the Great Basin (C Nevada).

Environment

Mountains in valleys, basins, and flats, 4600 ft. (1400 m) at Pine Butte Swamp (NC Montana), 6600 ft. (2000 m) at Birch Creek Fen (EC Idaho), 7200 ft. (2200 m) in the Laramie Basin (SE Wyoming), 9500 ft. (2900 m) in South Park (C Colorado), 6000 ft. (1800 m) at Monte Neva Hot Spring (C Nevada); *MAP* 17 in. (43 cm) at Pine Butte Swamp, 8 in. (20 cm) at Birch Creek Fen, 10 in. (26 cm) in the Laramie Basin, 11 in. (27 cm) in South Park, and less than 10 in. (26 cm) at Monte Neva Hot Spring; *habitat* seasonally moist to wet; *soils* alluvial, fine-textured, alkaline to highly alkaline; *parent material* carbonate rock formations.

Kelsey's phlox occurs in a variety of wetland communities, but most frequently in fens—wetlands fed by nutrient rich, calcareous groundwater. The soil surface of fens is sometimes white with accumulated deposits of sodium salts, high levels of which may inhibit the growth of some plants. Soil pH at sites where Kelsey's phlox occurs in EC Idaho averaged 8.9–9.6. Kelsey's phlox typically occurs at the margin of wetland habitat on low, relatively level benches above streams and fens, and on the tops and sides of hummocks within this habitat. These hummocks, likely the abandoned mounds of ant colonies (Lesica and Kannowski 1998), provide slightly raised places above the saturated soil zone. Kelsey's phlox also occurs in association with geothermal areas in SE Idaho.

Within the High Creek Fen wetland complex of C Colorado, Kelsey's phlox occurs in sites that are highest above the water table, yet wetted by capillary fringe. It occurs on hummocks of grass in these alkaline flats, where magnesium, potassium, and sodium salts accumulate to much higher percentages than in other communities within the fen.

In C Nevada, Kelsey's phlox is associated with unique geothermal areas, where it grows in the highly mineralized, saline clay soils on the sides and drainages of travertine hot spring mounds. These travertine mounds form when thermal waters pass through limestone and dissolve calcium carbonate, which is deposited at the mouth of the spring.

Associations

Primarily riparian shrubland, typically comprised of open stands of shrubby cinquefoil (*Dasiphora fruticosa* subsp. *floribunda*) with an herbaceous layer including Baltic rush (*Juncus balticus*), various sedges and grasses, and forbs such as few-flower shooting star (*Dodecatheon pulchellum*), Idaho blue-eyed grass (*Sisyrinchium idahoense*), and alpine meadowrue (*Thalictrum alpinum*). Kelsey's phlox is often locally abundant within its habitat, which occurs at the mesic ecotone between saturated riparian communities and dry, upland big sagebrush (*Artemisia tridentata*) steppe. At the northern edge of its range at Pine Butte Swamp (NC Montana), Kelsey's phlox occurs at the fringes of the dwarf-carr community type dominated by the low growing shrubby cinquefoil and bog birch (*Betula glandulosa*). It occurs elsewhere in W Montana in shrubland association of shrubby cinquefoil and rough fescue (*Festuca campstris*). At Birch Creek Fen and similar sites in EC Idaho, Kelsey's phlox grows on benches and hummocks in alkaline meadows along spring-fed streams, and to a lesser extent in adjacent upland habitat dominated by rubber rabbitbrush (*Ericameria nauseosa*). It occurs here with alkali primrose (*Primula alcalina*), a rare plant limited to Idaho and Montana (Cholewa and Henderson 1984), along with wetland specialists meadow milkvetch (*Astragalus diversifolius*) and park milkvetch (*A. leptaleus*).

PLATE 1. Phlox Mountain, Owl Creek Range, Wyoming

PLATE 2. *Phlox adsurgens*, Siskiyou County, California

PLATE 3. *Phlox albomarginata* subsp. *diapensioides*, Madison County, Montana

PLATE 4. *Phlox albomarginata* subsp. *diapensioides* associated with rock outcrop community, Rocky Mountains, Montana

PLATE 5. *Phlox alyssifolia*, Custer County, South Dakota

PLATE 6. *Phlox alyssifolia* associated with ponderosa pine savanna, Black Hills, South Dakota

PLATE 7. *Phlox amabilis*, Yavapai County, Arizona

PLATE 8. *Phlox andicola*, Garden County, Nebraska

PLATE 9. *Phlox austromontana*
subsp. *jonesii*, Washington
County, Utah

PLATE 10. *Phlox austromontana* subsp. *jonesii* associated with slickrock
habitat, Zion National Park, Utah

PLATE 11. *Phlox bifida*, Reynolds County, Missouri

PLATE 12. *Phlox caespitosa*, Ravalli County, Montana

PLATE 13. *Phlox caespitosa* habit

PLATE 14. *Phlox caespitosa* associated with ponderosa pine woodland, Bitterroot Mountains, Montana

PLATE 15. *Phlox colubrina*, Wallowa County, Oregon

PLATE 16. *Phlox condensata* subsp. *condensata*,
Summit County, Colorado

PLATE 17. *Phlox condensata* subsp. *condensata* habit

PLATE 18. *Phlox condensata* subsp. *condensata*
associated with subalpine conifer woodland,
Rocky Mountains, Colorado

PLATE 19. *Phlox condensata* subsp. *covillei*, Inyo County, California

PLATE 20. *Phlox condensata* subsp. *covillei* associated with subalpine bristlecone pine woodland, White Mountains, California

PLATE 21. *Phlox diffusa*, Tuolumne
County, California

PLATE 23. *Phlox diffusa* associated with subalpine parkland,
Sawtooth Range, Idaho

PLATE 22. *Phlox diffusa* associated with subalpine conifer woodland, Yosemite National Park, California

PLATE 24. *Phlox dispersa*, Inyo County, California

PLATE 25. *Phlox dispersa* associated with subalpine conifer woodland, Sierra Nevada, California

PLATE 26. *Phlox douglasii*, Jefferson County, Oregon

PLATE 27. *Phlox douglasii* associated with western juniper savanna, Columbia Plateau, Oregon

PLATE 28. *Phlox douglasii* associated with bunchgrass steppe, Columbia Plateau, Washington

PLATE 29. *Phlox drummondii* subsp. *drummondii*, Gonzales County, Texas

PLATE 30. *Phlox glaberrima* subsp. *interior*, Lawrence County, Missouri

PLATE 31. *Phlox glaberrima* subsp. *interior* associated with tallgrass prairie, Osage Plains, Missouri

PLATE 32. *Phlox gladiformis*, Garfield County, Utah

PLATE 33. *Phlox gladiformis* associated with montane conifer woodland, Red Canyon, Utah

PLATE 34. *Phlox griseola* subsp. *griseola*, Iron County, Utah

PLATE 35. *Phlox griseola* subsp. *tumulosa*, Beaver County, Utah

PLATE 36. *Phlox griseola* subsp. *tumulosa* associated with sagebrush shrubland, Great Basin, Utah

PLATE 37. *Phlox hoodii* subsp. *canescens,* Kittitas County, Washington

PLATE 38. *Phlox hoodii* subsp. *hoodii*, Scotts Bluff County, Nebraska

PLATE 39. *Phlox kelseyi*, Uinta County, Wyoming

PLATE 40. *Phlox kelseyi* habit

PLATE 41. *Phlox longifolia*, Las Animas County, Colorado

PLATE 42. *Phlox longifolia*, Lake County, Oregon

PLATE 43. *Phlox longifolia* associated with sagebrush steppe, Hart Mountain, Oregon

PLATE 44. *Phlox longifolia*, Lincoln County, Wyoming

PLATE 45. *Phlox missoulensis*, Granite County, Montana

PLATE 47. *Phlox multiflora*, Lincoln County, Wyoming

PLATE 46. *Phlox missoulensis* associated with montane grassland, Rocky Mountains, Montana

PLATE 48. *Phlox muscoides*, Fremont County, Wyoming

PLATE 49. *Phlox muscoides* associated with rock outcrop community, Wyoming Basin, Wyoming

PLATE 50. *Phlox nana*, Quay County, New Mexico

PLATE 51. *Phlox oklahomensis*, Comanche County, Kansas

PLATE 52. *Phlox ovata*, Augusta County, Virginia

PLATE 53. *Phlox pilosa* subsp. *fulgida*, Otoe County, Nebraska

PLATE 54. *Phlox pilosa* subsp. *longi-pilosa*, Greer County, Oklahoma

PLATE 55. *Phlox pilosa* subsp. *longipilosa* associated with mixedgrass prairie, Wichita Mountains, Oklahoma

PLATE 56. *Phlox pulcherrima*, Montgomery County, Texas

PLATE 57. *Phlox pulcherrima* associated with longleaf pine forest, Big Thicket, Texas

PLATE 58. *Phlox pulvinata*, Carbon County, Montana

PLATE 60. *Phlox pulvinata*, Albany County, Wyoming

PLATE 59. *Phlox pulvinata* associated with alpine fell-field, Bear Tooth Plateau, Montana

PLATE 61. *Phlox pulvinata* associated with alpine turf community, Medicine Bow Mountains, Wyoming

PLATE 62. *Phlox roemeriana*, Burnet County, Texas

PLATE 63. *Phlox roemeriana* associated with limestone glade, Edwards Plateau, Texas

PLATE 64. *Phlox speciosa* subsp. *lignosa*, Klickitat County, Washington

PLATE 65. *Phlox speciosa* subsp. *lignosa* habit

PLATE 66. *Phlox speciosa* subsp. *occidentalis*, Lewis and Clark County, Montana

PLATE 67. *Phlox subulata* subsp. *brittonii*, Rock-bridge County, Virginia

PLATE 68. *Phlox subulata* subsp. *brittonii* associated with shale barren, Allegheny Mountains, Virginia

PLATE 69. *Phlox tenuifolia*, Gilia County, Arizona

PLATE 70. *Phlox tenuifolia* associated with Sonoran desert scrub, Superstition Mountains, Arizona

PLATE 71. *Phlox viscida*, Wallowa County, Oregon

PLATE 73. *Phlox woodhousei*, Navajo County, Arizona

PLATE 72. *Phlox viscida* associated with scabland shrubland, Columbia Plateau, Oregon

Kelsey's phlox occurs in saline meadows and flats in the Laramie Basin (SE Wyoming). In C Wyoming, it occurs with meadow pussytoes (*Antennaria arcuata*), a rare plant known from Wyoming, C Idaho, and N Nevada in isolated islands of meadow communities surrounded by sagebrush steppe. In the High Creek Fen wetland complex of C Colorado, Kelsey's phlox is part of the halophytic flora of sodic peat communities dominated by sea milkwort (*Glaux maritima*) and curly bluegrass (*Poa secunda*). This is the harshest of the High Creek Fen communities, with vegetation cover less than twenty percent on these alkaline flats.

In C Nevada, Kelsey's phlox occurs on hot springs mounds with saline-tolerant plants including seashore saltgrass (*Distichlis spicata*) and shrubby seepweed (*Suaeda moquinii*). It is an associate of the rare Monte Neva Indian paintbrush (*Castilleja salsuginosa*), which is known from only two occurrences on travertine hot springs mounds in Nevada.

Cultivation

Kelsey's phlox is a rising star among the western phloxes. Selections made from wild populations in the Lemhi Mountains region of C Idaho began circulating among rock gardeners in the 1980s, and have proven both floriferous and relatively easy to grow. Given its occurrence in clayey, alkaline soils in the wild, Kelsey's phlox should prove widely adaptable as a garden plant, and its tolerance of seasonal wetness could help it survive in typically over-watered American landscapes. Grow in full sun. Cultivars are available.

Phlox longifolia
longleaf phlox
PLATES 41–44

There is an air to the sagebrush sea that eventually drowns all the senses. The vast sagebrush biome of western North America is a silvery, aromatic world unto itself, sovereign over the land and the lives of a host of creatures. Some, like sage grouse and pygmy rabbits, are obligates of *Artemisia* and can live nowhere else. Others are less entangled but still intimate associates. Longleaf phlox is one of them, a graceful, sometimes willowy wildflower that seems to follow these pungent shrubs wherever they go.

Although strongly allied with sagebrush, longleaf phlox is not entirely limited to its domain, also turning up in woodlands, mountain brush, and grasslands. This plant blankets the Intermountain West, as botanist Arthur Cronquist once observed, and can be so common you just stop noticing it. But I will never forget lying down with longleaf phlox in the meadows of Lemhi Pass, as I sought empathy with Lewis and Clark, who, upon ascending to this hopeful point on the Continental Divide, suddenly knew the Pacific Ocean was still a long, way away.

∽∾

Phlox longifolia Nuttall, J. Acad. Nat. Sci. Philadelphia 7: 41. 1834 [long-leaved; "foliis subulatis longissimis"] Upstanding suffrutescent perennial with erect-ascending flowering shoots (5–)10–40(–100) cm tall with ca. 5–10 leaf nodes below the inflorescence; *leaves* narrow to broadly linear to narrowly lanceolate to oblong-lanceolate, long-acuminate, the margin varying from thin and green to thick and pale, sparsely ciliate, surface glabrous to pilose, max. length 40–90(–100) mm (but some expressions with max. leaf length 15–40 mm) and width 1–2(–4) mm; *inflorescence* 3- to 18-flowered, rather lax, its herbage pubescent with fine hairs, glandless or gland-tipped, max. pedicel length (5–)10–40(–60) mm; *calyx* (8–)10–13(–15) mm long, united ⅜–⅝ its length, lobes linear-subulate with apex cuspidate, membranes conspicuously plicate to carinate; *corolla* tube (11–)13–17(–18) mm, lobes variable but often rather narrowly obovate (dimensions varying from 8 × 5 mm to 10 × 6 mm) with apex entire, short-apiculate, or erose to emarginate, hue light purple (sometimes deep purple), lilac, pink, or white, the eye sometimes striate; *stamens* borne on mid- to upper corolla tube, with some anthers exserted; *style* (8–)9–14(–18) mm long, united to tip which is free for 1–1.5 mm, with stigma placed among the anthers near corolla orifice or sometimes exserted; *flowering season* spring and early summer; April–July, depending on latitude and altitude.

∽∾

Taxonomic notes

The *Phlox longifolia* complex is taxonomically difficult, with as many as sixteen species and thirty-six infra-specific taxa named within this wide-ranging, variable, intergrading assemblage. Key morphological features that have been used to sort out this diversity include pubescence of inflorescence-herbage (glandular versus nonglandular), leaf length and width, and corolla tube length.

Wherry was the first to approach the task informed by both herbarium and field research, and in his early treatments interpreted *Phlox longifolia* as a single species with five (Wherry 1938) and then seven (Wherry 1941c, 1942) subspecies, but in his 1955 monograph divided the complex into two species—*P. longifolia* with glandless pubescence and *P. viridis* with conspicuously glandular pubescence, the former consisting of four subspecies and the latter of three subspecies. He regarded these seven taxa as a complex allied with but separate from a series of entities with elongate corolla tubes—*P. dolichantha, P. superba, P. stansburyi,* and *P. grayi.*

Subsequent workers have tended toward much more conservative schemes, even incorporating *Phlox stansburyi* and *P. grayi* into their interpretation of *P. longifolia*: Peabody (1979) recognized five varieties, Cronquist (1984) two varieties, and others (Welsh 2003; Wilken and Porter 2005) no subspecific taxa within *P. longifolia*. Wherry (1962) disagreed with such lumping, noting that field study "shows that most of these submerged taxa occur so frequently in pure stands and under specialized ecologic or geographic conditions as to merit some degree of nomenclatural recognition."

Wherry's observations may be meaningful, but distribution maps in his 1955 monograph show no discernable geographic segregation of these seven taxa, and no subsequent research has clarified the ecological associations of the members of this complex. As presently understood, the morphological variation represented by these taxa does not show clear correlation with geographic/ecological segregation. Pending further study, *Phlox longifolia* is interpreted here as a complex of up-standing long-styled phloxes with calyx membranes conspicuously carinate (raised lengthwise into a keel-like ridge) and with an average corolla tube length of 13–17 mm. This interpretation incorporates *P. grayi* (also treated as *P. longifolia* subsp. *brevifolia*) while segregating *P. longifolia* from *P. dolichantha* (corolla tube > 35 mm) and *P. stansburyi* (corolla tube 19–33 mm) (see discussion under *P. stansburyi*).

Careful field research throughout the range of the *Phlox longifolia* complex could demonstrate that some of the variation recognized by Wherry at the species or subspecies level corresponds geographically and ecologically to taxa of sagebrush (*Artemisia* subgenus *Tridentatae*) in western North America. Eleven species and a number of subspecies of sagebrush are recognized within the subgenus, with some taxa demonstrating patterns of niche differentiation and habitat partitioning correlated to soil properties such as elemental composition, texture, depth, and pH. Such a relationship appears exemplified by the strong association of a very dwarf (5–10 cm tall) expression of *P. longifolia* (interpreted by Wherry as *P. viridis* subsp. *compacta*) with dwarf sagebrush (*A. arbuscula*) in unique edaphic situations in the northern Great Basin.

Geography

Western interior of the United States, extending into southwestern Canada. The core range of the *Phlox longifolia* complex extends from the western slope of the Rocky Mountains westward to the Cascade–Sierra axis, and from the Columbia Basin southward into the Great Basin and Colorado Plateau, with disjunct occurrences on the eastern side of the Rocky Mountains on the Great Plains in SE Colorado and in the vicinity of Denver.

Environment

Plains, intermountain basins, dissected plateaus, hilly piedmont, and mountains at low to middle elevations, on upper slopes of rolling terrain and divides, mesas, and in canyons, 500–9000 ft. (150–2700 m); *MAP* 8–14 in. (20–36 cm); *habitat* dry to dry-mesic; *soils* loamy sand, sandy loam, loam, or clay loam, often stony, shallow to moderately deep; *parent material* varied, including shale.

Associations

Conifer woodland, shrubland, and grassland. Throughout its range, longleaf phlox is a relatively constant associate of juniper woodland and pinyon-juniper woodland. In the Rocky Mountains, these woodlands are dominated by Rocky Mountain juniper (*Juniperus scopulorum*) or Utah juniper (*J. osteosperma*). On the Columbia Plateau in C Oregon, longleaf phlox occurs in association with western juniper (*J. occidentalis* var. *occidentalis*) woodland. The composition of pinyon-juniper woodland varies, with singleleaf pinyon pine (*Pinus monophylla*) and Utah juniper codominant in the Great Basin, and two-needle pinyon pine (*P. edulis*) and Utah juniper codominant on the Colorado Plateau.

On the Columbia Plateau in E Washington, longleaf phlox is associated with ponderosa pine (*Pinus ponderosa*) woodland and savanna. In N Utah, it extends sparingly up into higher elevations where it occurs in quaking aspen (*Populus tremuloides*) / Engelmann spruce (*Picea engelmannii*) / subalpine fir (*Abies lasiocarpa*) forest. In the mountains of NE Nevada, longleaf phlox occurs in subalpine woodland comprised of open stands of limber pine (*Pinus flexilis*) and Intermountain bristlecone pine (*P. longaeva*).

Members of the *Phlox longifolia* complex occurs most frequently in shrubland and shrub-steppe dominated by sagebrush (*Artemisia* spp.). At the eastern limits of its range in the Middle and Northern Rocky Mountains, longleaf phlox occurs in relatively mesic meadows or parklands dominated by mountain big sagebrush (*A. tridentata* subsp. *vaseyana*) in association with Idaho fescue (*Festuca idahoensis*) or rough fescue (*F. campstris*). In NE Nevada, it occurs in mountain brush communities dominated by mountain big sagebrush with mountain snowberry (*Symphoricarpos oreophilus*). In montane areas of Utah, longleaf phlox occurs in oakbrush communities dominated by stands of Gambel oak (*Quercus gambelii*).

In the northern portion of its distribution on the Columbia Plateau, longleaf phlox is strongly associated with sagebrush steppe dominated by basin big sagebrush (*Artemisia tridentata* subsp. *tridentata*). Longleaf phlox occurs in each of the sagebrush steppe communities types identified by Daubenmire (1970) for the Columbia Plateau of Washington, which are differentiated by the species of codominant bunchgrass in association with basin big sagebrush, either Idaho fescue, blue-bunch wheatgrass (*Pseudoroegneria spicata*), curly bluegrass (*Poa secunda*), or needle-and-thread (*Hesperostipa comata*). In this region, longleaf phlox occurs most consistently in sagebrush steppe where bluebunch wheatgrass is the codominant. Longleaf phlox also occurs in relatively mesic sagebrush steppe dominated by three-tip sagebrush (*A. tripartita*) on the Columbia Plateau and on the Snake River Plain. It occurs to a lesser extent in shrub-steppe dominated by antelope bitterbrush (*Purshia tridentata*).

In the southern portion of its range, longleaf phlox occurs in Great Basin sagebrush shrubland which is dominated by basin big sagebrush but lacks the bunchgrass component and overall floristic diversity of sagebrush steppe. In this region it also occurs in desert shrubland where saltbush (*Atriplex* spp.) and black greasewood (*Sarcobatus vermiculatus*) are important species.

In xeric, shallow soil habitat in the central part of its range, longleaf phlox is a relatively constant component of dwarf-shrubland comprised of a low-growing species of sagebrush in codominance with a species of perennial bunchgrass. These communities may occur as large, expansive stands or as openings in a matrix with shrub-steppe or coniferous woodland. Longleaf phlox occurs most frequently in communities dominated by dwarf sagebrush (*Artemisia arbuscula*), but it also oc-curs with alkali sagebrush (*A. arbuscula* subsp. *longiloba*), and black sagebrush (*A. nova*). Idaho fescue, curly bluegrass, or bluebunch wheatgrass are the principal as-sociated grasses. Longleaf phlox occurs with dwarf sagebrush in N Nevada, and NE California. In SC Oregon a very dwarf (5–10 cm tall) expression of longleaf phlox is strongly associated with dwarf sagebrush, the phlox plants frequently growing within the shrub itself.

Longleaf phlox occurs in a variety of grassland communities, particularly in the northern and eastern portions of its range. At moderate elevations in Northern Rocky Mountains of Montana and Idaho, longleaf phlox occurs in communities dominated by Idaho fescue, being most common in the Idaho fescue / bluebunch wheatgrass community type. In the mountains of EC Idaho, longleaf phlox as-cends up to the alpine zone in a spike fescue (*Festuca kingii*) grassland community.

Longleaf phlox occurs in most of the steppe communities types identified by Daubenmire (1970) for the Columbia Plateau of E Washington, including blue-bunch wheatgrass / curly bluegrass, bluebunch wheatgrass / Idaho fescue, needle-and-thread / curly bluegrass, and long-awned purple three-awn (*Aristida purpurea* var. *longiseta*) / curly bluegrass associations. While it is an occasional component of the relatively mesic meadow-steppe of the Palouse region of SE Washington and adjacent Idaho and Oregon, it is more common in drier communities dominated by bluebunch wheatgrass. In NW Colorado, longleaf phlox occurs in Great Basin grassland dominated by bluebunch wheatgrass with Indian mountain-ricegrass (*Achnatherum hymenoides*).

Disjunct populations of longleaf phlox occur in E Colorado in shortgrass prairie dominated by blue grama (*Bouteloua gracilis*), with buffalograss (*Buchloe dactyloides*) or James' galleta (*Pleuraphis jamesii*) associated. Scattered flat sagebrush (*Artemisia bigelovii*) and winter-fat (*Krascheninnikovia lanata*) shrubs, along with tree cholla (*Opuntia imbricata*) cactus also occur in these grasslands. Longleaf phlox also occurs in association with the low shrub James' frankenia (*Frankenia jamesii*) in shallow soil habitat on outcroppings of Timpas shale in SE Colorado.

Natural history notes

A number of birds and mammals are essentially limited to sagebrush ecosystems in the American West, and longleaf phlox is an important food plant of several of these. It is a primary food for sage grouse (*Centrocercus urophasianus*) (Barnett and Crawford 1994; Klebenow and Gray 1968) and Townsend's ground squirrel (*Spermophilus townsendii*) (Rogers and Gano 1980), both characteristic of sagebrush steppe. It also is commonly consumed by rabbits, which have been observed to graze it into compact hemispheric cushions when the plants occur in openings between sagebrush shrubs (Daubenmire 1970).

Cultivation

References to longleaf phlox can be found in the literature of native plant horticulture and rock gardening, but it has not been widely cultivated. It has been offered on occasion by specialty nurseries, and has circulated in plant society seed exchanges. Longleaf phlox has potential for regional native plant horticulture, and some of the more dwarf expressions would be attractive subjects for rock garden use. Grow in full sun in dry to slightly moist, well-drained soils.

Phlox maculata
meadow phlox

The humidity of summer slightly blurs the vision in the Wapsipinicon Valley of northeast Iowa, the drenched air hanging like breath exhaled over ripening fields of corn and soybeans. It is quintessential Midwest farm country. Yet here and there, tucked in among the crops, are remnants of once-wild Iowa called fens. These spring-fed wetlands, too soggy to plow and plant, are coming into their prime at this time of year, glorious in the blue of lobelia, yellow of goldenrod, purple of joe-pye-weed, and pink of meadow phlox.

A really good fen can suck your shoes off, but what a small price for getting up close to one of these exuberant plant communities. Along with a suite of wetland generalists, a groundwater-cooled fen often harbors rare plants isolated from the boreal north, left behind in relict populations from glacier-pressed times. Because of its affinity for such fragile and often imperiled wetland habitats, meadow phlox is itself a rare plant in most parts of its range, the distribution of both phlox and fen often one and the same.

Phlox maculata Linnaeus, Sp. Pl. 1: 152. 1753 [spotted, of the stems; "caulis . . . *maculatus*"] Upstanding perennial herb from a slender shallow rhizome which at its apex turns up into a solitary flowering shoot 35–70 cm tall (75–125 cm tall in more fertile soils) with 10–15 spaced leaf nodes below the inflorescence, the nodes more numerous (18–35) and crowded in some population systems, stems typically streaked or spotted with purple; *leaves* linear below, widening upward, the longest near mid-stem and the widest just below the inflorescence, becoming lanceolate or often ovate, the uppermost truncate to cordate-clasping, max. length (45–)65–130(–140) mm and width 5–25 mm, upper surface glabrous to sparsely or rarely copiously pi-lose; *inflorescence* a cylindric panicle of small cymes, in vigorous clones becoming narrowly conical through stem-branching, 75- to 150- (up to 200-) flowered, pedi-cels often minutely pubescent, max. length 3–5(7) mm; *calyx* 5.5–9.5 mm long, united ½–⅝ its length, lobes triangular-subulate with apex sharp-cuspidate, usually glabrous outside but sparse-pilose on inner lobe surface, membranes flat to subpli-cate; *corolla* tube 18–25 mm long, lobes obovate to orbicular (dimensions varying from 8 × 7 mm to 11 × 11 mm), hue purple, pink, or white, sometimes with eye striae or a purple eye-ring; *stamens* with anthers included and placed just below co-rolla orifice; *style* 14–20(–23) mm long, united to tip which is free for 0.5–1 mm; *flowering season* begins late spring and continues into the middle of summer, with some population systems (sometimes identified as subsp. *pyramidalis*) flowering 3–6 weeks later, the flowering period sometimes extended in moist growing seasons.

Taxonomic notes

Wherry (1955) recognized *Phlox maculata* subsp. *pyramidalis*, which he separated from the nominate expression (subsp. *maculata*) by a suite of morphological char-acters, subsp. *pyramidalis* having stems crowded with more leaf nodes, narrower leaf shape, and a pyramidal (versus cylindrical) geometry to the inflorescence. Wherry also cited differences in flowering season, with subsp. *pyramidalis* bloom-ing about a month later than subsp. *maculata*. Finally, as mapped by Wherry, the subspecies show a level of geographic segregation, subsp. *maculata* occupying the northern part of the species range, subsp. *pyramidalis* the southern. While these differences are significant, Wherry noted "intermediates difficult to assign definitely to one or the other [subspecies] occur so frequently that they can not be segregated at the species level." Delineation of subspecies is particularly difficult in the central portion of the species range, particularly in the Appalachian Highlands, where meadow phlox may come into contact with sympatric occurrences of the allied *P. carolina* and *P. glaberrima*. As presently understood, the morphological variation represented by these subspecies does not show clear correlation with geographic/ecological segregation.

Geography

Eastern and central United States, concentrated in the Appalachian Highlands from New York south into Tennessee and Georgia, extending east onto the northern Atlantic Coastal Plain and west through the north-central Interior Plains to the lower Great Lakes region (N Illinois and N Indiana) thence to C Iowa and SE Minnesota, with outlying occurrences in the Ozark Plateaus in S Missouri. Reports of meadow phlox from several New England states and southeastern Canada likely represent garden escapes. Meadow phlox is strongly associated with fen habitat throughout its range. These unique wetlands (described below) have historically been geographically isolated, or are presently isolated due to habitat alteration and destruction. As a consequence, the actual distribution of meadow phlox is fragmented and it is a plant of conservation concern in a number of states.

Environment

Plains, dissected plateaus, hilly piedmont, and mountains, associated with fens, groundwater seeps, springs, river valleys, floodplain terraces, stream edge, and riverwash; *habitat* wet to hydric; *soils* deep, organic, and saturated (rarely inundated) most of the year; *parent material* typically calcareous bedrock or glacial deposits.

Meadow phlox is most often associated with fens—wetlands fed by nutrient-rich (minerotrophic) groundwater. Fens (and seeps) occur at stratigraphic and/or topographic breaks in the landscape that create hydrologic gradients causing groundwater to reach the land surface. Fens are often situated on slopes, where groundwater discharges out of a hillside, and are associated with groundwater that has been influenced by calcareous bedrock (limestone or dolomite) or glacial deposits and is rich in both calcium and magnesium bicarbonates. The high concentration of bicarbonates sometimes precipitates marl at the soil surface. Soils associated with fens are typically near-neutral in pH, but can range between 5.5 and 8.2. The relatively cool temperature of groundwater creates a cool zone in association with the fen.

Associations

Floodplain and riparian forest, swamps, marshes, fens, and riverwash communities. Meadow phlox occurs in a variety of wetland systems, but most often in those that develop in association with fens and calcareous seeps. These are small patch communities occurring within the matrix of larger ecological systems, and often transition into other wetland communities. Fens are distinguished from seeps when they are large enough to create gaps in tree canopies or develop an herbaceous layer dominated by sedges.

Meadow phlox occurs in forested fens, shrub fens, and prairie fens. Red maple (*Acer rubrum*) is the dominant tree of forested fens in which meadow phlox occurs. Shrub fens are a mosaic of shrubs with graminoid-dominated openings. Alders (*Alnus incana* and *A. serrulata*), common buttonbush (*Cephalanthus occidentalis*), dogwoods (*Cornus amomum* and *C. sericea*), northern spicebush (*Lindera benzoin*), eastern ninebark (*Physocarpus opulifolius*), and willows (*Salix candida, S. humilus*, and *S. sericea*) are typical of shrub fens in which meadow phlox occurs. Prairie fens develop in the context of tallgrass prairie and are dominated by typical fen sedges mixed with wet-mesic prairie grasses like big bluestem (*Andropogon gerardii*), yellow Indiangrass (*Sorghastrum nutans*), and prairie cordgrass (*Spartina pectinata*).

Throughout its range, meadow phlox occurs with a suite of characteristic herbaceous species. Some, like swamp milkweed (*Asclepias incarnata*), great blue lobelia (*Lobelia siphilitica*), and cattail (*Typha latifolia*), are wetland generalist species. Other associates of meadow phlox are more narrowly restricted to fen or calcareous seep habitat. The most important of these are the sedges, many of which form tussocks and create a hummocky microtopography in a fen. Some sedges occur with meadow phlox practically throughout its range, notably tussock sedge (*Carex stricta*) but also porcupine sedge (*C. hystericina*), bristly-stalk sedge (*C. leptalea*), shallow sedge (*C. lurida*), and prairie sedge (*C. prairea*). Three pteridophytes are frequent associates of meadow phlox: sensitive fern (*Onoclea sensibilis*), royal fern (*Osmunda regalis*), and eastern marsh fern (*Thelypteris palustris*). Commonly associated forbs include white turtlehead (*Chelone glabra*), spotted water-hemlock (*Cicuta maculata*), swamp thistle (*Cirsium muticum*), spotted joe-pye-weed (*Eupatorium maculatum*), common sneezeweed (*Helenium autumnale*), four-flower loosestrife (*Lysimachia quadriflora*), Riddell's goldenrod (*Oligoneuron riddellii*), stiff cowbane (*Oxypolis rigidior*), swamp lousewort (*Pedicularis lanceolata*), Virginia mountain-mint (*Pycnanthemum virginianum*), Canada burnet (*Sanguisorba canadensis*), New England aster (*Symphyotrichum novae-angliae*), and swamp aster (*Symphyotrichum puniceum*).

Meadow phlox occurs with a small number of plants that are fen specialists, some of which are geographically limited and considered species of conservation concern including marsh-marigold (*Caltha palustris*), small white lady's-slipper (*Cypripedium candidum*), queen-of-the-prairie (*Filipendula rubra*), fringed gentian (*Gentianopsis crinata*), grass-of-Parnassus (*Parnassia glauca* and *P. grandifolia*), and hairy valerian (*Valeriana edulis* var. *ciliata*).

Fens represent some of the rarest yet most biologically diverse wetlands in the eastern United States. Meadow phlox occurs in many of the best remaining fen communities, including the Cranberry Glades Botanical Area in West Virginia, one of the largest fens in the Appalachian Highlands, and Grasshopper Hollow in S Missouri, the largest fen complex in unglaciated North America. Hine's emerald dragonfly (*Somatochlora hineana*), one of the most endangered dragonflies in the United States, is associated with fens in which meadow phlox occurs.

Cultivation

The dramatic cylindrical to pyramidal shape of the meadow phlox inflorescence can be seen in a number of modern border phlox cultivars, as well as those figured in nineteenth-century gardening periodicals. Some of these are pure *Phlox maculata*, but other early summer (early June into July) flowering border phloxes may have *P. carolina* in their heritage. Cultivars of unknown parentage but suspected of being selections of *P. maculata* or of its hybrids with *P. carolina* should be identified with the *Phlox* Suffruticosa Group (see Appendix B). Meadow phlox can be used in herbaceous perennial gardens and borders, mixed borders, and rain gardens. Grow in full sun to partial shade in deep, evenly moist, well-drained soils, but tolerant of seasonally wet conditions. Meadow phlox is somewhat susceptible to foliar diseases (see *P. paniculata* species account for management practices).

Phlox missoulensis
Missoula phlox
PLATES 45, 46

The fresh snow was hard to reconcile with a flatlander's calendar. It was the middle of June, yet two inches had fallen overnight on the foothills of Montana's Anaconda Range. The rolling grassy ridges were dressed in a white that surely wouldn't last for long. But as morning dawned and the sun ascended, and the crystalline mantle melted away, flowery cushions of Missoula phlox came into view, bearing the crispness and prolonging the tonic of a high country summertime snow.

Endemic to the mountains of central and western Montana, Missoula phlox occurs on exposed slopes and ridge crests of moderate elevation. It was first collected on Waterworks Hill overlooking the town of Missoula, where it forms tight cushions on the gravelly summit. More typically it occurs in the beautiful rough fescue grasslands that ride montane ridges in this region. Here Missoula phlox grows a bit more lax, weaving in among the bunchgrasses and laying down groundcover in these short, showy meadows of larkspur and lupine, prairie smoke and locoweed.

~~~

Phlox missoulensis Wherry, Notul. Nat. Acad. Nat. Sci. Philadelphia 146: 7, figure 4. 1944 [of Missoula, Montana] Caespitose perennial herb 5–10 cm tall from a branched, spreading caudex, varying in habit from diffuse mat to condensed cushion in more xeric situations, the numerous erect-ascending flowering shoots 3–5 cm long with ca. 3 leaf nodes below the inflorescence; *leaves* linear to linear-lanceolate, thinnish but with subacerose tip, ciliate and surficially pilose with fine gland-tipped hairs, max. length 15–25 mm and width 1.5–2.5 mm; *inflorescence* 1-flowered, its herbage usually copiously glandular-pubescent, max. pedicel length 2–8 mm; *calyx* 9.5–13.5 mm long, united ⅜–½ its length, lobes subulate with moderate midrib and with apex cuspidate, membranes flat; *corolla* tube 10–13 mm long, lobes obovate to orbicular (average dimensions 8 × 6 mm), hue light blue to white, occasionally pink; *style* 5–7 mm long, united to tip which is free for ca. 1 mm, with stigma included; *flowering season* spring, May–June.

~~~

Taxonomic notes

Cronquist (1959) reduced this species to *Phlox kelseyi* var. *missoulensis*, a relationship and reduction with which Wherry (1962, 1965b) strongly disagreed. Campbell (1992) likewise treated it as a variant of *P. kelseyi*, but cited "extremely different habitats" plus "complete geographical isolation coupled with differences in ten morphological characters" in support of recognition of *missoulensis* as a distinct entity within the *P. kelseyi* complex. Field study affirms Wherry's interpretation as most consistent with the pronounced morphological and ecological differences between these two taxa (Locklear 2009).

Geography

Northern Rocky Mountains, endemic to the mountains of WC Montana, chiefly along or west of the Continental Divide where it occurs in the Anaconda, Garnet, Lewis and Clark, and Sapphire ranges, and on small mountains in the vicinity of Missoula (Waterworks Hill [also known as Urlin Hill], Mount Jumbo, and Dean Stone Mountain), with outlying occurrences east of the Continental Divide in the Little Belt Mountains.

Environment
Hilly piedmont and mountains at middle elevations, on exposed slopes and ridges that are windswept with negligible snow pack, 3600–6900 ft. (1100–2100 m), up to 8250 ft. (2500 m); *MAP* 13–15 in. (33–38 cm); *habitat* dry, typically water-stressed for a portion of the growing season; *soils* stony, shallow to moderately deep; *parent material* limestone, granite.

Associations
Primarily montane grassland dominated by rough fescue (*Festuca campestris*), with lesser amounts of other bunchgrasses associated, notably Idaho fescue (*F. idahoensis*) and bluebunch wheatgrass (*Pseudoroegneria spicata*). These meadowlike communities occur in areas otherwise forested by lodgepole pine (*Pinus contorta* var. *latifolia*), Douglas-fir (*Pseudotsuga menziesii*), and subalpine fir (*Abies lasiocarpa*), with the trees sometimes forming krummholz at the margins of the meadows. Associated forbs include Wyoming coral-drops (*Besseya wyomingensis*), Flathead larkspur (*Delphinium bicolor*), prairie-smoke (*Geum triflorum*), Pursh's silky lupine (*Lupinus sericeus*), white point-vetch (*Oxytropis sericea*), and American bistort (*Polygonum bistortoides*). Some of these grasslands provide habitat for small, unusual ferns known as moonworts, most notably peculiar moonwort (*Botrychium paradoxum*). Longleaf phlox *(Phlox longifolia)* and showy phlox (*P. speciosa*) sometimes co-occur with Missoula phlox in these grassland communities.

On drier, windswept sites with shallow, rocky soils, Missoula phlox occurs in sparsely-vegetated rock outcrop communities dominated with low-growing forbs including hairy false goldenaster (*Heterotheca villosa*), mountain douglasia (*Douglasia montana*), oval-leaf wild buckwheat (*Eriogonum ovalifolium*), Oregon bitter-root (*Lewisia rediviva*), and silvery ragwort (*Packera cana*).

Cultivation
Lincoln Foster collected seed of Missoula phlox at its type locality in 1966, and this species has been quietly circulating among rock gardeners ever since. Strains of Missoula phlox growing in meadows where they must compete with bunch-grasses may prove more adaptable to general garden culture than those occurring

on barren, rocky sites. In addition to its proven value as a rock garden plant, Missoula phlox could also have potential as a drought-tolerant landscape plant for the Mountain West. Grow in full sun in dry, well-drained soil.

Phlox multiflora
Yellowstone phlox
PLATE 47

The Medicine Wheel seems to have dropped from the sky. This mysterious circle of stone, eighty feet in diameter, with twenty-eight spokes radiating out from a central cairn, stands above timberline on a ridge in Wyoming's Bighorn Mountains. The Crow, Cheyenne, and other native peoples of the region revere the Wheel as a place of spiritual pursuit, but its creators are unknown to them. Carefully placed rocks declare purpose, and worn mountain trails reveal centuries of pilgrimage, but the meaning, emotion, and longings behind the work are lost in antiquity.

Whatever you might seek from the Medicine Wheel, your experience will be heightened in summer by the sweet fragrance of Yellowstone phlox, a wildflower so abundant here it perfumes the air. This mat-forming species is a familiar sight at middle elevations in the Rockies, from Colorado north into Montana, and from dry sage parkland up into the alpine. It often marks, like a living shard, ground roughed by the tread of our kind, from tourists seeking beauty in Yellowstone, to ancients seeking light at the Medicine Wheel.

༄ ༺ ༄

Phlox multiflora A. Nelson, Bull. Torrey Bot. Club 25: 278. 1898 [many-flowered; "numerous flowers"] Caespitose perennial herb from a diffusely branched, spreading caudex, ranging in habit from diffuse mats 8–20 cm tall, the numerous decumbent to erect-ascending flowering shoots 5–15 cm long, to more strongly condensed forms 4–8 cm tall in xeric situations with flowering shoots 2–4 cm long; *leaves* linear, thinnish and sometimes lax, at most sparingly ciliate and often glabrous, max. length 12–25 mm and width 1.5–2.5 mm (up to 30–45 mm long and 2–3 mm wide in more mesic habitat); *inflorescence* 1- to 3-flowered, its herbage glabrous to sparse-pilose with glandless or sometimes a few short gland-tipped hairs, max. pedicel length 7.5–15 mm, but ranging from 3–7 mm in xeric habitat to 18–35 mm in mesic habitat; *calyx* 8–13(–15) mm long, united ⅜–⅝ its length, lobes linear-subulate with moderate midrib and with apex cuspidate, membranes broad, flat, and deep-seated; *corolla* tube 9–15(–16) mm long, lobes obovate (average dimensions 8 × 6 mm), hue lilac to pink to white; *stamens* borne on upper corolla tube, with anthers mostly exserted; *style* 6–10(–12) mm long, united to tip which is free for 1 mm, with stigma placed among the anthers at or near corolla orifice; *flowering season* spring to early summer, June–August depending on altitude.

Taxonomic notes

Wherry (1955) recognized *Phlox multiflora* subsp. *depressa* and subsp. *patula*, which he separated from the nominate expression (subsp. *multiflora*) by differences in stature and size of floral parts, viewing them as a series running from the taller, open-growing subsp. *patula* through the mat-forming subsp. *multiflora* to the more reduced, cushion-forming subsp. *depressa*. These appear to represent a series of environmentally influenced expressions, with subsp. *patula* occurring in more mesic, protected canyon habitat on the east slope of the Rockies and subsp. *depressa* in more stressful habitat such as exposed escarpments in the Wyoming Basin and near-alpine situations in the mountains. As presently understood, the morphological variation represented by these subspecies does not show clear correlation with geographic/ecological segregation.

Geography

Chiefly Rocky Mountain System, concentrated in the Middle Rocky Mountains in Wyoming, extending into the Northern Rocky Mountains (E Idaho and W Montana) and south into the Wyoming Basin, the Southern Rocky Mountains (to C Colorado), and the northern edge of the Great Basin (S Idaho and N Nevada), with outlying occurrence in the Blue Mountain section of the Columbia Plateau (NE Oregon). Yellowstone phlox occurs with greatest abundance in the Absaroka, Bighorn, and Wind River mountain ranges of the Middle Rocky Mountains in Wyoming. In Colorado, Yellowstone phlox is most common in the intermountain basins of North Park, east of the Continental Divide, and Middle Park and the Gunnison Basin, west of the Continental Divide. The common name refers to the Greater Yellowstone Ecosystem, a region of the Middle Rocky Mountains in and around Yellowstone National Park encompassing NW Wyoming and adjacent parts of Idaho and Montana where Yellowstone phlox is a common element in the flora.

Environment

Intermountain basins, hilly piedmont, and mountains, primarily at middle elevations but up through timberline to alpine areas, on escarpments, bluffs, ledges, slopes, and ridges, 6000–12000 ft. (1800–3600 m) in Colorado, 7300–9900 ft.

(2200–3000 m) in Utah, and 6200–11000 ft. (1900–3300 m) in Wyoming; *MAP* 10–14 in. (25–36 cm); *habitat* dry to dry-mesic; *soils* medium- to coarse-textured, stony, shallow with bedrock at or near the surface; *parent material* varied, including limestone in Bighorn and Owl Creek mountains of Wyoming and quartzite in Lemhi Range of Idaho.

Associations

Conifer forest/woodland, shrubland, shrub-steppe, and grassland. Yellowstone phlox often occurs in large, conspicuous populations. It is often one of the most abundant herbs present in these communities and in some associations is the dominant forb and is recognized as a principal indicator species. Its prominence is attested to by place names like Phlox Creek in Yellowstone National Park and Phlox Mountain in the Owl Creek Range (NW Wyoming). Joseph Barrell (1969) called it "the universal phlox" of the Gunnison Basin (SW Colorado).

Yellowstone phlox is associated with conifer forest and woodland throughout its distribution, ranging from Utah juniper (*Juniperus osteosperma*) woodland and ponderosa pine (*Pinus ponderosa*) woodland at lower and drier elevations up through Douglas-fir (*Pseudotsuga menziesii*) forest and into subalpine woodland of whitebark pine (*Pinus albicaulis*). Yellowstone phlox occurs most frequently in mid-elevation forest codominated by lodgepole pine (*P. contorta* var. *latifolia*) and quaking aspen (*Populus tremuloides*), often at the interface between this community and openings or "parks" dominated by sagebrush steppe and/or grassland. It also occurs in association with open stands of limber pine (*Pinus flexilis*) on dry, rocky ridge-crests. In all of these communities, Yellowstone phlox is a component of an herbaceous layer comprised of graminoids including short sedge (*Carex rossii*), Geyer's sedge (*C. geyeri*), spike fescue (*Festuca kingii*), and prairie Junegrass (*Koeleria macrantha*), along with a diversity of forbs.

Yellowstone phlox is commonly associated with montane sagebrush steppe. These communities, sometimes referred to as "sage meadows," are dominated by big sagebrush (*Artemisia tridentata*), with Idaho fescue (*Festuca idahoensis*) typically the codominant bunchgrass. In the Middle Park region of Colorado, Yellowstone phlox is most common in sagebrush steppe dominated by mountain big sagebrush (subsp. *vaseyana*), but it also occurs with Wyoming big sagebrush (subsp. *wyomingensis*) and basin big sagebrush (subsp. *tridentata*). Typical forbs occurring with Yellowstone phlox in sagebrush steppe include arrow-leaf balsam-root (*Balsamoriza sagittata*), sulphur-flower wild buckwheat (*Eriogonum umbellatum*), prairie smoke (*Geum triflorum*), lupines (*Lupinus* spp.), prairie bluebells (*Mertensia lanceolata*), and mat beardtongue (*Penstemon caespitosus*). Yellowstone phlox also is a minor component of shrubland dominated by Saskatoon serviceberry (*Amelanchier alnifolia*) and black greasewood (*Sarcobatus vermiculatus*).

At middle elevations in the mountains of Wyoming and Montana, Yellowstone phlox occurs in montane grassland communities dominated by Idaho fescue. It is particularly abundant in Idaho fescue grassland in the Bighorn Mountains of Wyoming. In W Wyoming and adjacent Idaho, Yellowstone phlox occurs in unique montane tall forb meadows characterized by a luxuriant abundance of tall 40–120 cm forbs. Common associates in this community type include arrow-leaf balsam-root, hoary balsam-root (*Balsamoriza incana*), sulphur Indian paintbrush (*Castilleja sulphurea*), sticky geranium (*Geranium viscosissimum*), prairie smoke, fern-leaf lovage (*Ligusticum filicinum*), silvery lupine, and leafy Jacob's-ladder (*Polemonium foliosissimum*).

Yellowstone phlox is an important component of alpine forb meadows in the Absaroka Mountains of NW Wyoming, where it occurs in several of the more mesic alpine communities. In the mountain ranges of SW Montana and EC Idaho, Yellowstone phlox occurs at the lower edge of alpine habitat in turf communities dominated by blackroot sedge (*Carex elynoides*). At higher elevations and in more exposed alpine habitat Yellowstone phlox is replaced by alpine phlox (*Phlox pulvinata*).

Cultivation
Yellowstone phlox has been cultivated to a limited extent as a rock garden plant. The dark green, linear leaves contrast nicely with the crisp white flowers. The more open-growing expressions occurring in relatively mesic sage meadows and in protected canyons may prove more adaptable to general garden and landscape use. Grow in full sun in dry to slightly moist, well-drained soil.

Phlox muscoides
Shoshone phlox
PLATES 48, 49

Like some grizzled little troll from a fairy tale, Shoshone phlox always keeps interesting company—most of it dwarf; much of it in exile. This hoary, humped-back plant is a near constant presence in the wonderful rock plant communities of the Wyoming Basin, reigning amidst the mounds and mats, cushions and carpets that cling to wind-scoured escarpments and ridges. Many of these plants are the most compact, miniature species in their genus, and a number occur nowhere else, isolated here from both the norm and the world.

Dramatic and extreme reduction of leaf, stem, and inflorescence is required of plants banished to the cold desert barrens of the Wyoming Basin, a crucible melding, in Gretel Ehrlich's experience, "torrential beauty" with "absolute indifference." Perhaps it is telling that most of the companions of Shoshone phlox also have silvery foliage, as if the dross and alloy of easier living had been forged out of them. That plants even bother with such places makes you shudder a bit at the resolve to *be* and *beget* that throbs on this planet.

❧

Phlox muscoides Nuttall, J. Acad. Nat. Sci. Philadelphia 7: 42, plate 6, figure 2. 1834 [mosslike; "The whole plant depressed to the appearance of a hoary *Bryum* (a genus of mosses), no part scarcely rising half an inch from the ground!"] Caespitose perennial herb forming dense cushions to domed mounds 2–5 cm tall from a short-branched to spreading caudex, the numerous close-packed erect flowering shoots 1–3 cm long, the leaves so crowded and appressed as to give stems a quadrate (square) outline; *leaves* broad-subulate, thickish-margined and concave, cuspidate, basally arachnoid-tomentose but toward the tip glabrate, max. length 3–5 mm and width 0.5–1.5 mm; *inflorescence* 1-flowered, its herbage tomentose to sparsely pilose, max. pedicel length 0.5–1 mm; *calyx* 3–7 mm long, united ⅜–⅝ its length, lobes linear- to broad-subulate with thick margin and weak midrib and with apex cuspidate, membranes flat and deep-seated, obscured; *corolla* tube 4–8 mm (included in or somewhat exceeding the calyx), lobes elliptic (dimensions varying from 4.5 × 2.5 mm to 3 × 2 mm), hue white or rarely faint lavender; *stamens* borne on upper corolla tube, with anthers mostly exserted; *style* 2–6 mm long, united to tip which is free for ca. 1 mm, with stigma placed among the anthers at or near corolla orifice; *flowering season* spring to early summer, May–June.

❧

Taxonomic notes
Wherry (1941c, 1955, 1962) treated this taxon as *Phlox bryoides* (Nuttall 1848), and Cronquist (1959, 1984) as *P. muscoides*. Dorn, with much more field experience, settled on *P. muscoides* in his *Vascular Plants of Wyoming* (1988), noting, "[h]aving now seen the holotypes of *P. muscoides*, *P. bryoides*, and *P. caespitosa*, I must now agree with Cronquist and disagree with Wherry."

Geography
Chiefly Rocky Mountain System, concentrated in the Wyoming Basin and Middle Rocky Mountains (NW Colorado, SC Montana, and NE Utah/Wyoming),

extending into the Northern Rocky Mountains (EC Idaho and SW Montana) and the Southern Rocky Mountains (N Colorado), with outlying occurrences in the western Great Plains (NE Colorado, W Nebraska, and SE Wyoming), northern Great Basin (NE California, SC Oregon, Nevada, and Utah), and Colorado Plateau (S Utah). The common name commemorates the Shoshone people, whose ancestral homeland in the Wyoming Basin and Middle Rocky Mountains is the core of the range of this species and the provenance of its original collection.

Environment
Plains, intermountain basins, hilly piedmont, and mountains at low to middle elevations, on escarpments, bluffs, rims, rock ledges, and ridges, often subjected to very strong, prolonged winds, 3500–9000 ft. (1000–2700 m); *MAP* 8–12 in. (20–31 cm); *habitat* xeric; *soils* stony, with little organic matter, shallow with bedrock at or near the surface; *parent material* sandstone, limestone, siltstone, shale, slate, conglomerates, volcanic material.

The Wyoming Basin experiences some of the strongest prolonged winds of any region in North America, and the interaction of wind, aridity, harsh substrates, high summer temperatures, and cold winters (lows of −15 degrees F common from November through March, down to −50 degrees F) favor very low, ground-hugging plants, particularly those with cushion or matted growth forms familiar in alpine settings. Shoshone phlox always has a cushion-forming growth habit, but on extremely xeric sites it forms hard, dense mounds. Some of its companions are among the most morphologically reduced species of their genus.

Associations
Primarily rock outcrop communities dominated by herbaceous plants within a matrix of conifer woodland, shrubland, and grassland systems. These communities are characterized by a variety of dwarf, cushion-forming forbs. Bluebunch wheatgrass (*Pseudoroegneria spicata*) and curly bluegrass (*Poa secunda*) are important graminoids in these communities, with Indian mountain-ricegrass (*Achnatherum hymenoides*), spike fescue (*Festuca kingii*), prairie Junegrass (*Koeleria macrantha*), and needle-and-thread (*Hesperostipa comata*) often associated. Prairie sagebrush (*Artemisia frigida*), a low-growing subshrub, is often associated with occurrences of Shoshone phlox. Shoshone phlox is typically one of the most abundant herbs present in these communities and in some associations is the dominant forb and is recognized as a principal indicator species.

In the Middle Rocky Mountains, Shoshone phlox occurs in conifer-dominated woodland systems on escarpments, foothills, and the lower slopes of mountains.

These woodlands develop on shallow, rocky soils, often derived from limestone, and are characterized by open stands of limber pine (*Pinus flexilis*), Utah juniper (*Juniperus osteosperma*), or Rocky Mountain juniper (*J. scopulorum*). Shoshone phlox is an associate of a number of regional endemics in the sparse herbaceous layer of these woodlands, including Nuttall's false sagebrush (*Sphaeromeria argentea*) and the recently discovered species, Railroad Canyon wild buckwheat (*Eriogonum soliceps*) and Sheep Mountain twinpod (*Physaria eriocarpa*). At the far northwestern limits of its range on the Columbia Plateau in Oregon, Shoshone phlox occurs sparingly in western juniper (*J. occidentalis* var. *occidentalis*) woodland. At the southern limits of its range on the Colorado Plateau in Utah, Shoshone phlox occurs in woodland dominated by open stands of Colorado pinyon pine (*Pinus edulis*) and Utah juniper.

Shoshone phlox occurs in variety of shrublands, notably communities dominated by mountain big sagebrush (*Artemisia tridentata* subsp. *vaseyana*), Wyoming big sagebrush (*A. tridentata* subsp. *wyomingensis*), or curl-leaf mountain-mahogany (*Cercocarpus ledifolius*). In the Laramie Basin of SE Wyoming, Shoshone phlox occurs in openings in birchleaf mountain-mahogany (*C. montanus*) shrubland along with Laramie false-sagebrush (*Sphaeromeria simplex*), a rare local endemic.

On exposed, windswept sites in the Wyoming Basin, Shoshone phlox occurs in dwarf-shrubland dominated by black sagebrush (*Artemisia nova*). Shoshone phlox is often a dominant forb in these communities, accompanied by a suite of cushion plants that includes Hooker's sandwort (*Arenaria hookeri*), Sweetwater milkvetch (*Astragalus aretioides*), summer orophaca (*Astragalus hyalinus*), bun milkvetch (*Astragalus simplicifolius*), caespitose cat's-eye (*Cryptantha caespitosa*), single-stem wild buckwheat (*Eriogonum acaule*), low nailwort (*Paronychia sessiliflora*), rock-tansy (*Sphaeromeria capitata*), and sword Townsend-daisy (*Townsendia spathulata*). A number of rare regional endemics are associated with exposures of calcareous or otherwise unique geological formations within the Wyoming Basin, and Shoshone phlox is an associate of many of these including Owl Creek cat's-eye (*Cryptantha subcapitata*), Wyoming feverfew (*Parthenium alpinum*), stemless beardtongue (*Penstemon acaulis*), and Uinta greenthread (*Thelesperma pubescens*). Shoshone phlox also occurs in black sagebrush shrubland on the Colorado Plateau. In the northern Great Basin, disjunct occurrences of Shoshone phlox are associated with dwarf sagebrush (*Artemisia arbuscula*) shrubland.

Shoshone phlox occurs in dry, sparsely vegetated grassland formations, primarily mixedgrass prairie and shortgrass prairie. In the Middle Park intermountain basin of Colorado, it occurs in a bluebunch wheatgrass / prairie sagebrush community type. In the dry mountains of SW Montana and adjacent EC Idaho, Shoshone phlox ascends up into dry alpine tundra.

Shoshone phlox occurs in sparsely vegetated shale badland habitat in the Wyoming Basin, where it is an associate of several rare regional endemics including Porter's sagebrush (*Artemisia porteri*), Cedar Rim thistle (*Cirsium aridum*), Devil's Gate twinpod (*Physaria eburniflora*), and desert yellowhead (*Yermo xanthocephalus*), the latter an endangered species whose entire existence is limited to an area about 43 acres (17 hectares) in size. At eastern limits of its range in the Nebraska panhandle and adjacent Colorado and Wyoming, Shoshone phlox occurs on gravelly hilltops, bluff tops, rim-rock, and rock outcrops, particularly on exposed bluff tops and escarpments of the Arikaree sandstone formation. Typical associates include Hooker's sandwort, summer orophaca, tufted milkvetch (*Astragalus spatulatus*), few-flower wild buckwheat (*Eriogonum pauciflorum*), stemless four-nerve-daisy (*Tetraneuris acaulis*), matted prickly-phlox (*Leptodactylon caespitosum*), alpine bladderpod (*Lesquerella alpina*), plus regional endemics silky orophaca (*A. sericoleucus*), mountain cat's-eye (*Cryptantha cana*), slender parsley (*Musineon tenuifolium*), and plains nailwort (*Paronychia depressa*).

Cultivation

The sort of plant "that calls the water to the mouth" is how Reginald Farrer described Shoshone phlox in *The English Rock-Garden*. Most horticultural descriptions of this silvery, mound-forming plant are written from desire not experience, and have a tinge of resignation about them. That Shoshone phlox grows almost everywhere in the cold desert "bunneries" of the Wyoming Basin makes its beauty even more painful to bear. Shoshone phlox has been cultivated to a limited extent as a rock garden plant, and is sometimes offered under the name *Phlox bryoides*. Like the alpine phloxes, Shoshone phlox would be difficult to cultivate and keep in character except under the favorable conditions of a rock garden or alpine house. Grow in full sun in dry, well-drained soil.

Phlox nana
Santa Fe phlox
PLATE 50

It is country that made poets and troubadours out of humble unlettered *pastores*. Perhaps it was the loneliness of the Hispano shepherd's life, tending sheep, months on end, away from his village back in the Sangre de Cristo Mountains. Or maybe it was the magic of the landscape—a jumble of *cañon* and *mesa*, carved from bed-rock of smoldering reds and pinks. Whatever the reason, a shepherd's sojourn in the canyon country of northeast New Mexico marked him, and cracked open the window of his soul.

Santa Fe phlox adds its art to this romantic land, raising cerise-pink flowers above low clusters of grassy leaves. Ranging southward into Chihuahua, Mexico, it is primarily associated with grassland and pinyon-juniper woodland. In the mountains of Trans-Pecos Texas, however, Santa Fe phlox climbs up into forests of evergreen oaks, alligator juniper, and the theatrical, pink-barked madrone tree. Its eastern limit is the juniper-darkened escarpment of the Llano Estacado, a pres-ence known to wayfaring, wistful Hispano shepherds as *La Ceja*, the Eyebrow.

Phlox nana Nuttall, Proc. Acad. Nat. Sci. Philadelphia 4:10. 1848 and in J. Acad. Nat. Sci. Philadelphia (2) 1: 153. 1848 ["Dwarf"] Upstanding perennial herb form-ing open tufts of a few to multiple erect-ascending flowering shoots, these some-times suffrutescent and ranging from 10–30 cm tall with 6–12 close-set leaf nodes below the inflorescence for the nominate expression (*P. nana*), up to 60 cm tall with ca. 5 nodes in more southerly population systems (variant *P. mesoleuca*), all expres-sions varying markedly in stature and leaf size from year to year in relation to climatic conditions, the nominate expression (*P. nana*) arising from a taproot but more southerly expressions rhizomatous; *vesture* (particularly of the inflorescence-herbage) mostly glandular for nominate expression and for variants *P. mesoleuca* and *P. mexicana*, mostly glandless for variant *P. triovulata*; *leaves* linear or narrow-elliptic to lanceolate, the lowest narrow and short, at mid-stem longest and broad-est, the uppermost short but broad, mostly pubescent, max. length 25–45 mm and width 2–5 mm for nominate expression, 50–90 mm long for southern variants (*P. mesoleuca* and *P. triovulata*); *inflorescence* 2- to 6- (15-) flowered, max. pedicel length 3–30(–45) mm; *calyx* 10–16 mm long for nominate expression, 13–18 mm long for southern variants (*P. mesoleuca* and *P. triovulata*), lobes linear to linear-subulate with obscure midrib and with apex cuspidate, membranes ranging from flat to somewhat plicate for nominate expression and variant *P. mesoleuca* to distinctively plicate to somewhat carinate for variants *P. mexicana* and *P. triovulata*; *corolla* tube 12–18 mm long, pilose with gland-tipped or pointed hairs, rarely glabrous, lobes obovate or orbicular (dimensions varying from 12.5 × 9 mm to 14 × 11 mm) with apex entire or erose (with ragged margin), hue purple to lilac, pink, white, or exceptionally light yellow, the eye usually paled and whitish or yellowish, sometimes bearing deep-hued striae; *stamens* with anthers placed deep within corolla tube; *style* 2–3 mm long, united ¼–½ its length (6–8 mm long for variant *P. mexicana*), included; *flow-ering season* spring, but in years of adequate precipitation again in autumn or almost throughout the growing season.

Taxonomic notes

Following the description of *Phlox nana* (Nuttall 1848), two similar taxa were described from SW New Mexico—*P. triovulata* (Torrey 1858) and *P. mesoleuca* (Greene 1905). Wherry (1955, 1966) treated these as two distinct species separate from *P. nana*, but noted intermediates were common, "raising a question as to the desirability of maintaining them as distinct." Turner (1998a) recognized only *P. nana*, noting "my own field work in the Trans-Pecos region of Texas has convinced me that these several taxa, as recognized by Wherry [*P. mesoleuca* and *P. triovulata*], are not even worthy of varietal rank."

Compounding the situation are similar but poorly known taxa in Mexico. Brand (1907) described two unique color variants of *Phlox nana* (yellow- and purple-flowered) from Chihuahua based on collections made in 1887 by Cyrus Pringle. Maslin (1979) rediscovered populations of these color forms in Chihuahua and proposed full species rank as *P. lutea* and *P. purpurea*, but these names were never validly published. In addition, Wherry (1944) described *P. mexicana* from the vicinity of Durango, Mexico, but this entity is known only from two specimens.

Acknowledging Turner's substantial experience with the Texas flora in the field and in the herbarium, only *Phlox nana* is recognized here. Further study of the *P. nana* complex throughout its range is needed to resolve the taxonomy of this group.

Geography

Chiefly Basin and Range Physiographic Province, concentrated in the Mexican Highland and Sacramento sections (SE Arizona, S New Mexico, and W Texas), extending north into the southern Great Plains (E New Mexico) and the southern Rocky Mountains (NC New Mexico) and south into the highlands of northern Mexico.

The northern limits of the *Phlox nana* complex occur in the southern end of the Sangre de Cristo Range of the Southern Rocky Mountains in an area of NC New Mexico lying roughly between Santa Fe and Las Vegas, extending eastward onto the Great Plains in the canyons and tributaries of the Canadian and Pecos rivers as far as the Mescalero Escarpment—the west-facing "Caprock" edge of the Llano Estacado. The southern limits of the complex are the highlands of the Sierra Madre Occidental in the Mexican states of Chihuahua, Durango, and Sonora, with the southernmost occurrence of any species in the genus *Phlox* is represented by two collections from the "City of Durango" which Wherry treated as *P. mexicana*. Wherry observed that the ranges of *P. nana*, *P. triovulata*, and *P. mesoleuca* "interpenetrate," and distribution maps in his 1955 monograph show the distributions of these entities overlapping except at the northern limits of the complex where the material shows the attributes of classic *P. nana*.

Most of the range of the *Phlox nana* complex is characterized by islandlike mountain systems surrounded by desert basins. In the Trans-Pecos region of W Texas, Santa Fe phlox has been collected in the Chinati, Chisos, Davis, Dead Horse, Eagle, Franklin, Glass, Guadalupe (shared with New Mexico), Ord, Sierra Madera, and Sierra Vieja mountains. In S New Mexico it has been collected in the Black Range and the Capitan, Gallinas, Magdalena, Mimbres, Organ, Sacramento, and San Mateo mountains. At the western limits of its distribution, Santa Fe phlox has been collected in the Chiricahua and Swisshelm mountains (SE Arizona) and in the Sierra de los Ajos (NE Sonora, Mexico).

Environment

Plains, cuestas, dissected plateaus, hilly piedmont, and mountains at low to middle elevations, on upper slopes of rolling terrain and divides, escarpments, mesas, and ridges, and in canyons and on rocky creek beds, ranging 3500–10,000 ft. (1000–3000 m), 6000–7500 ft. (1800–2300 m) in NE New Mexico, 4400–7000 ft. (1300–2100 m) in W (Trans-Pecos) Texas, 5500–7200 ft. (1700–2200 m) in Chihuahua, Mexico, up to 10,000 ft. (3000 m) in certain mountain ranges; *MAP* 8–20 in. (20–50 cm); *habitat* dry to dry-mesic; *soils* stony, shallow with bedrock at or near the surface; *parent material* granite, rhyolite, sandstone, limestone. Wherry (1955, 1966) noted populations of *Phlox nana*, *P. mesoleuca*, and *P. triovulata* tend to occur in "discrete colonies differing in micro-environmental details," with *P. nana* occurring in "more sterile or drier places."

Associations

Conifer woodland, mixed conifer-oak woodland, and grassland. Throughout most of its range, the Santa Fe phlox complex is associated with low to mid-elevation mountain woodlands comprised of open stands of pines, junipers, and evergreen oaks, with an herbaceous layer of graminoids and forbs.

Santa Fe phlox occurs in pinyon-juniper woodland throughout much of its range in New Mexico, dominated by open stands of two-needle pinyon pine (*Pinus edulis*) in association with one-seed juniper (*Juniperus monosperma*). On drier sites these woodlands grade into juniper woodland and savanna, with singleseed juniper dominating in the northern part of the range of Santa Fe phlox, replaced by alligator juniper (*J. deppeana*) and redberry juniper (*J. coahuilensis*) in S New Mexico and Trans-Pecos Texas. At the northern limits of its range in the Sangre de Cristo Mountains of NC New Mexico, and in the Capitan and Sacramento mountains of SC New Mexico, Santa Fe phlox ascends up into montane conifer forest dominated by open stands of ponderosa pine (*P. ponderosa*).

In the mountains of Trans-Pecos Texas and across S New Mexico, Santa Fe phlox occurs in pine-oak forests and woodlands. Dominant species in these communities include gray oak (*Quercus grisea*), Emery's oak (*Q. emoryi*), alligator juniper, and Mexican pinyon (*Pinus cembroides*), along with Texas madrone (*Arbutus xalapensis*). In the mountains of SE Arizona and adjacent Sonora, Apache pine (*P. engelmannii*), Chihuahuan pine (*P. leiophylla* var. *chihuahuana*), and Arizona pine (*P. ponderosa* var. *arizonica*) become important, along with Arizona oak (*Q. arizonica*) and silver-leaf oak (*Q. hypoleucoides*). On the east slope of the Sierra Madre Occidental in Chihuahua, Mexico, Santa Fe phlox occurs in open oak woodlands and in mixed woodlands of oak, pine, and juniper, where Arizona oak, Emory's oak, gray oak, silver-leaf oak, and Chihuahuan oak (*Q. chihuahuensis*) are important species, along with alligator juniper, Mexican pinyon, Apache pine, and Chihuahuan pine. Big bluestem (*Andropogon gerardii*) and common wolf's-tail (*Lycurus phleoides*) dominate the herbaceous layer of these woodlands in Chihuahua, along with various species of grama (*Bouteloua* spp.) and muhly (*Muhlenbergia* spp.) grasses.

Santa Fe phlox is associated with grassland throughout most of its range, described variously as shortgrass prairie, plains grassland, Plains-Mesa Grassland, and Apacherian mixed shrub savanna. Blue grama (*Bouteloua gracilis*) is the dominant grass species throughout the range of Santa Fe phlox, both in grassland communities and in the herbaceous layer of woodlands and savannas. Santa Fe phlox typically occurs at the upper limits of grasslands at the interface with woodland of juniper, pinyon-juniper, or oaks, and in places where surfacing bedrock yields rocky, shallow soils that decrease the dominance of grass cover. In Trans-Pecos Texas, grassland supporting Santa Fe phlox occurs at middle elevations (4300–5500 ft. [1300–1700 m]) on mountain slopes, foothills, plateaus, high mesas, and in basins between mountains. At the southern limits of the range of Santa Fe phlox, these grasslands include a component of rosette plants and succulents characteristic of the Chihuahuan desert.

Shreve (1939) listed *Phlox mesoleuca* among the major associates of grasslands in Chihuahua, Mexico. These grasslands occur on elevated plains along the eastern front of the Sierra Madre Occidental, and are best developed above 5500 ft. (1700 m) elevation. Grama grasses dominate these grasslands, with blue grama accounting for at least 80 percent of the cover in approximately 60 percent of the grassland area. Maslin (1979) found a yellow-flowered form of Santa Fe phlox in the vicinity of Cusihuiriachic in central Chihuahua in November 1978, growing in a remnant of relatively undisturbed grassland habitat where the phlox plants occurred in association with small rock outcrops and "mingled with taller clumps of grasses." Kelaidis (1984), who accompanied Maslin, described the habitat as "prairie," adding "the thick grasses were filled with a rich variety of wildflowers," of which phlox was "the most common plant throughout the several-acre tract." The entity described by Wherry as *P. mexicana* is known from only two collections from the vicinity of Durango, Mexico, which is located in a region historically characterized as grassland dominated by grama grasses, notably blue grama, side-oats grama (*Bouteloua curtipendula*), and hairy grama (*B. hirsuta*) (Gentry 1957; Shreve 1942).

Cultivation

The "Mexican phlox" caused quite a stir when they began to trickle into American horticulture in the 1980s, especially selections with unique crimson and yellow flower coloration. But the fervor faded when it was discovered that Santa Fe phlox and its relatives, adapted to summer drought of the Southwest, were difficult to sustain under typical garden conditions. This was old news to European rock gardeners, who had been growing these plants for half a century. Two sage English alpinists, Graham Stuart Thomas and Will Ingwersen, summed up their experience in similar evaluations: "brilliant but frail"; "exciting but aggravating."

In September of 1978, T. Paul Maslin traveled to the state of Chihuahua, Mexico, in search of "the mythical yellow phlox" and rediscovered populations of phloxes with unique scarlet- and yellow-colored corollas. He returned in November of that year for another collecting trip, this time accompanied by Panayoti Kelaidis. Seed gathered on these and later expeditions were germinated and selections made from the resultant seedlings once they flowered. Crosses were subsequently made between these cultivars and other seedlings which in turn were given cultivar names. Siskiyou Rare Plant Nursery of Oregon was instrumental in making these cultivars available. Many of these cultivars have been grown at the Denver Botanic Gardens and have proven hardy to below zero degrees F.

Often collectively referred to as the "Mexican phlox," these selections are usually listed as cultivars of *Phlox mesoleuca* or *P. triovulata*, names applied to taller, more southern expressions of the first-described *P. nana*. Given the unsettled nomenclature of the *P. nana* complex, and the uncertain parentage of some of the cultivars selected from cultivated material, these are treated here under the name the *Phlox* Madrean Group (see Appendix B).

Santa Fe phlox can be used in containers, as a groundcover, or as a subject for rock gardens, wall gardens, and regional native plant horticulture. Grow in full sun to partial shade in dry, well-drained soil rich in loam and humus. Santa Fe phlox needs adequate water in summer but dryness in winter.

Phlox nivalis
trailing phlox

It doesn't take much breeze to make the Carolina Sandhills sing. The needle of the longleaf pine and the blade of the turkey oak are ever at the ready for the thrum and the pluck of the wind. Airy sigh and waxy clatter, scent of pitch and grind of sand, register lean and xeric on the senses. But beneath the musical, sky-infused canopy, stitched into a ticking of wiregrass clumps, springs an outrageous fancy-work of forbs. Running among them is trailing phlox, thin of leaf but generous of bloom, thriving on drought and flame.

"A pine fell in love with a place that belonged to lightning" is Janisse Ray's genesis account of longleaf pine country. Junkyard-born-and-raised daughter of the Georgia longleaf, her *Ecology of a Cracker Childhood* reveals a deep knowing and aching affection for these spare imperiled woodlands. The once-great longleaf pine ecosystem stretched from Virginia to Texas in a gigantic arc along the coastal plain, with trailing phlox tumbling along, part of the lively groundcover that brightens these fire-wrought forests, like "a comforter laid on the land."

Phlox nivalis Loddiges, Bot. Cab. 8: plate 780. 1823 validated by Sweet, Brit. Fl. Gard. 2: plate 185. 1827 [snowy, of the white corolla hue; "*Corolla* . . . of a snowy white" (Sweet 1827)] Caespitose suffrutescent perennial forming open tufts or diffuse mats from trailing-decumbent to prostrate nonflowering stems with persistent leaves, the numerous erect-ascending flowering shoots ranging from 8–15 cm tall for the nominate expression (subsp. *nivalis*) to 10–20 cm for subsp. *texensis* and 15–30 cm for subsp. *hentzii*, leaf nodes 3–5 (4–8 for subsp. *hentzii*) below the inflorescence, fascicles of smaller, nonflowering shoots present in leaf axils; *leaves* subulate to linear-subulate, ciliate, max. dimension 8–12 mm long and 1–2 mm wide (1.5–3 mm wide on flowering shoots) for subspp. *nivalis* and *texensis*, longer and wider for subsp. *hentzii* with leaves on sterile shoots 10–20 mm long and 1.5–3 mm wide and leaves on flowering shoots 12–25 mm long and 2.5–5 mm wide; *inflorescence* 3- or sometimes 6-flowered, its herbage pubescent, the some to many hairs tipped with small glands, max. pedicel length (3–)8–25(–30) mm for subsp. *nivalis*; *calyx* 6–10 mm long, united 7/16–5/8 its length, lobes linear-subulate with apex cuspidate, membranes flat to somewhat plicate; *corolla* tube 1–13(–17) mm long, lobes usually obovate (average dimensions 11 × 7 mm) with apex entire, erose (with ragged margin), or notched, the depth ranging from 1 (exceptionally 2) mm for subsp. *ni-valis* up to 3 or 4 mm for subspp. *hentzii* and *texensis*, hue deep purple to lilac, pink, less often white, the tube often deeper-hued, eye striae obsolete to deeper-hued to intense red; *stamens* with anthers included; *style* 1.5–4 mm long, united to tip which is free for 1–2 mm, with stigma included and placed below the anthers; *flowering season* spring, March–May, chiefly March and April toward the southern end of the overall range and April and May toward the northern end, sporadic flowers both earlier and later, sometimes in autumn in years of frequent alterations of cool and warm weather.

☙☜

Taxonomic notes

Considerable confusion has surrounded the identity of this species, particularly in horticultural literature, due to its similarity to *Phlox subulata* and the early mis-application of the Linnaean *P. setacea* by Curtis (1798b) to horticultural material that was in fact *P. nivalis*. Wherry (1929b, 1937, 1941b, 1955) clarified the relationship between these two species and affirmed the validity of *P. nivalis* as a name proposed by Loddiges (1823) and validated by Sweet (1827).

Key to the Subspecies of *Phlox nivalis*

1a. Pubescence mostly fine-glandular to glandless; endemic to a small area in SE Texas
.... **subsp. *texensis*** Lundell, Contr. Univ. Michigan Herb. 8: 77. 1942 |
Texas trailing phlox
1b. Pubescence mostly coarse-glandular; distributed in eastern United States 2
2a. Plant relatively compact, flowering shoots 3–15 cm tall; upper leaves linear-lanceolate, 4–12 mm long, 0.5–1.5 mm wide on sterile shoots, 1.5–3 mm wide on flowering shoots; pubescence of the inflorescence mostly with finely glandular tips; primarily of sandy soils on Atlantic and Gulf coastal plains . . . . **subsp. *nivalis*** | trailing phlox
2b. Plant relatively lax, flowering shoots usually 15–30 cm tall; upper leaves oblong-lanceolate, 6–20 mm long, 0.5–1.5 mm wide on sterile shoots, 2.5–5 mm wide on flowering shoots; pubescence of the inflorescence mostly with conspicuously glandular tips; plants of dry deciduous woodlands, primarily on heavier soils of the Appalachian Piedmont . . . .
. . . . **subsp. *hentzii*** (Nuttall) Wherry, Gen. Phlox 25. 1955 and in Baileya 4: 97. 1956 |
Piedmont trailing phlox

Geography

Southeastern and south-central United States along the Atlantic and Gulf coastal plains and southern Appalachian Piedmont. Collections of trailing phlox are reported from three parishes in Louisiana, but it is unclear whether these represent natural populations or escapes from cultivation. The nominate expression of trailing phlox (subsp. *nivalis*) occurs most abundantly in the Fall-line Sandhills region, a narrow belt of rolling hills at the interface of the Atlantic Coastal Plain with the Piedmont, running from C North Carolina south into E Alabama. The Fall-line Sandhills is recognized as a distinctive phytogeographic region that includes a number of local endemics and disjunct xerophytes.

Subspecies *hentzii* has the largest distribution of the three subspecies, ranging through the Appalachian Piedmont and Atlantic Coastal Plain from S Virginia south into C Florida and west into E Alabama.

Subspecies *texensis* is endemic to the Big Thicket region of the West Gulf Coastal Plain in SE Texas. Historical collections represent only about six definable populations from Hardin, Polk, and Tyler counties, with the largest remaining group of populations occurring on the Roy E. Larsen Sandyland Sanctuary in Hardin County, a property owned and managed by The Nature Conservancy. Texas trailing phlox was listed as an endangered species by the U.S. Fish and Wildlife Service in 1991.

Environment

Coastal plains and hilly piedmont; associated with upper slopes of gently rolling terrain, from sea level to 3000 ft. (900 m); *habitat* dry to dry-mesic; *soils* sandy to loamy; *parent material* variable.

The nominate expression of trailing phlox (subsp. *nivalis*) occurs in rolling hills and uplands with droughty sands, sandy clay loams, and loamy sands underlain by clays. These soils are classified as xeric or subxeric. It typically occurs in communities that are transitional between xeric and mesic habitat.

Subspecies *hentzii* occurs in uplands in dry, rocky, gravelly, loamy soils, sometimes underlain by impervious clays. It is sometimes associated granite and other more unusual rock types including diabase and gabbro.

Subspecies *texensis* occurs on sandy stream terraces laid down by early rivers and streams and by ancient longshore currents of the Gulf of Mexico. These soils contain little clay, are low in fertility, and drain water rapidly, creating a xeric environment in an area that receives 49 in. (125 cm) of annual precipitation. The xeric nature of the habitat is accentuated by the intense glare and reflected heat from the sand. Sites occupied by subsp. *texensis* have a sandy soil surface underlain by moisture-bearing clays or sandy-clay soils at depths of 1.5–6 ft. (0.5–2 m).

Associations

Pine- or oak-dominated forest/woodland/savanna, grassland, and rock outcrop communities. The nominate expression of trailing phlox (subsp. *nivalis*) is associated with longleaf pine (*Pinus palustris*) woodland and savanna characterized by open stands of longleaf pine with an understory of scrub oaks and an abundance of graminoids and forbs. The specific composition of the understory and herbaceous layer varies depending on the latitude, soil characteristics, and other factors. Harper (1906) listed trailing phlox among the plants occurring in "dry pine-barrens" in the Altamaha Grit region of the Coastal Plain in Georgia. Longleaf pine woodland and savannas were historically maintained by low intensity surface fires, and characteristic species of these plant communities are adapted to periodic fire. Wherry (1929b) noted trailing phlox "is fairly resistant to fire, rapidly producing new shoots on crowns from which the old stems have been burned away."

Subspecies *hentzii* is primarily associated with forest dominated by white oak (*Quercus alba*) with various other oaks and hickories in association. Pine species, primarily shortleaf pine (*Pinus echinata*) and loblolly pine (*P. taeda*), may also be common, along with tuliptree (*Liriodendron tulipifera*) and sweetgum (*Liquidambar styraciflua*). This subspecies also occurs in forests dominated by open stands of post oak (*Q. stellata*) and blackjack oak (*Q. marilandica*) in "Piedmont blackjack prairie," where it is a component of an herbaceous layer that is best developed in forest openings and along roadsides and dominated by grasses like little bluestem (*Schizachrium scoparium*).

Subspecies *hentzii* also is associated with remnant Piedmont prairies in North Carolina and South Carolina on shallow, clayey soils that often experience droughty growing seasons despite a yearly average of 43 in. (110 cm) of precipitation. These prairies may grade into sparsely vegetated glades underlain by gabbro bedrock. This subspecies is associated with several endangered plants in these imperiled prairie and glade communities, including smooth coneflower (*Echinacea laevigata*), Schweinitz's sunflower (*Helianthus schweinitzii*), Carolina birdfoot-trefoil (*Lotus unifoliolatus* var. *helleri*), and Georgia aster (*Symphiotrichum georgianum*), along with disjunct occurrences of downy phlox (*Phlox pilosa* subsp. *pilosa*), prairie rosinweed (*Silphium terebinthinaceum*), and other forbs characteristic of Midwestern tallgrass prairies. It also occurs in association with granitic outcrops and flatrock in the Piedmont of North Carolina, South Carolina, and Georgia. Zonation of vegetation occurs in these communities, from bare rock outward toward the adjacent forest, with trailing phlox typically occurring at ecotonal areas at the outward margin of the bare rock that are dominated by perennial herbs.

Subspecies *texensis* occurs in open woodlands with a canopy dominated by blue jack oak (*Quercus incana*), post oak, blackjack oak, and black hickory (*Carya texana*), with longleaf pine associated. This community has been variously described as oak-farkleberry sandylands, sandhill pine forest, and upland longleaf pine savanna. Lundell (1942) described the type locality and general habitat of subsp. *texensis* as "pine land." It often occurs in areas that are transitional between longleaf pine woodland and mixed forest of hardwood and pines, with the species composition of the overstory not as important as the percent canopy cover (ideal of 25–75 percent). The understory of these woodlands is comprised of scattered shrubs, such as farkleberry (*Vaccinium arboreum*), and an herbaceous layer of graminoids and forbs. Frequently associated herbaceous species include little bluestem, elegant blazing star (*Liatris elegans*), spotted beebalm (*Monarda punctata*), and slender rosinweed (*Silphium gracile*). Subspecies *texensis* is adapted to fire, which was common historically in longleaf pine savanna, and it appears that periodic fire is essential to its survival. Although the aboveground parts of the plant are destroyed by fire, underground parts are undamaged, and new growth appears within two weeks after a spring burn. Even flowering plants subjected to prescribed burns in April will resprout and flower again in May. See Warnock (1995) for valuable information on the ecology and conservation of *Phlox nivalis* subsp. *texensis*

Cultivation

For the better part of nearly two centuries both the botanical and horticultural identities of trailing phlox were intertwined with that of *Phlox subulata*, its immensely popular relative. Happily, a growing enthusiasm for native plants and regional horticulture in the southeastern United States has brought well-deserved attention to this species. Trailing phlox brings a different character to the garden than its cousin, being taller, forming more of a billowing mound than a mat, holding its flowers above its foliage, and coming into bloom about the time *P. subulata* is finishing. While it has obvious value as a rock garden plant, it also is well-suited as a groundcover in dry, partial shade situations. Grow in partial shade in dry, light, slightly acid, well-drained soil, although selections of subsp. *hentzii* should be tolerant of soils with higher clay content. Trailing phlox is less reliably hardy in northern gardens than *P. subulata* and requires some protection from winter cold; Wherry (1935b, 1946) encouraged American horticulturists to develop hardier strains of trailing phlox, noting stock obtained from northern portions of its range "can stand temperatures of 15 to 20 degrees below zero without injury." Cultivars are available. Horticultural hybrids are documented (see Appendix B).

Phlox oklahomensis
Oklahoma phlox
PLATE 51

Spring spreads slowly across the Gypsum Hills of Kansas and Oklahoma, a cool green ground fire burning through the prairie-straw. It will be May before color surges again from the crowns of little bluestem, sideoats grama, and other warm-season grasses. But a careful search in early April will reveal a vanguard of little wildflowers already at work on a fresh crop of seed—Easter daisy, windflower, bitterweed, and such. Blooming with them, in a few favored places, will be Oklahoma phlox.

A tufted, wiry little plant, Oklahoma phlox is a challenge to find in the Gypsum Hills and even more so in its historic range in the Flint Hills to the east. My first encounter began by scent rather than by sight, the spicy fragrance of Oklahoma phlox leading me to a colony that lay just over the crest of a grassy hill. The hunting is easier where the graceful Cimarron and its nimble, hardworking creeks have carved the country into mesas, buttes, and ravines, revealing the red earth and raising a backdrop against which phlox flowers shine like Venus at dawn.

Phlox oklahomensis Wherry, Notul. Nat. Acad. Nat. Sci. Philadelphia 146: 2, figure 1. 1944 [of the state of Oklahoma] Caespitose suffrutescent perennial forming colonies of discrete crowded tufts 8–15(–20) cm tall from a rhizomatous caudex, the erect-ascending flowering shoots 4–8(–12) cm long with 3–5 leaf nodes below the inflorescence; *leaves* linear to lanceolate, ciliate and the upper pilose, max. length 30–60 mm and width 2–5 mm, leaves longest at mid-stem, nonflowering shoots occur in leaf axils at flowering time, these elongating and bearing rather large persistent leaves; *inflorescence* 3- to 6- (9-) flowered, its herbage pubescent with hairs which vary from nearly all fine-glandular to all glandless, max. pedicel length 8–20 mm; *calyx* 7–10 mm long, united ⅜–⅝ its length, lobes linear-subulate with apex cuspidate, membranes flat to somewhat plicate; *corolla* tube 8–12 mm long, lobes variable but usually narrowly obovate (average dimensions 8 × 6 mm) with apex notched 1–3 (rarely 4) mm deep, hue lilac, pink, or lavender, sometimes ranging to white, the eye sometimes bearing a pair of deep-hued striae at base of each corolla lobe; *stamens* borne on upper corolla tube, one anther occasionally exserted; *style* 1.5–3 mm long, united ca. ½ its length, with stigma included and placed below the anthers; *ovary/capsule* with 2 ovules/seeds per locule; *flowering season* early spring, late March through early May, peaking mid-April, one of the earliest forbs to bloom in communities in which it occurs.

ᑍᑌᑎ

Geography

South-central Interior Plains, limited to a small area in S Kansas and NW Oklahoma. The distribution of Oklahoma phlox is bi-centric, with one area of occurrence in the Gypsum Hills (also called the "Red Hills") of SC Kansas and NW Oklahoma and the other in the Flint Hills of SE Kansas, separated by a distance of about 120 miles (90 km). Oklahoma phlox is reported from Kay County in NE Oklahoma (Taylor and Taylor 1980), just south of the Kansas-Oklahoma state line in the Flint Hills region, but may have been extirpated there. Never a common plant, Oklahoma phlox is more frequent and abundant in the Gypsum Hills than the Flint Hills. Oklahoma phlox is one of the relatively few plants endemic to the Great Plains.

Wherry (1955) reported Oklahoma phlox from the Ozark Plateaus section of the Interior Highlands in NW Arkansas, but Marsh (1960) determined these plants to be a local variant of *Phlox bifida*, which he described as *P. bifida* subsp. *arkansana*. Cypher (1993) concurred with Marsh in her dissertation on the *P. bifida* complex. Reports persist in Texas floristic literature that Oklahoma phlox is known historically from NC Texas but ecological and biogeographic relationships favor the interpretation that this plant represents a disjunct occurrence of *P. bifida* (see *P. bifida* species account), and Oklahoma phlox should be excluded from the flora of Texas.

Environment

Plains, cuestas, and hills, on upper slopes of rolling to dissected terrain and the rims of small canyons and ravines, 980–1500 ft. (300–450 m); *MAP* 25–38 in. (64–97 cm); *habitat* dry; *soils* medium-textured, stony, shallow to moderately deep with bedrock at or near the surface; *parent material* in the Gypsum Hills sandstone, shale, gypsum, and dolomite of the Ogallala, Rush Springs, and Marlow formations, in the southern Flint Hills limestone of the Chase and Council Grove groups.

Populations of Oklahoma phlox are most frequently encountered along canyon rims and in association with rock outcroppings, perhaps because this habitat is less accessible to grazing cattle and less impacted by fire. Oklahoma phlox appears to grow best in areas of low to moderate grazing. Observations following a 1996 wildfire in the Gypsum Hills showed little negative impact on Oklahoma phlox, with many robust populations observed throughout the burned area (Springer and Tyrl 2003). Fire may improve Oklahoma phlox habitat by eliminating eastern red-cedar trees (*Juniperus virginiana*), a troublesome invasive species in rangelands.

Field observations suggest that Oklahoma phlox is not threatened in the Gypsum Hills, but that it may be declining in the southern Flint Hills due to local range management practices of spring burning, which historically takes place annually during the peak flowering season. Comparative assessment of Oklahoma phlox populations in the Gypsum Hills after a twenty-year period (1982–2002) indicates relative stability of the species in this region (Springer and Tyrl 2003).

Associations

Grassland, occurring in tallgrass prairie in the eastern portion of its range and mixedgrass prairie in the western. Little bluestem (*Schizachyrium scoparium*) along with sideoats grama (*Bouteloua curtipendula*) and blue grama (*B. gracilis*) dominate mixedgrass prairie in the Gypsum Hills, while big bluestem (*Andropogon gerardii*) becomes more important in the upland tallgrass prairie of the Flint Hills. Eastern red-cedar is often prevalent where Oklahoma phlox occurs in the canyons and dissected terrain of the Gypsum Hills. Notable associates of the early spring flora include Carolina anemone (*Anemone caroliniana*), stemmed four-nerve-daisy (*Tetraneuris scaposa*), and silky Townsend-daisy (*Townsendia exscapa*) in the Gypsum Hills, and prairie pleatleaf (*Nemastylis geminiflora*) in the Flint Hills. See Springer (1983) and Springer and Tyrl (1989, 2003) for valuable information on the ecology, reproductive biology, and conservation of Oklahoma phlox.

Cultivation

Oklahoma phlox is a small but showy plant well-suited to the culture and context of a rock garden or raised bed. With its tufted growth habit, fine-textured foliage, and deeply cleft flowers, it can be used in a manner similar to the better-known. *Phlox bifida*, with the advantage of being more tolerant of dry, sunny conditions. Oklahoma phlox has not been widely cultivated, even though Wherry (1955) declared it "worthy of more horticultural attention." Its early flowering season, beginning late March and peaking mid-April, would make Oklahoma phlox one of the first plants to bloom in the garden, but would also limit its contributions to a fairly narrow window of the growing season, a situation that could be accommodated by planting it with low-growing, nonaggressive companion species such as grasses that would provide cover as it drifts off into dormancy. Grow in full sun in dry, well-drained soil.

Phlox opalensis
Opal phlox

A trip across the desert scrublands of southwest Wyoming is a journey most people just want to be done with. Even Charles Christopher Parry, enthusiastic and effusive botanical explorer of the nineteenth century, was uncharacteristically negative toward this land, lacing words like *monotonous* and *forbidding*, even *repulsive*, into the narrative of his travels in the region. Hardly chamber-of-commerce language, but how much affection can you work up for a place where the dominant plants are named *greasewood* and *saltbush*, and the air smells of selenium?

Yet, as Parry discovered, a closer look at what appear to be barren clay hills will reveal communities of hunkered-down plants with the gemlike charm of alpines. Opal phlox occurs here, taking the edge off the raw landscape with a profusion of white blossoms in the spring. Discovered in 1990 near the town of Opal, known only from this corner of Wyoming and adjacent Utah, this mat-forming phlox often occurs in large attention-catching colonies, luring you into the badlands to discover cryptic gardens spending themselves on the sky.

༄

Phlox opalensis Dorn, Vasc. Pl. Wyoming, ed. 2. 304. 1992 [of Opal, Wyoming] Caespitose perennial herb forming loose mats under 7 cm tall from a branched, spreading caudex, the numerous decumbent to semi-erect flowering shoots 5–10 cm long with elongate internodes, densely covered with crinkly multicellular hairs; *leaves* oblong to lance-subulate, mostly lanceolate, 2–10 mm long, 0.5–1 mm wide, mucronate, papillate, pubescent mostly toward base with crinkly multicellular hairs; *inflorescence* 1-flowered; *calyx* 6–8 mm long, lobes lance-subulate to narrowly triangular with apex spinulose, membranes more or less carinate, obscured by crinkly multicellular hairs; *corolla* tube 10–12 mm long, lobes 6–7 mm long, limb 12–16 mm in diameter, hue white, sometimes pink; *stamens* with anthers ca. 1 mm long; *style* 6–9 mm long; *flowering season* spring, mid-April to late June or early July depending on seasonal moisture conditions.

༄

Geography
Rocky Mountain System, nearly endemic to the Wyoming Basin where it occurs in the Green River and Bridger basins (SW Wyoming), then extends into the adjacent foothills of the Uinta Mountains of the Middle Rocky Mountains (NE Utah and SW Wyoming).

Environment

Intermountain basins, on dissected terrain (gullies and dry washes), flats, benches, badlands, hill slopes, 5500–7700 ft. (1700–2300 m); *MAP* 8–10 in. (20–25 cm); *habitat* xeric; *soils* fine-textured, often with a surface of chert or reddish sandstone gravel that accounts for more than 50 percent of the total cover, alkaline with high concentrations of salts, may have selenium present; *parent material* light-colored clays and shales derived from the Green River or Bridger formations.

Associations

Sparsely vegetated barrens and openings within a matrix of desert shrubland. Gardner's saltbush (*Atriplex gardneri*) and rubber rabbitbrush (*Ericameria nauseosa*) are dominant shrubs, with Wyoming big sagebrush (*Artemisia tridentata* subsp. *wyomingensis*), black greasewood (*Sarcobatus vermiculatus*), bird's-foot sagebrush (*A. pedatifida*), and smooth woody-aster (*Xylorhiza glabriuscula*) also important. Gardner's saltbush and smooth woody-aster are indicative of seleniferous soils. Vegetative cover is low (15–35 percent) in these communities. Frequently associated forbs include Hooker's sandwort (*Arenaria hookeri*), ground milkvetch (*Astragalus chamaeleuce*), oval-leaf wild buckwheat (*Eriogonum ovalifolium*), mat penstemon (*Penstemon caespitosus*), hollyleaf clover (*Trifolium gymnocarpon*), and several dwarf composites including thrift mock goldenweed (*Stenotus armeriodes*), western aster (*Machaeranthera grindelioides*), and Nuttall's false sagebrush (*Sphaeromeria argentea*). Opal phlox occurs with the rare Wyoming endemic large-fruited bladderpod (*Lesquerella macrocarpa*) in the Upper Green River Basin. See Fertig (1996) for valuable information on the ecology of Opal phlox.

Cultivation

Opal phlox has not been reported in cultivation, although seed has been available. It is a showy species, with potential as a drought-tolerant landscape plant for the Mountain West. Given its tolerance of heavy clay soils, Opal phlox may prove more adaptable to cultivation than most western cushion-forming phloxes. Grow in full sun.

Phlox ovata
Allegheny phlox
PLATE 52

Spring is slipping into summer all along the Appalachian Highlands, and the woods are closing up behind. In the dappled days of April, the forest floor was flush with trilliums, hepaticas, dogtooth violets, and the like. Now, in June, the show shifts to sunny meadows and roadsides, where the muted tones of mayapple season give way to colors from the loud side of the spectrum. Fire pink, Indian paintbrush, and golden star are part of this spirited second wave, as is Allegheny phlox, its light-stoked petal-alchemy flaming in wavelengths of violet.

In the phlox-rich flora of Appalachia, this species takes up the baton after creeping, timber, and Cherokee phlox have had their runs. Seen most often at roadsides, the natural niche of Allegheny phlox is grassy meadowlike habitat where tree cover is scant, sometimes over surfacing bedrock. I recall a beautiful stand of this phlox in the mountains of Virginia, in a colorful sun-drenched Allegheny glade made for a moment even more glorious by the king-sized slapdash swoop of a pileated woodpecker across the scene.

Phlox ovata Linnaeus, Sp. Pl. 1: 152. 1753 [egg-shaped, of the leaves; "foliis ovatis"] Upstanding perennial herb with 1 or 2 short (5–10 cm long) decumbent nonflowering shoots tipped with persistent leaves, normally rooting only at a sub-terminal node, from which arises a solitary flowering shoot 25–50(–65) cm tall with 4 (rarely 3 or 5) leaf nodes below the inflorescence; *leaves* variable in outline and size, those on nonflowering shoots and low on flowering ones elliptic to oblong and long-petioled, those higher up oblong to ovate and sessile, the largest 50–150 mm long and 12–50 mm wide, margin smooth or sometimes minutely ciliolate, surfaces glabrous; *inflorescence* mostly 15- to 30-flowered, its herbage glabrate to closely pubescent with minute hairs, a few of which may be obscurely gland-tipped, max. pedicel length (4–)5–15(–20) mm; *calyx* (7–)9–11(–13) mm long, united ½–⅝ its length, lobes elongate-triangular with moderately strong midrib and with apex blunt to sharp-cuspidate, membranes narrow, flat to subplicate; *corolla* tube (12–)16–24 mm long, lobes obovate (average dimensions 8 × 7 mm to 14 × 12 mm), hue bright to dull purple, pink, or rarely white, sometimes paler in the eye with a medial purple stripe at the base of each corolla blade; *stamens* elongate, 1 or 2 anthers sometimes exserted; *style* (12–)14–19(–21) mm long, united to tip which is free for 1 mm, with stigma sometimes exserted; *flowering season* late spring, early May through June, into early July at higher elevations.

Taxonomic notes

A challenging nomenclatural issue has plagued Allegheny phlox, precipitated by a paper (Reveal et al. 1982) that called into question the validity of the Linnaean name *Phlox ovata*, the next available name being *P. latifolia* (Michaux 1803). The issue was reviewed and processed through a series of eight nomenclatural papers and ultimately resolved in favor of *P. ovata* (see Locklear 2010), but floristic and ecological workers in the eastern United States have mostly used *P. latifolia* for this species since 1982.

Geography

Chiefly Appalachian Highlands, concentrated in the mountains of the Blue Ridge, Valley and Ridge, and Appalachian Plateaus physiographic provinces, extending east locally onto the Appalachian Piedmont, with outlying occurrences west in the northeastern Interior Plains in NW Ohio and E Indiana. Allegheny phlox does not appear to be a common species anywhere within its range, but occurs at greatest frequency in the Allegheny Mountains of Virginia and West Virginia, hence the common name.

Environment

Primarily mountains but also hilly piedmont, 1000–4500 ft. (300–1400 m); *habitat* dry-mesic; *parent material* ultramafic rock (rich in magnesium), limestone, dolomite. Allegheny phlox occurs in the driest habitat of any of the upstanding, long-styled eastern phloxes, most of which are associated with mesic to wet habitat such as floodplains, stream banks, and fens. Wherry (1935b), the first botanist to gain a clear taxonomic and ecological picture of Allegheny phlox, described its "favorite habitat" as "the moderately acid litter of an upland oak forest." Wherry (1932b) observed Allegheny phlox "occasionally enters alluvial meadows and rarely tracts underlain by calcareous rocks. . . . In succession it comes in fairly early, reaches its best development at an intermediate stage, and dies out as climax forest conditions are approached."

Associations

The community ecology of Allegheny phlox has not been delineated with much precision in the past, due in part to confusion with other eastern upstanding long-styled phloxes. Allegheny phlox is primarily associated with dry open forest and woodland dominated by species of oak, with hickories and pines also important. White oak (*Quercus alba*) is the dominant tree throughout these communities. In the Allegheny Mountains of Virginia, associated forbs include scarlet Indian paintbrush (*Castilleja coccinea*), green-and-gold (*Chrysogonum virginianum*), wild crane's-bill (*Geranium maculatum*), bowman's-root (*Porteranthus trifoliatus*), sundial lupine (*Lupinus perennis*), and fire-pink (*Silene virginica*). On the driest sites these woodlands grade into barrens or glades dominated by an herbaceous flora.

In the southern Blue Ridge Mountains, Allegheny phlox occurs in barrens over outcrops of ultramafic rock. This rare community type is characterized by open stands of pitch pine (*Pinus rigida*) and white oak, with an herbaceous layer dominated by the grasses northern dropseed (*Sporobolus heterolepis*) and big bluestem (*Andropogon gerardii*). In SW Virginia, Allegheny phlox occurs in dolomite barrens that support open woodland, shrub thickets, and small grassy openings. Chinquapin oak (*Quercus muhlenbergii*) is the most constant and abundant tree in these barrens, with an herbaceous layer dominated by little bluestem (*Schizachyrium scoparium*), prairie ragwort (*Packera plattensis*), and glade wild quinine (*Parthenium auriculatum*). The endangered plant, smooth coneflower (*Echinacea laevigata*) is sometimes associated with these communities.

Cultivation

Though one of the first of the phloxes to make the transition from the wilds of America to the gardens of Europe, the horticultural run of Allegheny phlox was a short one, and it seems to have faded into the background by the early 1800s. Some of this can be blamed on the taxonomic fog that has enveloped this species for over 200 years, but some of it is likely due to its preference for drier habitat and leaner soils, making it relatively short-lived in the typical flower border. Allegheny phlox does have potential as a border phlox for drier landscape situations and its late spring to early summer (early June into July) flowering season would be a valuable contribution to the floral progression of any garden or landscape. It can be used in herbaceous perennial gardens and borders, mixed borders, and regional native plant horticulture. Grow in full sun to partial shade in dry to slightly moist, well-drained soil.

Phlox paniculata
summer phlox

There is more verve and variety packed into the LaRue-Pine Hills Ecological Area than any other place in Illinois. Hard by the Mississippi, the air heavy with big-river mood and musk, its richness is due to a unique juxtaposition of creepy swamp and towering limestone bluffs. The preserve is thick with reptiles and amphibians, and is notorious for its twice-a-year road-closing snake migrations that find cottonmouths, rattlesnakes, and other slithering things on the move between summer habitat in LaRue Swamp and rocky hibernating dens at the base of the bluffs.

If you walk "Snake Road" in July or August, your anxiety will be eased by the cheerful presence of summer phlox, growing between the lotus-choked swamp and the pine-topped bluffs. A plant of low woods, this our tallest phlox ranges across the eastern United States from the Ozarks to the Appalachians. With large, showy heads of butterfly-plundered flowers, summer phlox is a fountain of color and animation in the serotinal forest. It is a favorite fueling stop of migrating cloudless sulphurs, south-bound on chartreuse wing to wintering grounds of their own.

❧❧

Phlox paniculata Linnaeus, Sp. Pl. 1: 151. 1753 [with flowers in panicles; "corymbis compositis"] Upstanding perennial herb from a thick short rhizome which from irregularly spaced nodes sends up multiple stout flowering shoots 75–200 cm tall with 15–40 leaf nodes below the inflorescence, lower leaves being opposite but upper leaves becoming subopposite, stems glabrous to soft- or rough-pubescent, sometimes red-streaked; *leaves* narrow-elliptic below, becoming elliptic-lanceolate to ovate upward, sessile or short-petioled, thinnish with conspicuous areolate veins, the margin serrulate and ciliolate, surfaces glabrous or the lower fine-pilose and the upper sometimes bristly, especially toward margin, max. length (attained a few nodes below the inflorescence) 100–150 mm and width 25–50 mm; *inflorescence* a multiply-compound panicle of small cymes, the flowers numbered in the hundreds, its herbage glabrous to fine-pubescent, the hairs only rarely gland-tipped, max. pedicel length 4–8 mm; *calyx* 6–10 mm long, united ⅜–⅝ its length, lobes narrow-subulate with apex aristate with awn 1–2 mm long, membranes subplicate; *corolla* tube (16–)18–24(–26) mm long, conspicuously pilose or rarely glabrous, lobes narrow-obovate to orbicular (dimensions varying from 7 × 5 mm to 12 × 12 mm) with apex entire or emarginate, hue dull to bright purple, pink or occasionally white, the eye sometimes moderately purple-striate; *stamens* with anthers cream-white, one or two often exserted; *style* 15–25 mm long, united to tip which is free for 1–1.5 mm, with stigma sometimes exserted; *flowering season* summer and autumn, early July to late September, occasionally into early winter.

❧❧

Geography

Eastern and central United States, in the Appalachian Highlands, Interior Highlands, and intervening portions of the Interior Plains. Due to the ease with which summer phlox has escaped cultivation over the years, it is difficult to determine its historic natural range with precision. Wherry (1933) considered summer phlox "undoubtedly native" from "northern Georgia to Arkansas, northeastern Kansas, northern Indiana, and central New York." The association of summer phlox with relatively intact, high quality natural habitat, rather than roadsides or disturbed areas, gives some assurance of its native occurrence in a particular area.

Environment

Dissected plateaus, hilly piedmont, and mountains, along river valleys, floodplain terraces, stream edge, gravel wash, riverscour, and other riparian habitat where periodic flooding creates openings in the forest canopy, and at the base of bluffs, 250–2500 ft. (75–750 m); *habitat* mesic; *soils* alluvial, mineral rich. Wherry (1933) described the "typical habitat" of summer phlox as "a thinly wooded alluvial flat where the vegetation has reached an intermediate stage of succession." More than any other species in the genus, summer phlox has escaped cultivation and occurs in a variety of culturally disturbed habitats including roadside ditches, railroad tracks, abandoned homesteads, old cemeteries, fencerows, and degraded woodlands.

Associations

Primarily floodplain and riparian forest and with riverwash communities. While the species composition of the regional upland forest types (such as oak-hickory and beech-maple) differs across the range of summer phlox, the associated floodplain and riparian forests of the eastern United States are typically comprised of the same suite of species that includes silver maple (*Acer saccharinum*), river birch (*Betula nigra*), sycamore (*Platanus occidentalis*), eastern cottonwood (*Populus deltoides*), black willow (*Salix nigra*), and American elm (*Ulmus americana*). Summer phlox also occurs at the lower edges of mesic upland hardwood forests where these intersect floodplain and riparian forest. Sugar maple (*A. saccharum*), tuliptree (*Liriodendron tulipifera*), and American beech (*Fagus grandifolia*) are often components of these forests.

Frequently associated forbs in these forested communities include tall bellflower (*Campanulastrum americanum*), large tick-treefoil (*Desmodium glutinosum*), great blue lobelia (*Lobelia siphilitica*), and woodland pinkroot (*Spigelia marilandica*), along with asters, sedges, and goldenrods. In low forests in Illinois and Indiana, summer phlox is a common associate of purple fringeless orchid (*Platanthera peramoena*), and the similar flower color, inflorescence size, and blooming time of these two species may lead to them being confused at a distance.

Throughout its range, summer phlox occurs in riverwash communities associated with gravel bar and other open stream edge habitat. Scouring floods in the spring and drought stress on the well-drained, course-textured substrate in summer reduce tree cover, resulting in shrubby and/or prairielike herbaceous vegetation dominated by grasses like big bluestem (*Andropogon gerardii*) and little bluestem (*Schizachyrium scoparium*). Along the Jacks Fork and Current rivers in the Ozarks of SC Missouri, summer phlox occurs in all but the earliest stages of succession on gravel bars in communities dominated by shrubs including Ozark witch-hazel (*Hamamelis vernalis*) and pale dogwood (*Cornus amomum* subsp. *obliqua*) (Witherspoon 1971).

Cultivation

Summer phlox has a very long history in horticulture (see chapter 2), and is one of the most widely grown herbaceous perennials in the world. Its horticultural attributes are legion: bold clusters of showy flowers, rich fragrance, and towering height. But it is the flowering season (July–September) of summer phlox that has earned it greatest appreciation, carrying the garden through the dog days of summer, holding down the fort until the asters and blazing stars can come on board.

Summer phlox has been the subject of more intensive selection and breeding efforts than any other species in the genus, and growers have pursued a long list of "improvements" over the years. Regarding the flowers, individually termed "pips," the aim has been large, round, flat corollas in strong, clear colors. Regarding the inflorescence, the goal has been an architecture that yielded a triangular or pyramidal shape similar to that of lilac, with a compact, well-branched "truss" capable of holding the flower cluster together in the wind and rain. The flowering season has been extended by selecting strains that bloomed a bit earlier or later than the norm. Dramatic variation in height has also been achieved. English horticulturists Harmer and Elliott (2001) presented a height classification scheme for summer phlox cultivars ranging from "short" (under 3 ft. [90 cm]) to "medium" (3–4 ft. [90–120 cm]) to "tall" (over 4 ft. [120 cm]).

Late summer (July–September) flowering border phlox cultivars have been informally classified under the name "Decussata," derived from *Phlox decussata* (Pursh 1813), an early and widely used synonym for *P. paniculata*. These represent selections of *P. paniculata* or crosses derived from different strains or cultivars of the species. Early references to the Decussata Group postulated hybridization involving *P. paniculata* with *P. maculata*, but Wherry (1933, 1935b, 1955) provided evidence refuting this, noting little trace of the distinctive characters of *P. maculata* in most late summer–flowering phlox cultivars or the seedlings derived from them. In addition, analyses of the chromosome numbers of dozens of late summer–flowering cultivars showed that each possess the standard compliment of fourteen chromosomes (Flory 1931; Meyer 1944), providing further evidence of nonhybrid origin. Finally, most of the late summer–flowering cultivars produce abundant viable seed, which would not be expected from inter-species hybridization. The correct nomenclature for horticultural selections of late summer–flowering phloxes is to treat them as cultivars of *P. paniculata*. Thus, the cultivar 'Cool of the Evening' is correctly rendered *Phlox paniculata* 'Cool of the Evening'.

The Dutch have been the most serious breeders of summer phlox in recent years, aiming at unnatural flower sizes and colorations and for very short plants that can be used at the front of the border or in containers. The "Feelings" series developed by Rene van Gaalen features plants with highly reduced corollas, the result being that the calyces are the ornamental feature of the flowers.

The Hardy Plant Society, an international group of nursery professionals and serious gardeners headquartered in the United Kingdom, seeks to discover, propagate, and reintroduce some of the older, rarer cultivars of summer phlox. Ironically, cultivars with smaller flowers have become desirable, as they harken back to Victorian-era cottage gardens.

In the United States, some of the more recently introduced summer phlox cultivars have been selected for their resistance to powdery mildew (*Erysiphe cichoracearum*) and other foliar diseases. Seldom the result of intentional breeding programs, most have been discovered in gardens or landscapes where they are the "last man standing" among their diseased kin, or hardy "pass-along plants" that have circulated among gardeners for years. Don Jacobs, proprietor of Eco-Gardens in Decatur, Georgia, has breed superior wild forms of summer phlox to European cultivars, resulting in selections with sturdy stems, lower flowers, and greater mildew resistance.

German nurseryman Georg Arends crossed *Phlox paniculata* with *P. divaricata*, naming the resultant hybrid *P. arendsii*. Arends (1912) described the parentage as "*P. decussata* [crossed] with a hybrid from *P. canadensis* and *P. Laphamii*." Correctly rendered *P. ×arendsii* (see Appendix B), the resultant plants are short, sturdy, and can be used in the front of the garden border in full sun. The Arendsii hybrids have not proven to be exceptionally hardy, but have been used in breeding programs to develop new generations of shorter statured border phlox (Harmer and Elliott 2001). Coen Jansen of the Netherlands has been an important breeder, releasing nine cultivars in the 1990s resulting from crosses of *P. ×arendsii* with cultivars of *P. paniculata*. The "Spring Pearl" series was introduced in the late 1990s by the Jacob Th. De Vroomen nursery of the Netherlands, from crosses between *P. ×arendsii* and a pink cultivar of *P. paniculata*. Arendsii hybrids and their cultivars must be propagated asexually as no viable seeds are produced.

Summer phlox can be used in herbaceous perennial gardens and borders and mixed borders. The impact (color and fragrance) of summer phlox is most effective when plants are grouped in masses. Combine summer phlox with lower growing plants in the foreground since leaves on the lower portions of the stems can look "rather tatty" by flowering season. The flowers of most cultivars are highly attractive to butterflies and the large panicles of summer phlox make excellent cut flowers.

Summer phlox grows best in full sun (preferred) to light shade and in deep, loamy, evenly moist, well-drained soil enriched with organic matter. Dividing clumps every three or four years and replanting in enriched soil will maintain vigor. The roots of summer phlox are fibrous and shallow and benefit from a light mulch, which reduces the need for overhead watering and the incidence of foliar disease.

Summer phlox is susceptible to powdery mildew disease caused by the fungus, *Erysiphe cichoracearum*. This appears to be more of a problem in North America than in Europe, and susceptibility varies by cultivar. Consult regional evaluations of mildew resistance such as Hawke (1999) for the Chicago region and Bir (2003) for North Carolina for cultivar recommendations. Management practices include (1) improving air circulation by selective thinning the number of stems within a plant and adequate spacing (3–6 ft. [1–2 m]) from other plants, (2) planting in full sun, (3) eliminating or reducing overhead watering, and (4) removing infected leaves and stems each autumn. Leaf spot and leaf blight foliar diseases can also weaken plants, but the management practices above will also reduce incidence. A stem-borne nematode (*Ditylenchus dispsaci*) can stunt and distort growth, but using clean nursery stock propagated from root cuttings and removal of stems and plant debris in fall reduce incidence.

Phlox pattersonii
Coahuila phlox

To someone with Missouri roots, there is an Ozark-feel to the Sierra Madre Oriental. While nearly a thousand miles and a host of different plant communities intervene between the Ozark Highlands and the mountains of northeast Mexico, the two areas have a number of tree, shrub, forb, fern, and moss species in common. These shared plants hint at ancient comings and goings, brokered by another rocky highland, the Edwards Plateau of Texas.

Coahuila phlox is one piece of this biogeographic puzzle. Known only from three small mountain ranges in the Mexican states of Coahuila and Nuevo Leon, this robust plant bears a strong resemblance to the Ozark subspecies of downy phlox. Coahuila phlox dwells in oak-shaded canyons that are a mesic safe haven from the hostile thornscrub that laps at the foot of the mountains. Whether Coahuila phlox is the stranded offspring of Ozark downy phlox, or its ancient progenitor, depends on whether the Sierra Madre is *Ozarkian*, or the Ozarks *Madrean*.

⤫⤬⤫

Phlox pattersonii Prather, Pl. Syst. Evol. 192: 62, figure 2. 1994 [after Thomas Frederick Patterson (b. 1950), collector of the type] Upstanding suffrutescent perennial with erect-ascending flowering shoots to 40 cm tall arising basally or from decumbent nonflowering stems, with (9–)15–24 leaf nodes below the inflorescence, pubescent with hairs eglandular near base, glandular above, lower leaves being opposite but upper leaves alternate; *leaves* usually linear-oblong below, becoming lanceolate-elliptic above, sessile, 13–35 mm long and 2–20 mm wide, glandular-pubescent (rarely eglandular) with evenly spaced hairs on both surfaces, acute, mucronate; *inflorescence* 2- to 10- (14-) flowered, bracts lanceolate, 2–6 mm wide, 7–24 mm long, attenuate, pedicels glandular-pubescent, 3–12 mm long; *calyx* 10–15 mm long, united ¼–½ its length, lobes linear, glandular-pubescent, long-attenuate; *corolla* tube 11–15(–18) mm long, scarcely expanded above, externally pubescent, lobes orbicular (average dimensions 8–11[–15] mm long and almost as wide) with apex cuspidate, hue blue or lavender; *stamens* with anthers included; *style* 2–3 mm long, united ca. ⅜ its length, with stigma deeply included and placed below the anthers; *flowering season* late June to mid-August.

⤫⤬⤫

Geography
Endemic to the Sierra Madre Oriental in northeastern Mexico where it has been collected at the northern end of the mountain system in the states of Coahuila and Nuevo Leon. Occurrences are known from isolated, narrow mountain ranges located along the eastern front of the Sierra Madre Oriental, specifically the Sierra Santa Rosa in Coahuila and the Sierra Gomas and Sierra Lampazos in Nuevo Leon.

Environment
Hilly piedmont and mountains at low to middle elevations, in relatively protected areas along open ridges and in arroyos and canyons, often on north-facing slopes, 3600–7200 ft. (1100–2200 m); *habitat* dry-mesic; *soils* stony, humus-rich, shallow; *parent material* limestone. Prather (1994) described Coahuila phlox growing in Bustamante Canyon in the Sierra Gomas "in humus-rich soil in the steep, dry bed of a seasonal stream in deep shade." Muller (1939) noted development of a "deep humus layer" as characteristic of montane mesic forest in the Sierra Madre of Nuevo Leon, adding this layer was also found in the densely wooded arroyos of the montane low forest.

Associations

Information about the ecological associations of Coahuila phlox is derived from herbarium specimens cited by Prather (1994) in his original description of the species, interpreted in light of vegetation types described by Cornelius Muller (1939, 1947) for Nuevo Leon and Coahuila. In general, Coahuila phlox ranges from shrub-dominated, chaparral-like communities on the lower slopes of mountains up into mixed temperate woodlands of oak and pine. These mountains are essentially surrounded by Tamaulipan thornscrub dominated by acacias (*Acacia* spp.) and honey mesquite (*Prosopis glandulosa*), which occupies much of the Gulf Coastal Plain in Coahuila and Nuevo Leon.

Coahuila phlox is associated with mixed woodland comprised of open stands of low trees dominated by oaks, with pines and junipers often associated. Muller described this community as "montane low forest." At its lowest elevations, this woodland is dominated by plateau live oak (*Quercus fusiformis*) along with other trees and shrubs typical of escarpment woodlands of the Edwards Plateau of Texas. At higher elevations, this woodland grades into what Muller described as "montane mesic forest," with a different, more diverse suite of oak species (*Q. canbyi*, *Q. laceyi*, and others) complemented by pines (*Pinus cembroides*, *P. teocote*, and others).

Within the context of these mixed woodlands, and the drier shrubland described below, Coahuila phlox also occurs in more densely forested communities that develop in the shelter of arroyos and canyons. Here it occurs with tree species characteristic of more mesic habitat, including bigtooth maple (*Acer grandidentatum*) and eastern hop-hornbeam (*Ostrya virginiana*).

At lower elevations, Coahuila phlox occurs in what Muller described as "piedmont scrub" for Nuevo Leon and "piedmont shrub" for Coahuila—plant communities dominated by open stands of large shrubs and small trees. Low, shrubby forms of plateau live oak dominant this community type. Two specimens cited by Prather state association with slimleaf-rosewood (*Vauquelinia corymbosa*), a shrubby species which Muller noted as characteristic of this community type. Two specimens also indicate the intriguing association of Coahuila phlox with stands of an arborescent palm, *Brahea bella*—"Palm Canyon" near Muzquiz in Coahuila and a "*Quercus-Vauquelinia-Ptelea-Palma* association" in a large arroyo in the Sierra Gomas in Nuevo Leon. Both Muller and Baker (1956) noted the occurrence of this palm in open forests on valley floors and lower slopes on the east flank of the mountains in Coahuila and Nuevo Leon.

Cultivation

Coahuila phlox has not been reported in cultivation, although seed has been available. It has potential as a drought-tolerant landscape plant for the southwestern United States. Grow in partial shade in dry to moist, well-drained soil rich in humus.

Phlox pilosa
downy phlox
PLATES 53–55

A dally with a good tallgrass prairie will leave you wobbly. I remember wading into a great one on a breezy June morning in Kansas. Pale purple coneflowers dominated the scene, weaving and bucking like boxers. White spires of larkspur and indigo trusses of baptisia rocked along. Squadrons of monarchs and regal fritillaries cruised the tossing flowery canopy. And adding to the ruckus—amid the roilings of rose, spiderwort, coreopsis, and leadplant—were flashes of pink from the clustered blossoms of downy phlox.

If a little patch of tallgrass can make you dizzy, what would a million acres do to you? Sadly, the tallgrass prairie is a dim shadow of its former self and such questions, as Aldo Leopold lamented, are "never to be answered, and perhaps not even asked." Yet remnants of various sizes remain from the Gulf Coast to Manitoba, and where these are found so too is downy phlox, mixing it up with the general spangle of prairie forbs or occurring en masse and turning grassy swales into rills and washes of lavender, pink, and magenta.

❧❧❧

Phlox pilosa Linnaeus, Sp. Pl. 1: 152. 1753 [beset with straight, soft, spreading hairs; "foliis . . . villosis"] Upstanding perennial herb with a few simple or sparingly branched erect-ascending flowering shoots 25–50(–75) cm tall with 6–12 leaf nodes (10–18 for subsp. *longipilosa*) below the inflorescence, sometimes accompanied by erect nonflowering shoots, leaves mostly opposite (but uppermost 7 or 8 pairs alternate for subsp. *longipilosa*); *vesture* (particularly of the inflorescence-herbage) mostly glandular for nominate expression (subsp. *pilosa*) and for subsp. *longipilosa* and subsp. *ozarkana*, mostly glandless for subsp. *deamii* and subsp. *fulgida*, and glabrous or nearly so for subsp. *sangamonensis* and for populations along Gulf Coastal Plain formerly treated as *P. pilosa* subsp. *detonsa*, hairs long (2–4 mm) and jointed for subsp. *longipilosa*, mostly under 1 mm for all other expressions; *leaves* linear below, widening to linear-lanceolate and lanceolate upward (uppermost leaves of subsp.

ozarkana ovate-lanceolate to ovate with cordate-clasping bases), the largest close to the inflorescence, 4–80(–120) mm long and 3–9(–12) mm wide (10–20[–30] mm wide for subsp. *ozarkana*); *inflorescence* (12-) 24- to 49- (ca. 100-) flowered, at first compact (especially so for subsp. *deamii*) but becoming open-paniculate, max. pedicel length 4–8(–12) mm; *calyx* (6–)8–12(–15) mm long, united ⅜–⁷⁄₁₆ its length, lobes subulate with apex aristate with awn 0.5–2 mm long (lobe longer and attenuate with awn to 3 mm in subsp. *longipilosa* and in southern population systems), membranes flat to moderately plicate; *corolla* tube 8–12(–16) mm long, usually pilose with glandless or sparingly glandular hairs (sometimes glabrous, especially in Mississippi valley portion of Gulf Coastal Plain), lobes oblanceolate to obovate (dimensions varying from 9.5 × 6.5 mm to 12 × 8 mm) with apex obtuse, rounded, apiculate, or (rarely) erose to emarginate, hue purple to pink to white, the eye often paled and bearing deep-hued striae; *stamens* with anthers included; *style* 1.5–3(4) mm long, united ¼–⅝ its length, with stigma included; *flowering season* spring to early summer, May–July.

Key to the Subspecies of *Phlox pilosa*

1a. Stems per plant numerous, with 10–18 leaf nodes; uppermost leaves and bracts alternate, sometimes as many as 7 or 8 being so; inflorescence stems and calyx densely pilose with long jointed hairs mostly 2–4 mm long; calyx lobes apically bearing relatively long, somewhat twisted awn; endemic to granite outcrops in Wichita Mountains of SW Oklahoma . . . . **subsp. *longipilosa*** (Waterfall) Locklear, J. Bot. Res. Inst. Texas 3: 647. 2009 | Kiowa downy phlox

1b. Stems per plant few, averaging 6–14 leaf nodes; uppermost leaves and bracts mostly opposite; hairs of upper stems and calyx less than 1 mm long; calyx lobes aristate . . 2

2a. Plants 15–30 cm tall; inflorescence somewhat compact, its pubescence glandless; open woodlands in S Indiana, W Kentucky, and NW Tennessee . . . . **subsp. *deamii*** D. A. Levin, Brittonia 18: 145. 1966 | Deam's downy phlox

2b. Plants more than 25 cm tall; inflorescence open, with bracts scattered throughout 3

3a. Stems, leaves, and calyx glabrous or nearly so; endemic to C Illinois . . . . **subsp. *sangamonensis*** D. A. Levin & D. M. Smith, Brittonia 17: 264. 1965 | Sangamon downy phlox

3b. Stems, leaves, and calyx downy phlox pubescent 4

4a. Upper leaves of stems ovate or ovate-lanceolate, broadest and somewhat heart-shaped at base, at least the upper leaves with gland-tipped hairs; stems sometimes branched, with gland-tipped hairs ....
.... **subsp. *ozarkana*** (Wherry) Wherry, Gen. Phlox 49. 1955 | Ozark downy phlox

4b. Upper leaves of stems mostly linear to narrowly oblanceolate or linear-lanceolate, rounded or truncate but not noticeably broader at base; stems usually simple, hairs either gland-tipped or glandless 5

5a. Inflorescence-herbage pubescent with gland-tipped hairs or glabrous (in some southern populations); calyx lobes reflexed .... **subsp. *pilosa*** | downy phlox

5b. Inflorescence-herbage pubescent with lustrous, glandless hairs ....
.... **subsp. *fulgida*** (Wherry) Wherry, Gen. Phlox 49. 1955 and in Baileya 4: 97. 1956 | Dakota downy phlox

Geography

Eastern and central United States, extending into southern Canada. The downy phlox species complex has the largest area of distribution of any of the eastern phloxes. The nominate expression of downy phlox (subsp. *pilosa*) has the largest distribution within the species complex, from the Atlantic Coastal Plain and Appalachian Piedmont west through the Interior Highlands to the edge of the southern Great Plains, and from the lower Great Lakes region south through the Interior Plains to the Gulf Coastal Plain. Downy phlox is largely absent from the Appalachian Highlands.

Subspecies *deamii* is endemic to a small area of the Interior Low Plateaus region (S Indiana, W Kentucky, and NW Tennessee). Subspecies *pilosa* occurs within range, but within this region of sympatry, populations of subsp. *deamii* are more frequent and have more individuals per population (Levin and Smith 1966).

Subspecies *fulgida* extends the range of the *Phlox pilosa* complex to the north and west, its distribution concentrated in the north-central Interior Plains, extending as far north as southeastern Manitoba, Canada, and west to the eastern portions of Nebraska and the Dakotas.

Subspecies *longipilosa* is endemic to the Wichita Mountains region of SW Oklahoma, remote from the rest of the *Phlox pilosa* complex. It is known at present only from the Quartz Mountains at the western edge of the Wichita Mountains in Greer and Kiowa counties. A single historical collection is known from the Wichita Mountains National Wildlife Refuge in Comanche County, and it also is reported from Washita County, Oklahoma.

Subspecies *ozarkana* has a bi-centric distribution, with the largest area of occurrence in the Interior Highlands (N Arkansas, SE Kansas, S Missouri, and E

Oklahoma) extending south to the adjacent upper portion of the West Gulf Coastal Plain (Arkansas and Louisiana), and the other in the southern Appalachian Highlands centered in N Alabama and adjacent NW Georgia and S Tennessee.

Subspecies *sangamonensis* is endemic to the watershed of the Sangamon River in the eastern Interior Plains, in a small, glaciated region of C Illinois (Champaign and Piatt counties) of less than 100 square miles.

Environment

The nominate expression of downy phlox (subsp. *pilosa*) occurs on gently to moderately steep rolling hills and plains. Soils tend to be sandy or relatively shallow, formed over a variety of parent material including sandstone, shale, chert, limestone, and dolomite.

Subspecies *deamii* is associated with soils that posses an impermeable or slowly permeable clay layer (fragipan) at or just below the surface. These poorly drained soils are often wet in spring but dry rapidly in summer or during drought, creating a dry-wet (xerohydric) fluctuating environment that is stressful for plants. Plant communities associated with such soils are termed "flatwoods," and have an open tree canopy and poorly developed understory.

Subspecies *fulgida* is associated with soils that are loamy, moderately- to well-drained and often derived from glacial till or loess parent material. Its prairie habitat is typically classified as relatively mesic, and in the drier western edge of its range in Nebraska and the Dakotas it tends to be associated with swales and lower terrain. It sometimes occurs in drier habitat, such as xeric "goat prairies" in Wisconsin and upland slopes in the Loess Hills bordering the Missouri River Valley in W Iowa. Where subsp. *pilosa* occurs in proximity in N Illinois, subsp. *fulgida* is characteristic of soils having a deep A horizon, high availability of soil moisture, high organic content, and good internal drainage, while subsp. *pilosa* is associated with soils having a thinner A horizon, lower availability of soil moisture, and less organic matter.

Subspecies *longipilosa* occurs on moderately steep to steep slopes of relatively small, talus-sided mountains at elevations of 1700–2000 ft. (500-600 m) in soils derived from granitic bedrock and often containing granite fragments and cobbles. It is usually more common on north slopes or areas near the base of slopes where moisture is more available.

Subspecies *ozarkana* is associated with rocky soils derived from surfacing bedrock of a variety of geological formations, including dolomite, limestone, sandstone, or chert in Missouri, novoculite in the Ouachita Mountains of Arkansas and Oklahoma, and nepheline syenite in C Arkansas.

Subspecies *sangamonensis* is associated with habitat along a few miles of the Sangamon River, including floodplain and bluff tops.

Associations

Members of the downy phlox species complex are associated with conifer woodland/savanna, mixed conifer/hardwood forest/woodland, oak-dominated forest/woodland/savanna, grassland, and rock outcrop communities.

Downy phlox occurs in longleaf pine (*Pinus palustris*) woodland and savanna on the Atlantic Coastal Plain in Georgia and the Gulf Coastal Plain in Florida, Alabama, and Texas. These forested communities are dominated by open stands of longleaf pine with an understory of clonal oaks and an herbaceous layer of wiregrass (*Aristada stricta*) and other perennial grasses along with a high diversity of forbs. William Bartram wrote of downy phlox in the account of his 1744 journey across the upper Florida peninsula when he described longleaf pine "groves" with openings "ornamented with a variety and profusion of herbaceous plants and grasses," including "almost endless varieties of the gay Phlox . . . that enamel the swelling green banks."

Downy phlox is associated with oak-dominated communities throughout most of its range. It typically occurs in drier, upland landscape positions where the canopy structure is fairly open. In N Illinois and adjacent Indiana, Wisconsin, and Michigan, downy phlox occurs in dry soil oak savanna dominated by open stands of black oak (*Quercus velutina*). These savannas often occur in mosaic with sand prairie, a sandy-soil variant of tallgrass prairie dominated by little bluestem (*Schizachryium scoparium*) and other dry-soil adapted prairie species. In E Kansas, E Oklahoma, and NC Texas, downy phlox occurs in post oak (*Q. stellata*) and blackjack oak (*Q. marilandica*) woodland in a mosaic with tallgrass prairie termed "Cross Timbers." On the Appalachian Piedmont of South Carolina, downy phlox occurs in a community termed "Piedmont blackjack prairie," comprised of open stands of post oak and blackjack oak.

Subspecies *deamii* occurs in post oak flatwoods and barrens, an unusual and rare forest type dominated by open, savanna-like stands of post oak, sometimes in nearly pure stands. Post oak barrens are natural openings surrounded by post oak flatwoods. The herbaceous layer of both flatwoods and barrens is sparse, with poverty oatgrass (*Danthonia spicata*) dominating, along with various moss and lichen species. The graminoids and forbs of the herbaceous layer of post oak flatwoods and barrens is different from that of limestone barrens and glades in the region, being comprised of both xeric prairie and southern wetland/bottomland species. The apparent rarity of subsp. *deamii* and its unique habitat warrants further inventory and conservation measures.

Subspecies *fulgida* occurs in oak and pine savanna on the eastern edge of its distribution. Bur oak (*Quercus macrocarpa*) is the dominant species of deep-soil savannas in which this subspecies occurs in Kansas, Missouri, and Iowa. On drier, sandier soils in Minnesota and Wisconsin, it occurs in savanna dominated by open stands of black oak. Northward, black oak is replaced by northern pin oak (*Q. ellipsoidalis*), also called Hill's oak, which co-occurs with jack pine (*Pinus banksiana*) in pine-oak barrens. At the northern limits of its range in Minnesota and Wisconsin, subsp. *fulgida* occurs in pine barrens dominated by open stands of stunted jack pine in a matrix of grasses and forbs dominated by little bluestem.

Subspecies *ozarkana* occurs in oak-hickory forest and woodland dominated by open stands of white oak (*Quercus alba*) and/or black oak, with post oak, blackjack oak, and black hickory (*Carya texana*) in association. Drier forests and woodlands have shortleaf pine (*Pinus echinata*) in association or dominating. It also occurs in savannas dominated by post oak in the Missouri Ozarks and by shortleaf pine in Ouachita Mountains of Arkansas and Oklahoma. These trees occur as scattered individuals in a matrix of prairie grasses and forbs, dominated by little bluestem, with yellow Indiangrass (*Sorghastrum nutans*) and big bluestem (*Andropogon gerardii*) associated. Subspecies *ozarkana* also occurs to some extent in association with glades, typically not on the open glade but at the edges or at the interface with adjacent savanna vegetation.

Downy phlox is associated with a variety of prairie formations throughout its range. In the southwestern portion of the tallgrass prairie ecosystem (NW Arkansas, SE Kansas, SW Missouri, and E Oklahoma), it is associated with mesic to dry-mesic tallgrass prairie dominated by big bluestem, little bluestem, yellow Indiangrass, and sideoats grama (*Bouteloua curtipendula*). In a study of seventy-seven remnant tallgrass prairies in this region, downy phlox occurred in 22 percent of the prairies, making it one of the most important forb species (Eyster-Smith 1984). Downy phlox is associated with the "Cherokee Prairie" of C Arkansas, a local phase of tallgrass prairie developed on soils derived from acidic sandstones and shales, and the "Grand Prairie" of E Arkansas and NE Louisiana, a formation of tallgrass prairie associated with Pleistocene alluvial terraces along the Mississippi River.

From Indiana into S Michigan and across Ohio, downy phlox is a component of scattered occurrences of tallgrass prairie and oak openings in an eastward extension of the tallgrass prairie known as the "Prairie Peninsula." These communities exist only as tiny remnants, primarily in old cemeteries and along railroad right-of-ways. Big bluestem is the dominant grass in the mesic prairies in which downy phlox occurs. A zone of intergradation occurs at the southern end of Lake Michigan, with subsp. *pilosa* tending to occur in upland sites dominated by little bluestem and subsp. *fulgida* occurring in more mesic sites dominated by big bluestem. Subspecies *fulgida* replaces subsp. *pilosa* in tallgrass prairies from C Illinois westward.

In EC Texas and on the Gulf Coastal Plain, downy phlox occurs in association with Blackland prairie. These grasslands develop on blackland soils—dark, humus-rich, clayey soils (Vertisols) formed over deposits of calcareous parent materials such as limestone, chalk, and marl. Blackland prairie in Texas is dominated by little bluestem and yellow Indiangrass, and typically has a more diverse forb component than tallgrass prairie to the north. In Alabama, Arkansas, Louisiana, and Mississippi, downy phlox is associated with blackland prairies that occur as islands in pine and hardwood forests that develop on more acidic soils. These small prairies, often associated with upland slopes, are dominated by little bluestem and grade into oak savanna and woodland.

Along the Gulf Coastal Plain of Texas and SW Louisiana, downy phlox occurs in association with coastal prairie. Similar to tallgrass prairie, coastal prairie occurs on mesic upland sites as opposed to the low, wet prairie of the coastal plain. Less than 1 percent of the original coastal prairie survives, and its floristic composition can only be surmised from historical accounts and tiny remnants. Extant remnants of coastal prairie in Texas are dominated by little bluestem, brown-seed paspalum (*Paspalum plicatulum*), and yellow Indiangrass. In Louisiana, the dominant species are switchgrass, little bluestem, big bluestem, and yellow Indiangrass. Coastal prairie in SW Louisiana is called "Cajun Prairie" in reference to the French Acadians (locally known as "Cajuns") who settled in the region in the mid-eighteenth century. Downy phlox is a common species in the few remaining remnants of Cajun Prairie

Subspecies *fulgida* occurs throughout the northern and central portions of the tallgrass prairie. In Manitoba, Minnesota, North Dakota, South Dakota, and Wisconsin, this grassland is dominated by big bluestem, with porcupine needlegrass (*Hesperostipa spartea*) and northern dropseed (*Sporobolis heterolepis*) the other major species. To the south, the latter two grasses are replaced by little bluestem and Indiangrass. In C Illinois, it occurs in hill prairies on the upper slopes of loess bluffs along rivers. Subspecies *fulgida* was among plant species observed to revegetate mounds of bare soil produced by the activities of pocket gophers (*Geomys bursarius*) in tallgrass prairie in NW Iowa (Wolfe-Bellin and Moloney 2001). This subspecies of downy phlox is an associate of several imperiled tallgrass prairie endemics—Mead's milkweed (*Asclepias meadii*), small white lady's-slipper orchid (*Cypripedium candidum*), prairie bushclover (*Lespedeza leptostachya*), and western prairie white-fringed orchid (*Platanthera praeclara*).

Historical accounts attest to the prominence of downy phlox (particularly subsp. *fulgida*) in the tallgrass prairie prior to settlement and agricultural conversion. Charles Simmons Harrison (1906) recalled riding on horseback over the prairies of Minnesota in the 1850s where he encountered enormous stands of downy phlox:

> One day I drove through a garden of thousands of acres of wild Phloxes. I can never forget the scene. All around I was greeted with those happy, smiling faces, and all the air was incense-laden. Far as the eye could reach I was surrounded by those great masses of loveliness. I was in raptures.

Albert W. Herre (1940) described the tallgrass prairie of C Illinois in the 1870s:

> One of the most marvelous sights of my whole life . . . was the prairie in spring. Unfading are my memories of that waving rippling sea of lavender when the 'wild sweet William', a species of *Phlox* . . . was in full flower. It stretched away in the distance farther than the eye could reach, while I sat entranced in the rear of the wagon.

Subspecies *longipilosa* occurs in mixedgrass prairie dominated by little bluestem along with big bluestem, hairy grama (*Bouteloua hirsuta*), blue grama (*B. gracilis*), and sideoats grama. Small soapweed yucca (*Yucca glauca*) and forbs such as downy Indian paintbrush (*Castilleja sessiliflora*) and hairy false goldenaster (*Heterotheca villosa*) are typical of this plant community. This grassland community occurs within the context of open woodland dominated by post oak or plateau live oak (*Quercus fusiformis*) with an herbaceous layer dominated by little bluestem.

Subspecies *sangamonensis* is associated with tallgrass prairie remnants, woodland borders and openings, and abandoned fields. Most of the plant communities in the Sangamon River Basin have been altered by human activity and are in various stages of succession, but this subspecies appears to have been associated primarily with mesic to wet-mesic tallgrass prairie dominated by yellow Indian grass, switchgrass, and big bluestem. Streamside forest vegetation in the vicinity of the type locality is dominated by silver maple (*Acer saccharinum*) in the most frequently flooded habitat, grading into common hackberry (*Celtis occidentalis*) and shingle oak (*Quercus imbricaria*) on higher ground, and ultimately white oak in habitat that is free of flooding.

Downy phlox is associated with a variety of herbaceous plant systems in the southeastern United States, most of which occur as small patch communities on shallow, rocky soils within a matrix of forest or woodland. These plant communities have been referred to as glades, barrens, prairies, patch prairies, prairie barrens, and xeric limestone prairies. Downy phlox occurs most frequently in herbaceous plant communities dominated by little bluestem.

In North and South Carolina, downy phlox occurs in association with remnant Piedmont prairies, which are thought to have occurred historically as pockets of prairie vegetation in the forested Carolina Piedmont. Downy phlox is a component of grassy openings locally referred to as barrens in the Coosa Valley of NW Georgia and adjacent Alabama, and in the Big Barrens region of the Interior Low Plateaus in Kentucky and Tennessee. It occurs in similar communities in S Indiana. These communities have been termed "deep-soil barrens" to distinguish them from xeric limestone prairies and limestone glades associated with surfacing bedrock.

Downy phlox occurs in glade and rock outcrop communities in the northeastern and southeastern United States, typically in the context of dry, open oak and/or pine woodland. These small patch herbaceous plant communities occur in association with shallow stony soils and surfacing bedrock, and are dominated by little bluestem and other grasses and forbs characteristic of Midwestern tallgrass prairies. Most have been degraded by agricultural practices, quarrying, and by encroachment of woody plants, and exist only as vulnerable remnants. These herbaceous plant communities occur in association with a variety of geological substrates, including diabase (Pennsylvania and Virginia), gabbro (South Carolina), limestone (Alabama, S Illinois, S Indiana, and Tennessee), and sandstone (Arkansas, S Illinois, and W Louisiana). In the Bluegrass Region of C Kentucky, downy phlox occurs in association with xeric limestone prairies and "cat-prairies" that develop over calcareous substrates including limestone, dolomite and shale. Downy phlox occurs in association with the notable Ketona dolomite glades of C Alabama and the Catahoula sandstone glades of Louisiana and E Texas.

Natural history notes

Downy phlox has a unique relationship with phlox flower moth (*Schinia indiana*). The larvae of this rare and poorly known day-flying noctuid moth (Family Noctuidae) feed on developing seed capsules and flowers and adults nectar at the flowers (Balogh 1987; Hardwick 1958; Kwiat 1908; Swengel and Swengel 1999). The reddish purple coloration of the forewings of adults provides camouflage when this moth is perched on the flowers, and appears most effective when the moth is resting in a mix of buds, flowers, sepals, pedicels, and shadows within the flower cluster rather than on the dorsal surface of a fully opened host flower. This unique moth may be an imperiled species due to the highly fragmented nature of the tallgrass prairie remnants where downy phlox occurs. The type locality of phlox flower moth is in NW Indiana, where subsp. *pilosa* was historically common in sandy prairies. The most recent field studies of phlox flower moth have been conducted in N Iowa, W Minnesota, and NW and C Wisconsin, well within the range of subsp. *fulgida*. The apparently obligatory relationship of phlox flower moth with downy phlox is unique because all other known phlox pollinators are generalists and visit a variety of species.

Cultivation

Though downy phlox is the most geographically widespread of all of the eastern phloxes, it has not attained parallel prominence in the garden world. At issue are not its beauty or hardiness, but how best to use it. A preference for relatively dry habitat does not make downy phlox a good fit for the traditional flower bed or border, but it has been grown in rock garden settings. Wherry saw considerable horticultural potential in subsp. *ozarkana*, and today this robust, showy subspecies is gaining recognition as an outstanding garden plant. Interest in native plant horticulture, particularly in the Midwest, has made the prairie expressions of downy phlox more available in recent years. Subspecies *deamii* is also worthy of horticultural trial, combining compact clusters of showy flowers with tolerance of clayey soils. A unique local strain of downy phlox native to the prairies of the Chicago region was a popular garden subject for a period of time and deserves reintroduction. Described as *Phlox argillacea* (Clute and Ferriss 1911), it appeared in Liberty Hyde Bailey's *Manual of Cultivated Plants* in 1924 and received the Royal Horticultural Society Award of Merit in 1927. Downy phlox should be grown in full sun to light shade in dry to slightly moist, well-drained soil. Cultivars are available.

Phlox pulcherrima
Big Thicket phlox
PLATES 56, 57

No tree can match the bearing of a beech. Lordly creatures with muscled trunks of pewter-gray, there is an owly, old-soul character to American beech that imparts a kindhearted melancholy to any stand of woods. Held to a supporting role in the forests of the Appalachian Highlands, beech takes on greater celebrity in the Big Thicket of East Texas. Here it is the much-admired sovereign of the beech-magnolia-loblolly ecosystem, a beautiful forest with a smooth southern disposition, given a Texas twang by brassy patches of Big Thicket phlox.

There is richness to the ecology of the Big Thicket that sets it apart from other regions of Texas, with plant communities as dissimilar as cypress sloughs and arid sandylands occurring side-by-side. It's not the rarity of the flora and fauna that sets the Thicket apart, but the complex diversity of the whole—a place where a roadrunner could come eye-to-eye with an alligator. Big Thicket phlox is one of the few endemic plants here, a tall, showy wildflower that forms bright, vigorous colonies of color in forest openings and along roadsides in spring.

❧

Phlox pulcherrima (Lundell) Lundell, Field & Lab. 13: 12. 1945. Basionym: *P. pilosa* Linnaeus subsp. *pulcherrima* Lundell, Contr. Univ. Michigan Herb. 8: 79. 1942 [most beautiful, of the flowers; "The beauty of its large rose-pink corollas is scarcely equaled by any other form"] Upstanding perennial herb with solitary to a few simple erect-ascending flowering shoots up to 60 cm tall, accompanied by shorter erect nonflowering shoots, pubescent with short eglandular recurved hairs, glabrescent, internodes elongate above, with 7–10 leaf nodes below the inflorescence, lower leaves opposite; *leaves* linear or narrowly linear-lanceolate, up to 125 mm long, 3–6 mm wide, sessile, chartaceous, attenuate to the spinescent apex, sparsely pubescent, glabrate, persistently ciliate at the base with coarse hairs, upper surface obscurely papillate, the midvein impressed; *inflorescence* of cymes within terminal corymb, usually few-flowered, foliaceous bracts spreading or reflexed, those at base lanceolate, slightly wider than the leaves, pedicels 5–18 mm long, pubescent with short recurved eglandular hairs; *calyx* up to 13 mm long, lobes subulate or linear with apex conspicuously awn-tipped; *corolla* tube up to 21 mm long, glabrous outside, short villous within at base, lobes obovate (dimensions 16 × 15 mm) with apex apiculate or rounded, glabrous, hue rose-pink with darker reddish eye; *stamens* with upper anthers barely included; *style* 2–3 mm long, united ca. ⅜ its length, with stigma deeply included; *flowering season* spring, April–May; tetraploid, $2n = 28$ (Levin 1968; Smith and Levin 1967).

❧

Taxonomic notes

Wherry (1955, 1966) treated this taxon as *Phlox pilosa* subsp. *pulcherrima*, but acknowledged full species rank was "worthy of consideration." His concept included populations from the Interior Low Plateaus region of the south-central United States later recognized as *P. pilosa* subsp. *deamii* (Levin 1966). Subsequent workers have affirmed full species status (Ferguson 1998; Levin 1968; Turner 1998a). See also Levin and Schaal (1970).

Geography
West Gulf Coastal Plain, endemic to E Texas but concentrated in the Big Thicket region of SE Texas. Ferguson (1998) considered *Phlox pulcherrima* "an East Texas endemic."

Environment
Coastal plains on upper slopes of gently rolling terrain, from slightly above sea level to 500 ft. (150 m); *MAP* 40–60 in. (102–152 cm); *habitat* mesic; *soils* sandy loam.

Associations
Big Thicket phlox was historically associated with woodland and savanna dominated by longleaf pine (*Pinus palustris*). Of the twelve specimens cited by Lundell (1942) in his original description of the species, ten were described as occurring "in pine land." Wherry (1966) likewise described its habitat as "grassland and open pineland." The term "pine land" as used of the Big Thicket or "Pineywoods" region of East Texas refers to plant communities dominated in early settlement times by open stands of longleaf pine. Harper (1920) studied plant communities near Kountze, in the vicinity of the type locality of Big Thicket phlox, and described longleaf pine as "practically the only tree on the uplands there." Lundell (1942) described Tyler County, Texas, as "the center of distribution" for Big Thicket phlox, and added, "it occurs throughout in pine lands."

The historic upland longleaf pine community of the Big Thicket is a pyro-edaphic climax with open-canopied stands of longleaf pine over a graminoid-dominated herbaceous layer. Shortleaf pine (*Pinus echinata*) is sometimes codominant with longleaf pine. Major understory trees include blue jack oak (*Quercus incana*), post oak (*Q. stellata*), and blackjack oak (*Q. marilandica*), with sweetgum (*Liquidambar styraciflua*) and dogwood (*Cornus florida*) increasing in unburned or previously logged stands. With complete fire exclusion for long periods, longleaf pine savanna will progress to a mixed hardwood-lobbly pine (*P. taeda*) forest. American beech (*Fagus grandifolia*), in particular, increases when fire is suppressed in the Big Thicket.

During the past 100 to 200 years, the open longleaf pine woodlands of the Big Thicket region have been replaced by more closed forests dominated by hardwoods or a mixture of pines and hardwoods, likely derived from forests along creeks and streams in the region. This is evidenced by land survey records for Hardin County, Texas, from the mid- to late 1800s that show beech and magnolia (*Magnolia grandiflora*) as the dominant species of narrow strips of forest along streams (Schafale and Harcombe 1983), and by the observations of Harper (1920) who noted small waterways in the vicinity of the type locality of Big Thicket phlox were bordered by strips of "hammock-like vegetation" dominated by loblolly pine and magnolia, with sweetgum, beech, and several species of oaks also associated. Most extant occurrences of Big Thicket phlox are associated with these mixed forests that appear to have been derived from more distinctive forests and woodlands that existed prior to settlement.

Cultivation

Though a Texan, Cyrus Lundell was not exaggerating when he declared his discovery "one of the finest," and bestowed on it the name *pulcherrima*—"most beautiful." Wherry (1955) concurred, calling it "strikingly beautiful, well deserving of the epithet." Big Thicket phlox has not been reported in cultivation, at least not under the name *Phlox pulcherrima*, but with its tall frame and large clusters of showy flowers it has outstanding potential as a plant for perennial gardens and borders. Southern horticulturists looking for the next great garden phlox would do well to spend some time poking around the Big Thicket. Grow in partial shade in moist to slightly dry, well-drained soil.

Phlox pulchra
Alabama phlox

The early days of June find the Alabama Cumberlands opulent in oakleaf hydrangea. A rustic, big galoot of a shrub with flowers in heavy ivory panicles, it seizes the stage from the genteel dogwood like a dobro taking lead on a fiddle tune. Its blooming marks the transition to summer and brings a touch of hankering, sad-song sweetness to these rocky, wooded hills. Joining along is a suite of second-shift woodland-edge wildflowers that includes, in a scattering of places, Alabama phlox.

Discovered by Edgar Wherry in the watershed of the Black Warrior River, Alabama phlox bears clusters of pink flowers on stems a foot-and-a-half tall. Known to some as "Wherry phlox," the man himself proposed "Alabama phlox," a name which remains fitting since this species has never been found outside the state. A revisiting of historic collection sites, many altered by logging or swamped by the virulent kudzu vine, suggest Alabama's striking namesake phlox also may be an imperiled species.

*Phlox pulchra (*Wherry) Wherry, Gen. Phlox 101. 1955. BASIONYM: *P. ovata* Linnaeus var. *pulchra* Wherry, Bartonia 16: 37. 1935 [beautiful, of the flowers; "Because of the striking beauty of its flowers"] Upstanding perennial herb from a branching rhizome from which arise clusters of flowering shoots 25–50 cm tall with 6 (rarely 5 or 7) leaf nodes below the inflorescence, accompanied by a few ascending nonflowering shoots 10–20 cm tall; *leaves* varying, those lower on flowering shoots narrowly elliptic and short-petioled, upper leaves broadly elliptic and sessile, fine ciliate and pilose to glabrate, max. length 30–60 mm and width 10–20 mm; *inflorescence* 12- to 36-flowered, its herbage glabrous or, when the stems and leaves are pilose, the bracts moderately but the calyx scarcely so, max. pedicel length 10–20 mm; *calyx* 8.5–12.5 (usually 10) mm long, united ½–⅝ its length, lobes broadly subulate with obscure midrib and with apex sharp-cuspidate, membranes moderately narrow and plicate; *corolla* tube 20–24 mm long, lobes obovate to orbicular (dimensions varying from 12 × 10 mm to 15 × 14 mm) with apex entire and obtuse, rarely erose, hue lilac, lavender, pink, or rarely white, often in pastel shades; *stamens* with some anthers exserted; *style* 20–25 mm long, united to tip which is free for 1 mm, with stigma sometimes exserted; *flowering season* late spring to early summer, late May–July.

Geography

Chiefly Appalachian Highlands at the southern limits of the Appalachian Plateaus (Cumberland Plateau section) and Valley and Ridge physiographic provinces in C Alabama, with outlying occurrence on the East Gulf Coastal Plain in S Alabama. Alabama phlox is endemic to Alabama, with most populations occurring in the central part of the state. Know historically from Autauga, Bibb, Blount, Butler, Jackson, Lawrence, and Shelby counties, recently collected in Bibb and Pike counties. Alabama phlox appears to have been eliminated in some places where it occurred historically by competition from kudzu (*Pueraria montana*) and Japanese honeysuckle (*Lonicera japonica*), both introduced invasive vines.

Environment

Dissected plateaus and hilly piedmont, also coastal plains, on upper slopes of rolling to dissected terrain, bluffs, and in ravines, 350–450 ft. (107–137 m); *MAP* 54–59 in. (137–150 cm); *habitat* dry-mesic; *soils* sandy loam, stony, with bedrock near the surface; *parent material* sandstone, acidic shales. Wherry (1935a) described habitat at the type locality for Alabama phlox as "ravines" and "open woods," and described (1955) its general habitat as "moderately acid soils at margins of or openings in woodland." Herbarium specimens from various collectors bear labels describing the habitat as "ravines," "rocky woods," "shaley woods," and "shaley bluffs." Most known populations of Alabama phlox are located at roadsides. It appears to occur naturally in canopy gaps.

Associations

Primarily forest dominated by white oak (*Quercus alba*), with chestnut oak (*Q. prinus*) associated on drier sites and tuliptree (*Liriodendron tulipifera*) on more mesic sites. Loblolly pine (*Pinus taeda*) is also frequently associated. The understory consists of flowering dogwood (*Cornus florida*), sourwood (*Oxydendrum arboreum*), and bigleaf magnolia (*Magnolia macrophylla*), with a shrub stratum characterized by oakleaf hydrangea (*Hydrangea quercifolia*) and mapleleaf viburnum (*Viburnum acerifolium*). Alabama phlox occurs to a limited extent south of the Cumberland Plateau in longleaf pine (*P. palustris*) woodland.

Cultivation

Edgar Wherry discovered Alabama phlox and, in describing it, chose the name *pulchra*—"[b]ecause of the striking beauty of its flowers." He also noted its "considerable horticultural promise." Alabama phlox has lived up to Wherry's epithet and expectations, and is now recognized as one of the best phloxes for American gardens. Although limited to Alabama, this upstanding, medium-height phlox with showy heads of flowers has proven hardy from Pennsylvania to Minnesota and Nebraska, and rated as one of the best border phlox in the evaluation program of the Chicago Botanic Garden.

The cultivar 'Morris Berd' is the most widely-known selection of Alabama phlox, but is sometimes listed in the trade as a cultivar of *Phlox glaberrima*. Mr. Morris Berd, for whom the cultivar was named, confirmed the identity of this plant (pers. comm.). According to Mr. Berd, the cultivar originated from a "good color form of *P. pulchra*" that Wherry had asked a cooperator in a southern state to send Berd, who frequently grew plants for Wherry in his garden at Media, Pennsylvania. Berd later distributed stock of this plant to other gardeners, and was surprised to one day find his name associated with it. Additional cultivars are available.

Alabama phlox can be used in herbaceous perennial gardens and borders, mixed borders, and regional native plant horticulture. Grow in partial shade in moist to slightly dry, well-drained, somewhat acid soil.

Phlox pulvinata
alpine phlox

PLATES 58–61

It is the lexicon of the alpine: *fell-field*; *krummholz*; *felsenmeer*; *scree*. Hearty words ringing with the rush of dazzling light, ripping winds, frost-riven rock, and fervent colors. It is the stony, short-of-breath language of life above timberline, a vocabulary you can use up on the Beartooth Plateau. Straddling the Wyoming-Montana line, the Beartooth is a sky-scraping tableland that sports magnificent expanses of alpine tundra in a mosaic of subtly different plant communities. But from rock face to sedge-mantled meadow, one plant ties it all together—alpine phlox.

While other species of phlox are true ecological alpines, the label is best applied to the most widespread of the highlanders. Alpine phlox spans a far-flung empire, from the Middle and Northern Rockies westward across the mountains of the Great Basin to the Sierras of California. It occurs in a wide variety of alpine habitat where it is a common and sometimes dominant aspect of the flora, perfuming the thin air and compelling oxygen-deprived ecologists to name entire plant communities after it.

❧

Phlox pulvinata (Wherry) Cronquist, Vasc. Pl. Pacific NorthW. 4: 135. 1959. BASIONYM: *P. caespitosa* Nuttall subsp. *pulvinata* Wherry, Notul. Nat. Acad. Nat. Sci. Philadelphia 87: 10. 1941 [having the appearance of a cushion, of the plant growth habit; "Habit pulvinate"] Caespitose perennial herb forming cushions 2.5–7.5 cm tall from a branched, spreading caudex, the numerous decumbent to semi-erect flowering shoots 1–5 cm long; *leaves* linear-subulate, the margin slightly to strongly thickened and tip cuspidate, ciliate with coarse hairs, surficially pilose to glabrous, the uppermost sometimes glandular, max. length 6–12 mm and width 0.75–1.75 mm (leaves elliptic-lanceolate to oblong and width 2–3 mm for variant treated as *P. caespitosa* subsp. *platyphylla*); *inflorescence* 1- to 3-flowered, its herbage glandular-pubescent, max. pedicel length 0.5–3(–5) mm; *calyx* 5–10 mm long, united ⅜–⅝ its length, lobes linear-subulate with moderate midrib and with apex cuspidate, membranes flat; *corolla* tube 7–14 mm long, lobes obovate (average dimensions 7 × 4 mm), hue lavender to white, sometimes with bluish sheen; *stamens* borne on upper corolla tube, with some anthers exserted; *style* 2.5–5 mm long, united to tip which is free for 0.5–1 mm, with stigma located among anthers and near corolla orifice; *flowering season* alpine spring and summer, June–August.

Taxonomic notes

Three names have been misapplied to *Phlox pulvinata*: *P. caespitosa*, derived from abbreviation of the basionym, *P. caespitosa* subsp. *pulvinata* (see *P. caespitosa* species account for circumscription of this species); *P. sibirica*, derived from abbreviation of *P. sibirica* subsp. *pulvinata* (Weber 1973), a synonym that sometimes appears in the floristic and ecological literature of Colorado; *P. variabilis* (Brand 1907), which Wherry (1955) considered a distinct species but which Cronquist (1959) and most subsequent works have considered synonymous with *P. pulvinata*.

Geography

Chiefly Rocky Mountain System, with the core of the range in the Northern and Middle Rocky Mountains (Idaho, Montana, and Wyoming). Apart from this roughly contiguous area, alpine phlox occurs in the Southern Rocky Mountains (Colorado and S Wyoming), the Blue Mountain section of the Columbia Plateau (NE Oregon), the Sierra Nevada (California), on isolated mountains in the Great Basin (California and Nevada), and on high-elevation plateaus of the Colorado Plateau (Utah).

Environment

Mountains at high elevations, on exposed slopes and ridges subject to desiccating winds and intense solar radiation, associated with rock ledges and pavements, fell-fields, talus, and scree slopes, 9400–12,000 ft. (2800–3600 m) in the Southern Rocky Mountains of Colorado, 8500–12,000 ft. (2600–3600 m) in the Absaroka Mountains of Wyoming, 8200–10,000 ft. (2500–3000 m) in the Bitterroot Mountains of Montana, and 10,000–13,000 ft. (3000–3900 m) in the White Mountains of California; *MAP* varied across distribution, mostly as snow in winter, with high winds limiting amount and duration of accumulation; *habitat* dry to dry-mesic; *soils* minimally developed, stony, with little organic matter, shallow with bedrock at or near the surface, sometimes subject to soil frost activity, deeper and better developed on moderate slopes and more protected sites; *parent material* granite, quartzite, basalt, serpentine, limestone, dolomite. Alpine phlox appears to have generalized habitat preferences which enable it to be successful in a diversity of habitats.

Associations

Primarily subalpine conifer woodland and with dry alpine communities. Subalpine woodland occurs at the upper limits of tree growth and is comprised of open stands of conifers, the trees typically stunted in size and deformed by wind into krummholz. Alpine plant associations occur above treeline and are comprised of dwarf shrubs, graminoids, and perennial herbs. Alpine phlox is typically one of the most abundant herbs present in these communities and in some associations is the dominant forb and is recognized as a principal indicator species.

Whitebark pine (*Pinus albicaulis*) is the most common subalpine tree species throughout the range of alpine phlox, particularly in the Middle and Northern Rocky Mountains, but also in the Sierra Nevada. In the Front Range of the Southern Rocky Mountains, from Rocky Mountain National Park in Colorado north through the Medicine Bow Range of Wyoming, alpine phlox occurs in subalpine woodland dominated by Engelmann spruce (*Picea engelmannii*), Rocky Mountain subalpine fir (*Abies bifolia*), and limber pine (*Pinus flexilis*). In the Bitterroot Mountains of Montana and Idaho, alpine phlox occurs in subalpine woodland dominated by subalpine larch (*Larix lyallii*).

On limestone substrates occurring at moderate, nonalpine elevations in the Absaroka Mountains of Wyoming and Montana, alpine phlox is associated with unique "wind-blast" cushion plant communities that develop on exposed, rocky ridges within the context of limber pine woodlands. These communities are comprised of a suite of regional limestone endemics that includes aromatic pussytoes (*Antennaria aromatica*), Jones' columbine (*Aquilegia jonesii*), snow Indian paintbrush (*Castilleja nivea*), Howard's forget-me-not (*Eritrichium howardii*), one-flower kelseya (*Kelseya uniflora*), and shoshonea (*Shoshonea pulvinata*).

Above treeline throughout its range, alpine phlox is associated with a variety of dry-mesic alpine plant associations including grassland, turf communities, and fell-field communities. In the Middle and Northern Rocky Mountains, alpine phlox occurs in high elevation grasslands dominated by Idaho fescue (*Festuca idahoensis*). These grasslands occur on gentle to steep slopes at the lower reaches of the alpine zone. On soils derived from dolomite and volcanic substrates in the mountains of EC Idaho, alpine phlox is a major component of a dry alpine grassland community dominated by spike fescue (*F. kingii*). In the Wallowa Mountains of Oregon, alpine phlox occurs in alpine grassland comprised of rough fescue (*F. campestris*) and Idaho fescue.

Alpine phlox is a common, often major component of turf communities in the Rocky Mountains dominated by dwarf, fibrous-rooted graminoids. Typical graminoids include kobresia (*Kobresia myosuroides*), blackroot sedge (*Carex elynoides*), rock sedge (*C. rupestris*), purple reedgrass (*Calamagrostis purpurascens*), shortleaf fescue (*Festuca brachyphylla*), and timberline bluegrass (*Poa glauca* subsp. *rupicola*). Characteristic forbs in these communities include Ross' avens (*Geum rossii*), old-man-of-the-mountain (*Tetraneuris grandiflora*), northern yellow point-vetch (*Oxy-*

tropis campestris), skunk polemonium (*Polemonium viscosum*), American bistort (*Polygonum bistortoides*), and mountain meadow cinquefoil (*Potentilla diversifolia*).

In fell-field habitat, graminoids become less prominent, being replaced by cushion- and mat-forming shrubs and forbs of which alpine phlox is often the most common and abundant species. Forbs and shrubs typical of these communities in the Rocky Mountains include alpine sagebrush (*Artemisia scopulorum*), eight-petal mountain-avens (*Dryas octopetala*), arctic forget-me-not (*Eritrichium nanum*), alpine stitchwort (*Minuartia obtusiloba*), dense spike-moss (*Selaginella densa*), moss campion (*Silene acaulis*), and whip-root clover (*Trifolium dasyphyllum*). In the Sweetwater Mountains of California, alpine phlox occurs in fell-field communities with dagger-pod (*Anelsonia eurycarpa*), Mt. Hood pussy-paws (*Cistanthe umbellata*), oval-leaf wild buckwheat (*Eriogonum ovalifolium* subsp. *nivale*), Gordon's ivesia (*Ivesia gordonii*), Mount Rose lupine (*Lupinus caudatus* subsp. *montigenus*), and bottlebrush squirrel-tail (*Elymus elymoides*).

On limestone and dolomite substrates in the mountains of SW Montana and adjacent portions of Wyoming and Idaho, fell-field communities in which alpine phlox occurs are enriched by a suite of regional endemics that include snow Indian paintbrush, mountain douglasia (*Douglasia montana*), Howard's forget-me-not, and Hayden's clover (*Trifolium haydenii*).

Cultivation

Of all the high altitude phloxes, alpine phlox is the one least likely to break the gardener's heart. The most widespread and ecologically adaptable of the group, alpine phlox has enjoyed at least some success in cultivation given the accommodating conditions of a rock garden. It is sometimes offered under the name *Phlox caespitosa*. Grow in full sun in dry, well-drained soil.

Phlox pungens
Beaver Rim phlox

Beaver Rim is a jarring, hawk-hurling escarpment, carved from the aching desert heart of Wyoming. Only cushion plants and a scattering of limber pine cling to its rocky brink, giving you a clear view of the Wind River Range to the west. Stand for long on these austere, wind-hammered cliffs, and your eye begins to yearn for respite, finding it afar in the snow-capped mountains, or at hand in the bracing white flowers of Beaver Rim phlox.

A mat-forming species with prickly leaves, Beaver Rim phlox is a rare plant known only from central Wyoming. First encountered on the escarpment, it occurs elsewhere within the Wind River and Green River basins, often in the company of other cold desert specialists. It makes an impressive display in the watershed of the Little Popo Agie River, where erosion has exposed the siltstones and shales of the Chugwater Formation, and brilliant white patches of Beaver Rim phlox light up the Mars-red landscape.

Phlox pungens Dorn, Vasc. Pl. Wyoming 299. 1988 [sharp, of the leaf tip; "leaves very stiff and pungent"] Caespitose perennial herb forming cushions under 7 cm tall from a short-branched caudex, the numerous sub-erect flowering shoots 3–6 cm long, internodes pubescent with short, mostly straight often gland-tipped hairs; *leaves* very stiff and pungent, mostly lanceolate or lance-linear, 4–8 mm long, 1–1.5(–2) mm wide, pubescent, often glandular, ciliate, punctate, the margins and dorsal midrib usually strongly thickened, gradually tapering into a sharp terminal bristle; *inflorescence* 1-flowered; *calyx* 6–9 mm long, lobes sharply spinulose-tipped, densely glandular-pubescent; *corolla* tube 8–12 mm long, lobes 4–6 mm long, hue white to bluish; *style* 6–9 mm long; *flowering season* mid-May through June.

Geography
Rocky Mountain System, endemic to the Wyoming Basin in WC Wyoming where it occurs in the Wind River and Green River basins.

Environment
Intermountain basins and hilly piedmont, on dissected terrain, badlands, escarpments, bluffs, and rock ledges and outcrops, 6000–7400 ft. (1800–2200 m); *MAP* less than 10 in. (25 cm); *habitat* xeric; *soils* medium-textured, stony, shallow to moderately deep with bedrock at or near the surface; *parent material* in Wind River Basin sandstones and limestones of the Nugget, Splitrock, and White River formations and siltstones and shales of the Chugwater Formation, in Green River Basin marlstones, shales, and limey sandstones of the Green River Formation.

Associations
Sparsely vegetated barrens and openings within a mosaic of conifer woodland and shrubland. Limber pine (*Pinus flexilis*) and/or Utah juniper (*Juniperus osteosperma*) are the dominant trees, with scattered shrubs intermixed including Wyoming big sagebrush (*Artemisia tridentata* subsp. *wyomingensis*), black sagebrush (*A. nova*), and antelope bitterbrush (*Purshia tridentata*), along with various bunchgrasses.

Beaver Rim phlox can be locally abundant and sometimes occurs in large, conspicuous populations. Associated forbs in the Wind River Basin include the rare Wyoming endemics Cedar Rim thistle (*Cirsium aridum*), Fremont's bladderpod (*Lesquerella fremontii*), Devil's Gate twinpod (*Physaria eburniflora*), and desert yellowhead (*Yermo xanthocephalus*), the latter an endangered species. Associates in the Green River Basin include the endemic Big Piney milkvetch (*Astragalus drabelliformis*).

Cultivation

Beaver Rim phlox has not been reported in cultivation, although seed has been available. It has potential as a rock garden plant. Grow in full sun in dry, well-drained soil.

Phlox richardsonii
Arctic phlox

There is arctic, there is alpine, and there is arctic-alpine. Known to Russian ecologists as the *goltsy*, the alpine zone of an arctic mountain is one of the most brutal places on the planet. To survive here above the Arctic Circle a plant must be both *cryophyte* and *xerophyte*, able to endure cold and dry in the extreme. Arctic phlox qualifies on both counts, taking on the goltsy in the company of other cushion- and mat-forming stout-hearts.

Arctic phlox ranges from the Bering Strait region of northwest Alaska eastward across the Brooks Range and other interior mountains of Alaska and the Yukon Territories, to the Arctic seacoast of Canada. It is also known from the Asiatic side of the Bering Strait on the Chukotka Peninsula. I have never been to the Arctic, but I have seen this plant in its prime, shipped to me via express mail by a naturalist in Nome, Alaska, the flowers still fragrant and the roots still clinging to stones from a place called Anvil Mountain.

❧❧

Phlox richardsonii Hooker, Fl. Bor.-Amer. (Hooker) 2: 73, tab. 160. 1838 [after Sir John Richardson (1787–1865), collector of the type] Caespitose perennial herb forming loose to dense mats 3–12 cm tall from a short-branched caudex, the flowering shoots 2–6 cm long; *leaves* linear to subulate (thick-textured and fleshy for subsp. *richardsonii*); *vesture* ranging from basally arachnoid-hairy for subsp. *richardsonii* to copiously ciliate in subsp. *alaskensis*, max. length 5–15 mm and width 1–2.5 mm; *inflorescence* 1- to 3-flowered, its herbage pubescent with mixed glandular and glandless hairs, max. pedicel length 1–6 mm for subsp. *alaskensis* and subsp. *richardsonii*, (4–)8–25 mm for subsp. *borealis*; *calyx* tube length ranging from 5–7 mm for subsp. *richardsonii* to 7–10 mm for subsp. *alaskensis* to 9–12 mm for subsp. *borealis*, united ⅜–⅝ its length, lobes linear with weak midrib with apex cuspidate, membranes narrow, flat; *corolla* tube length ranging from 6–11 mm for subsp. *richardsonii* to 8–12 for subsp. *alaskensis* and subsp. *borealis*, lobes broadly obovate (dimensions ranging from 6 × 5 mm for subsp. *richardsonii* to 9 × 6 mm for subsp. *alaskensis* to 14 × 13 mm for subsp. *borealis*), hue lilac, lavender, or white; *style* ranging from 4–8 mm long for subsp. *alaskensis* and subsp. *richardsonii* to 7–12 mm long for subsp. *borealis*, united to tip which is free for 1 mm; *flowering season* arctic spring and summer, late June into July.

∽∾∽

Taxonomic notes

The identity of *Phlox richardsonii* has long been entangled with that of *P. sibirica* (Linnaeus 1753). Wherry (1955) separated the phloxes of the North American Arctic (recognized by him as *P. borealis*, *P. richardsonii* subsp. *alaskensis*, and *P. richardsonii* subsp. *richardsonii*) from the Eurasian *P. sibirica* on the basis of morphological characters. Locklear (2009) affirmed these plus noted significant ecological differences, with *P. sibirica* associated with mid-elevation, graminoid-dominated steppe communities, primarily in the cold, dry interior plateaus of Middle Asia, and *P. richardsonii* occurring almost entirely above the Arctic Circle in association with alpine dwarf-shrubland and alpine herb communities.

Key to the Subspecies of *Phlox richardsonii*

1a. Longest pedicels (4–)8–25 mm long; corolla lobe 8–14 mm long; styles 7–12 mm long; associated with coastal areas of the Bering Strait . . . .
. . . . **subsp. *borealis*** (Wherry) Locklear, J. Bot. Res. Inst. Texas 3: 647. 2009 |
Bering Sea Arctic phlox
1b. Longest pedicels 1–6 mm long; corolla lobe 6–9 mm long; styles 4–8 mm long 2
2a. Leaves flattened, thinnish, 8–15 mm long; pubescent with rather coarse hairs; calyx 7–10 mm long; corolla limb diameter up to 2.2 cm; associated with higher interior foothills and mountains of Alaska, Yukon, and the Northwest Territories . . . .
. . . . **subsp. *alaskensis*** (Jordal) Wherry, Gen. Phlox 127. 1955 and in
Baileya 4: 98. 1956 | Brooks Range Arctic phlox
2b. Leaves awl-shaped, thickish, 5–10 mm long; pubescent with fine, arachnoid hairs; calyx 5–7 mm long; corolla limb diameter ca. 1 cm; associated with the Arctic seacoast
. . . . **subsp. *richardsonii*** | Arctic phlox

Geography

The distribution of the Arctic phlox species complex ranges from the Bering Sea region of Alaska east to northwestern Canada, mostly north of the Arctic Circle, with a few highly limited occurrences reported from Asia. The nominate expression of Arctic phlox (subsp. *richardsonii*) is known from relatively few stations in northwestern Canada and Alaska. The core of its range is along the Arctic seacoast of Canada, from Cape Bathurst in the District of Mackenzie of the Northwest Territories, east to Clifton Point, a mainland promontory on the coast of Nunavut Territory, with isolated occurrences west to the Arctic coast of Alaska. It also occurs on Banks Island of the Canadian Arctic Archipelago.

Subspecies *alaskensis* occurs in the higher interior foothills and mountains of Alaska, Yukon, and the Northwest Territories. It occurs most abundantly in the Brooks Range (N Alaska), but also in the Alaska Range (C Alaska) and in the Mentasta, Nutzotin, and Wrangell mountains (SC Alaska). It also occurs in the Ogilvie Mountains and Richardson Mountains of the Yukon Territory.

Subspecies *borealis* is distributed along the coast of the Seward Peninsula (NW Alaska), in the vicinity of the Bering Strait. It also has been reported (as *Phlox alaskensis*) on the Asiatic side of the Bering Strait in the far southwestern portion of the Chukotka Peninsula of the Russian Far East (Balandin and Razzhivin 1980; Razzhivin 1986, 1994). On American side of the Strait, occurrences of subsp. *borealis* are known from coastal areas of Norton Sound and Kotzebue Sound, and from Cape Krusenstern, Cape Lisburne, Cape Prince of Wales, Cape Sabine, and Cape Thompson.

Environment

Plains, hilly piedmont, and mountains at low to middle elevations, on escarpments, bluffs, slopes, ridges, rock outcrops, fell-fields, talus, and scree slopes, also terraces, cutbanks, and slopes of stream valleys, 2000–5000 ft. (600–1500 m) for subsp. *alaskensis*, from near sea level up to 2000 ft. (600 m) for subsp. *borealis*, from near sea level up to 200 ft. (60 m) for subsp. *richardsonii*; *habitat* dry to xeric; *soils* minimally developed, stony, with little organic matter, shallow with bedrock at or near the surface, sometimes subject to cycles of freeze-and-thaw (cryopedogenic) activity that result in "patterned ground"; *parent material* primarily calcareous rock formations such as limestone, dolomite, marbleized limestone, but also basalt, rhyolite, sandstones, shale.

Associations

Alpine dwarf-shrubland and alpine herb communities. All three subspecies of

Arctic phlox are associated with dwarf-shrubland dominated by species of prostrate, mat-forming shrubs, primarily eight-petal mountain-avens (*Dryas octopetala* subsp. *octopetala*). Arctic willow (*Salix arctica*) is sometimes associated with mountain-avens and occasionally replaces it as the dominant shrub in communities in which Arctic phlox occurs. These shrubland systems typically have a strong component of dwarf, fibrous-rooted graminoids in association, with major species including alpine sedge (*Carex glacialis*), rock sedge (*C. rupestris*), bulrush sedge (*C. scirpioidea*), shortleaf fescue (*Festuca brachyphylla*), and kobresia (*Kobresia myosuroides*). Mosses and lichens are common associates, particularly fruticose lichens of the genus *Cladonia* and *Cetaria*. In more exposed, xeric habitat, lichens may become more common than graminoids. In the vicinity of Cape Bathurst on the Canadian Arctic seacoast, the nominate expression of Arctic phlox (subsp. *richardsonii*) occurs in dwarf shrub tundra dominated by entire-leaved mountain-avens (*D. integrifolia*) and arctic willow. On Banks Island, it occurs in association with dwarf shrub tundra dominated by arctic willow.

All three subspecies of Arctic phlox occur in alpine herb communities—sparsely vegetated associations dominated by dwarf, cushion-forming forbs, with woody plants absent or nearly so. In contrast to shrub tundra, these herbaceous communities occur in association with more xeric habitat such as cliffs, bluff tops, rock outcrops, or unstable slopes, often on limestone or dolomite bedrock. Associated graminoids are similar to those found in dwarf-shrubland. Alpine herb communities are open with much bare rock between individual plants.

Arctic phlox occurs with a characteristic suite of dwarf, cushion-forming plants in both dwarf-shrubland and alpine herb communities. Common associates include arctic forget-me-not (*Eritrichium nanum*), stitchworts (especially *Minuartia obtusiloba*), pointvetchs (especially *Oxytropis nigrescens*), saxifrages (especially *Saxifraga bronchialis* and *S. oppositifolia*), northern spike-moss (*Selaginella sibirica*), and moss campion (*Silene acaulis*). Subspecies *alaskensis* occurs with the rare species Muir's fleabane (*Erigeron murii*) and Calder's bladderpod (*Lesquerella calderi*). Subspecies *borealis* occurs with several endemic Beringian species including Bering Sea wormwood (*Artemisia senjavinensis*), Bering Sea douglasia (*Douglasia beringensis*), Walpole's poppy (*Papaver walpolei*), and matted starwort (*Stellaria dicranoides*).

Arctic phlox is typically one of the most abundant herbs present in these communities and in some associations is the dominant forb. Richardson (1828) noted the strong fragrance of Arctic phlox near the type locality at Cape Bathurst, commenting "the air was perfumed by numerous tufts of a beautiful phlox." In her book *North to the Rime-Ringed Sun* (1934), famous explorer Isobel Wylie Hutchinson noted Arctic phlox on mountains near Nome, Alaska, "where it carpets the ground with lovely magenta patches in late June and early July." Johansen (1924) described Arctic phlox as the "dominant plant" on the top of bluffs near Cape Bathurst, and Kuc (1996) described a *Phlox richardsonii*–dominated tundra formation restricted to coastal areas of Banks Island. Kuc also noted association of Arctic phlox with the human-altered vegetation of villages and former Eskimo camps on the island.

Cultivation

The name "*Phlox borealis*" has circulated in Minnesota horticulture for a number of years, linked to plants cultivated at the Minnesota Landscape Arboretum that were introduced to the collections by Dr. Leon Snyder. The true identity and original source of these plants is presently unknown. Plants under this name have also been cultivated in rock gardens in Newfoundland, Canada, and in Europe. Grow in full sun in dry, well-drained soil.

Phlox roemeriana
golden-eye phlox
PLATES 62, 63

There is a clang and clatter to the flora of Hill Country that somehow, in the end, achieves song. Beloved by Texans for its clear-running streams, cedar-scented woodlands, and stony, squared-jawed character, this is the geographic and spiritual heart of the Lone Star State. It is also a center of plant diversity for the North American continent. Drawing instrumentation and style from Chihuahuan desert, Big Thicket bayou, and Madrean evergreen woodland, with riffs and licks of local origin tossed in, Hill Country feels like *Western Swing* sounds.

As if Texans needed more trigger for brag and swagger, Hill Country also has a phlox all its own. Golden-eye phlox is endemic to this region, occurring in colonies that sheet rocky limestone glades and woodland openings with a thin, soft pink in the early spring. It's a diminutive plant, one of only three annuals in the genus, yet its flowers are among the largest of any phlox. Each gold-bossed bloom is a tender work of organic verse, rewarding close inspection with a glint of daybreak, whiff of clove, and a little music of the spheres.

ᏬᎦᏬ

Phlox roemeriana Scheele, Linnaea 21: 752. 1848 [after Karl Ferdinand Roemer (1818–1891), collector of the type] Annual herb with simple erect stem, varying in stature in relation to amount of precipitation, 5–35 cm tall; *leaves* at base of plant opposite, up the stem becoming subopposite, passing into alternate bracts below the inflorescence, lower leaves elliptic-oblanceolate, upper lanceolate, margins ciliate, surfaces pilose to glabrate, max. length 20–50 mm and width 2–10 mm; *inflorescence* a 3- to 15-flowered sympodium, its herbage bearing coarse glandless hairs, max. pedicel length 15–35 mm; *calyx* 11–17 mm long, united ⅜–½ its length, lobes linear, foliaceous with scarcely any midrib and with apex subaristate, membranes flat; *corolla* tube 9–13 mm long, lobes obovate (average dimensions 14 × 11 mm) with apex apiculate (corolla tube is equaled or exceeded by the calyx, an unusual situation in *Phlox*), hue bright purple, varying to lilac or rarely to white, usually with conspicuous yellow eye bordered by white; *style* 1.5–2 mm long, united ¼–½ its length, with stigma included and placed below the anthers; *ovary/capsule* with multiple (3–5) ovules/seeds per locule; *flowering season* spring, late March to early May.

❧❧

Geography
Chiefly Edwards Plateau section of the Great Plains in C Texas, extending into the Rolling Plains and Cross Timbers regions of the Central Texas section of the Great Plains in NC Texas; endemic to Texas.

Environment
Cuestas and dissected plateaus, on upper slopes of rolling terrain and divides, on escarpments, mesas, bluffs, glades, and rock ledges, 500–2500 ft. (150–750 m); *MAP* 25–30 in. (64–76 cm); *habitat* xeric; *soils* stony, shallow with bedrock at or near the surface; *parent material* limestone. Golden-eye phlox habitat is characterized by shallow soil depth, high light levels, and low water availability during the summer, which appears to limit the establishment of woody species in favor of grasses and annual forbs.

Associations
Sparsely vegetated glades and barrens within a mosaic of forest, woodland, and savanna. Plateau live oak (*Quercus fusiformis*) and Ashe's juniper (*Juniperus ashei*) dominate the xeric woodlands and savannas in which golden-eye phlox occurs on the Edwards Plateau, with post oak (*Q. stellata*) and blackjack oak (*Q. marilandica*) characterizing occurrences in the Cross Timbers region. The herbaceous layer of these woodlands is dominated by grasses typical of mixedgrass prairie including little bluestem (*Schizachyrium scoparium*) and sideoats grama (*Bouteloua curtipendula*). Texas prickly-pear (*Opuntia engelmannii* var. *lindheimeri*), a shrubby cactus, is sometimes associated in drier sites. Suppression of natural fire on the Edwards Plateau in the last century has resulted in an increase in woody vegetation, par-

ticularly Ashe's juniper, at the expense of savanna and grassland. Associated forbs include a number of other annuals, many of which are endemic to Texas including Texas Indian paintbrush (*Castilleja indivisa*), Texas heron's-bill (*Erodium texanum*), Texas yellow-star (*Lindheimera texana*), Texas bluebonnet (*Lupinus texensis*), and Drummond's skullcap (*Scutellaria drummondii* var. *edwardsiana*). Golden-eye phlox can be locally abundant and often occurs in large, conspicuous populations. Wherry (1955) noted, "[i]n average seasons the plants grow in profusion, carpeting gravelly tracts with gorgeous sheets of color; when rainfall is deficient, however, they are sparse."

Natural history notes

A crab spider (*Misumenops celer*) is commonly found on golden-eye phlox flowers, where it has been observed to create a "bower" by webbing together two of the five corolla lobes of the phlox flower (Ott et al. 1998). This sit-and-wait predator hunts for insects that visit the flowers including butterflies, moths, and bees, relying on stealth rather than a web to catch its prey. The bower may act as a hiding place for the spider, since it occupies the structure after constructing it.

Cultivation

"A little dandy" is how flamboyant Texas plantsman Carroll Abbott described golden-eye phlox in his *Knowing and Growing Texas Wildflowers*. Abbott called this species an "excellent rock garden plant," but whether his praise was based on actual experience or perceived potential is unclear since seed of this elegant annual seems impossible to come by. Wild populations of golden-eye phlox are maintained at the wonderful Lady Bird Johnson Wildflower Center in Austin, Texas. Grow in full sun in dry, well-drained soil.

Phlox sibirica
Siberian phlox

The Altay Mountains of southern Siberia are said to be so beautiful, so heavy with transcendence that Russian writers have given up looking for words that work there. One of them, Valentin Rasputin, could only describe the landscape as a place of "divine sounds made visible." These highlands are homeland to Siberian phlox, a shrubby species that occurs here in dry rocky habitat. The sole Eurasian species in the genus, Siberian phlox breaks out of the Altay and across a swath of the Earth's surface that would have won an admiring nod from Ghengis Khan.

Siberian phlox ranges from the Ural Mountains almost to the Pacific Ocean, but the heart of its distribution is the mountainous world of central Asia where Kazakhstan, Mongolia, and Russian Siberia meet. It is most often associated with bunchgrass steppe communities, typically at an interface with forests of larch or pine. The region around Lake Baikal in southern Siberia is another stronghold, where it occurs in the surrounding mountains and on the mysterious, steppe-cloaked island of Olkhon, one of the many reputed resting places of the Khan himself.

∽⌒∾

Phlox sibirica Linnaeus, Sp. Pl. 153. 1753 [of the region of Siberia] Upstanding suffrutescent perennial, sparingly and diffusely branched, forming tufts 8–15 cm tall from a taproot, the erect-ascending flowering shoots 5–10 cm long with ca. 5 leaf nodes below the inflorescence; *leaves* linear, long-accuminate, sometimes falcate, thinnish, fine-ciliate and sparse-pilose, max. length 30–60 mm and width 1.5–3 mm; *inflorescence* (1-) 3- to 6-flowered, its herbage pubescent, the hairs varying from all gland-tipped to all glandless, max. pedicel length 20–40 mm; *calyx* 8–13 mm long, united ⅜–⅝ its length, lobes linear with obscure midrib and with apex aristate, membranes somewhat plicate; *corolla* tube 10–12 mm long, slightly flaring upward, lobes obovate (average dimensions 9 × 6 mm) with apex entire or erose to emarginate, hue pink; *style* 7–10 mm long, united to tip which is free for 1–1.5 mm; *ovary/capsule* typically with 2 ovules/seeds per locule; *flowering season* spring to early summer.

∽⌒∾

Geography

The core distribution of Siberian phlox is centered in the mountains and cold dry interior plateaus of Middle Asia, where it occurs in the Altay Mountains of southern Siberian Russia and adjacent Kazakhstan and Mongolia, in the Sayan and Transbaikalian mountain systems in the vicinity of Lake Baikal in southern Siberia, and in the highlands around Lake Khubsugul in northwestern Mongolia. From this center, the range of Siberian phlox approaches Europe to the west in the Ural Mountains and the Pacific Ocean to the east in the Chukotka Peninsula. These far-flung occurrences are isolated and appear to be relicts of a once more contiguous range. Siberian phlox is not reported from China, but may occur in the extreme northwest or northeast sectors along the border shared with Mongolia.

The westernmost occurrences of Siberian phlox are found on the eastern slope of the Ural Mountains, a range running north to south almost from the Arctic Ocean to the Caspian Sea. The Urals have traditionally been considered the boundary between Europe and Asia. Siberian phlox occurs in small, scattered populations in the southern part of the Northern Urals and the northern part of the Middle Urals. The southwestern limit of Siberian phlox is in the vicinity of Lake Balkhash in eastern Kazakhstan.

To the east of the mountain systems of Middle Asia, Siberian phlox occurs in *Dauria* (sometimes rendered *Dahuria*), a region encompassing northeastern Mongolia, the Chita region of Siberian Russia, and northeastern China. This region includes the basin of the Selenge River, Transbaikalia (the region south and east of Lake Baikal), the basin of the Nercha River, the steppe east of Hentiyn Nuruu and west of the Lesser Khingan mountains.

To the northeast, Siberian phlox is known from mountainous regions in the Russian Republic of Sakha (formerly Yakutia), where it is associated with landforms influenced by the Lena, Yana, Indigirka, and Kolyma rivers. Siberian phlox ranges farther north and east in a few isolated occurrences in the continental (western and central) parts of the Chukotka Peninsula of the Russian Far East. Yurtsev (1972) listed Siberian phlox among the plant species "characteristic of the mountains of the Yakutia sector" below the Arctic Circle, but noted it occurs only in "isolated relict sites" of steppe vegetation in northern Chukotka, a region otherwise characterized by tundra.

Environment

Hilly piedmont and mountains, primarily at middle elevations, on rock ledges and outcrops, talus, and scree slopes, also terraces and slopes of stream valleys, 1200–6200 ft. (360–1900 m) in the Altay Mountains of southern Siberia; 5300 ft. (1600 m) at Lake Khubsugul in northern Mongolia; *MAP* 9–16 in. (22–40 cm) in the Altay Mountains of southern Siberia, 12–16 in. (30–40 cm) in northern Mongolia, 10 in. (26 cm) in Yakutia in northeastern Siberia; *habitat* xeric to dry-mesic; *soils* stony, shallow with bedrock at or near the surface; *parent material* granite, slate, metamorphic limestone. Siberian phlox is often described as a *petrophilous* (rock-loving) species in Russian floristic and ecological literature, and its habitat is variously described as rock outcrops, stony slopes, scree, and cliffs.

Associations

Conifer forest/woodland, shrubland, grassland, and rock outcrop communities. Descriptions of the ecology of Siberian phlox invariably link it to graminoid-dominated herbaceous plant associations, using general descriptions like dry meadows, meadow slopes, steppe, steppe slopes, and forb steppes. *Steppe* is a treeless plant association dominated by perennial bunchgrasses and forbs. Steppe is most strongly developed in the uplands of Middle Asia, most notably the Altay mountain system, which is considered the center of origin and evolution of the modern Eurosiberian steppe flora.

In central Yakutia, Siberian phlox occurs in herb-rich forests on rocky uplands or steep south-facing slopes, dominated by open stands of Scotch pine (*Pinus sylvestris*) or Cajander larch (*Larix cajanderi*) in combination with steppe species (Ermakov et al. 2002). In this same region, Siberian phlox occurs in association with a unique landform locally called an "alase" (a woodless area) (Mirkin et al. 1985), a depression formed by thermal disturbance of permafrost, with steep slopes culminating in a flat bottom with a central lake. While the vegetation of this sub-arctic region is forest (taiga) dominated by larch, Scotch pine, and broadleaf white birch (*Betula platyphylla*), the slopes of an alase are vegetated with various steppe formations comprised of species typical of the steppes of Mongolia. In this habitat, Siberian phlox occurs in a mesic variant of steppe dominated by tundra fescue (*Festuca lenensis*), *Artemisia commutata*, and sulphur pasqueflower (*Pulsatilla flavescens*), which correlates to what is termed (in Soviet phytocoenology) "meadow rich-herb steppe." Mirkin et al. (1985) described this as the *Pulsatilletum flavescentis* association, and noted the "very colourful phenology" of this species-rich steppe which "becomes yellow, rose and blue when *Pulsatilla flavescens, Phlox sibirica* and *Myosotis sylvatica* bloom. . . . It reminds one of a bright carpet in July."

In the Altay and Western Sayan Mountains of southern Siberia, Siberian phlox occurs in mid-elevation, nonforested communities (Ermakov et al. 2006), notably "shrubberies" dominated by threelobe spirea (*Spirea trilobata*) on south-facing slopes surrounded by zonal meadow steppes. Siberian phlox is relatively common in the herbaceous layer of these communities, along with *Artemisia gmelinii, Sedum hybridum, Orostachys spinosa*, and *Thymus serphyllum*. In this same region, Siberian phlox occurs in rock-crevice and scree communities dominated by succulent herbs including *Sedum hybridum*, and *Hylotelephium ewersii*, along with *Dracocephalum nutans*.

Siberian phlox occurs in "mountain steppe" in the mountain ranges of north-central Mongolia and around Lake Baikal in southern (Baikalian) Siberia. Characteristic graminoids include *Stipa krylovii*, tundra fescue, and *Poa attenuata*. Prairie Junegrass (*Koeleria macrantha*) and needleleaf sedge (*Carex duriuscula*), two graminoids that also occur in North American prairies, are also associated with these mountain steppe communities. Dwarf subshrubs of the genus *Artemisia* are also characteristic including prairie sagebrush (*A. frigida*), another species with American connections. In the same region, Siberian phlox occurs in "meadow steppe," also referred to as "forest steppe." These associations occur in the context of Siberian larch (*Larix sibirica*) forest, either as openings in the forest or at the transition between steppe and forest. Meadow steppe occurs under somewhat more mesic conditions than mountain steppe, and has a more diverse forb component, hence the name "forb steppe" that is also used for this association. The American counterparts to this Asian ecological system would seem to be Northern Rocky Mountain Subalpine-Upper Montane Grassland or Rocky Mountain Subalpine-Montane Mesic Meadow, which occur at middle elevations in the mountains of Idaho, Montana, and Wyoming. These forb-rich meadows are dominated by Idaho fescue (*Festuca idahoensis*), with Yellowstone phlox (*Phlox multiflora*) a frequent and showy component.

In the continental western and central portions of the Chukotka Peninsula of the Russian Far East, Siberia phlox is a rare associate of isolated stands of "tundra-steppe." These plant communities are comprised of a codominance of species characteristic of both alpine tundra and steppe, and occur as isolated, presumably relict sites within the larger matrix of mesic to hygric tundra, often on the dry, upper slopes of river valleys. Siberian phlox occurs in tundra-steppe dominated by avens (*Dryas grandis*), typical of alpine tundra, along with graminoids and forbs characteristic of Mongolian-Siberian steppe. In an intriguing parallel, Hood's phlox (*Phlox hoodii*), a common species in the mixedgrass prairies of the northern Great Plains of North America, occurs in small, isolated steppelike associations in interior Alaska and the Yukon Territory that are analogous to those of the Russian Far East.

In the Ural Mountains of western Russia, Siberian phlox occurs in unique plant communities associated with cliffs and massive rock outcrops. These communities, associated with river valleys, are comprised of species that are rare in this otherwise forested region, many of which have affinities with the steppe communities of Siberia and are subjects of conservation concern. One such site in the Northern Urals is even named "Siberian Phlox Cliff" (Knyazev et al. 2007).

Cultivation

With its ornamental character and vast distribution, Siberian phlox is surely cultivated somewhere by somebody, but reports are lacking in western horticultural literature. It would make a fine rock garden plant. Grow in full sun to light shade in dry, well-drained soil.

Phlox speciosa
showy phlox
PLATES 64–66

It was the fall of 1809 and in a cabin along the Natchez Trace in Tennessee a man lay dying of self-inflicted wounds. The troubled soul was Meriwether Lewis, who only three years earlier had been hailed as hero of the one of the greatest enterprises in American history. But those heady days were past, and the captain of President Jefferson's celebrated Corps of Discovery was apparently overwhelmed at being a better woodsman, hunter, and botanist than businessman, socialite, and politician.

In *Undaunted Courage*, Steven Ambrose respectfully wonders about the stream of images coursing through the Captain's mind as he contemplated ending his life. Perhaps, mixed with that which caused such despair were scenes of the great beauty Lewis witnessed in the untrammeled reaches of the young United States. If so, surely he revisited early May along the Kooskooskee, the Clearwater River of present-day Idaho, where he did his most productive botanical collecting, and discovered one of the most beautiful wildflowers of the American West—showy phlox.

~~~

Phlox speciosa Pursh, Fl. Amer. Sept. (Pursh) 149. 1813 [showy] Upstanding suffrutescent perennial or subshrub, varying in habit from the nominate expression (subsp. *speciosa*) which is upstanding with a few erect-ascending stems averaging 40 cm tall and slightly woody at the base to more strongly woody subshrubs ranging from medium stature* (averaging 25 cm tall) and sparingly branched (subsp. *nitida* and subsp. *occidentalis*) to low (mostly under 25 cm) and densely branched (subsp. *lignosa*); *vesture* (especially inflorescence-herbage) glandular-pubescent for nominate expression and all subspecies except subsp. *nitida* which is glabrous (densely glandular-pubescent expressions having a musky scent when rubbed); *leaves* of non-flowering shoots uniform in outline (linear or oblong), those of flowering shoots exhibiting progression in leaf outline and passing to lanceolate upwards, dimensions ranging from 18–35 mm long and 2–3 mm wide for subsp. *lignosa* to (25–)30–55 (–75) mm long and (2–)3–8(–10) mm wide for other subspecies; *inflorescence* (3-) 6- to 18-flowered, max. pedicel length (8–)12–35(–60) mm; *calyx* 8–13 mm long (10–15[–19] mm long for subsp. *nitida*), united ⅜–⅝ its length, lobes linear with obscure midrib and with apex cuspidate, membranes flat to somewhat plicate; *corolla* tube 9–15 mm long, lobes obovate to orbicular (average dimensions 10 × 7 mm but ranging from 9 × 5.5 mm for subsp. *lignosa* to 17 × 10 mm in some colonies of subsp. *nitida*) with apex entire and obtuse to emarginate or with a notch 1–2(–4) mm deep, hue varying from purple to pink and to white, with the eye striae varying in intensity from conspicuous to obsolete; *stamens* borne on mid- to upper corolla tube, with some anthers exserted; *style* 1.5–4 mm long, united ¼–½ its length, with stigma deeply included and placed below the anthers; *flowering season* spring, April–June.

Key to the Subspecies of *Phlox speciosa*

1a. Herbage glabrous, except inside calyx lobes . . . . **subsp.** *nitida* (Suksdorf)
        Wherry, Proc. Acad. Nat. Sci. Philadelphia 90: 134. 1938 | Klamath showy phlox
1b. Herbage pubescent, at least terminally 2
2a. Leaves thick-textured, glandular-pubescent to glabrescent; plants densely branched
     with woody tissue well-developed; associated with rain-shadowed western edge of
     Columbia Plateau . . . .
       . . . . . . **subsp.** *lignosa* Brand, Pflanzenr. (Engler) 4 (250): 74. 1907 | Yakima showy phlox
2b. Leaves thin-textured, pubescent to glabrous; plants relatively tall and open-growing
    3
3a. Height averaging 25 cm; leaves firm, short-acuminate, the largest 25–50 mm long and
     3–9 mm wide; associated with Cascade–Sierra axis and Northern Rocky Mountains . . .
       . . . . . . **subsp.** *occidentalis* (Durand ex Torrey) Wherry, Proc. Acad. Nat. Sci. Philadelphia
                                              90: 133. 1938 | Sierra showy phlox
3b. Height averaging 40 cm; leaves lax, long-accuminate, the largest 30–60 mm long and
     2–7 mm wide; associated with Palouse region of Idaho and Washington . . . .
     . . . . **subsp.** *speciosa* | showy phlox

Geography

Pacific Northwest and northwestern interior of the United States and extending into southwestern Canada. The overall distribution of the showy phlox species complex is bi-centric, with a main area of occurrence centered in Washington in association with the Columbia Plateau and adjacent portions of the Northern Rocky Mountains and the Northern and Middle Cascade Mountains; and a second in California and S Oregon associated with the Southern Cascade Mountains, the Sierra Nevada, and California Coast Ranges. The nominate expression of showy phlox (subsp. *speciosa*) is concentrated on the northern Columbia Plateau and adjacent Northern Rocky Mountains, where it occurs most abundantly in the Palouse region of W Idaho and E Washington.

     Subspecies *nitida* occurs most abundantly in the Klamath Mountains section of the Pacific Mountain System (NW California and SW Oregon), extending south into the California Coast Ranges and remote north to the Middle Cascade Mountains (Oregon and Washington).

Subspecies *occidentalis* is the most wide-ranging of the subspecies, occurring in the mountains along the Cascade–Sierra axis from the eastern slope of the Northern and Middle Cascade Mountains south through the Southern Cascade Mountains and the Sierra Nevada of California, with outlying occurrence in the Northern Rocky Mountains in W Montana. It is most common on the western slope of the northern and central Sierra Nevada.

Subspecies *lignosa* has the smallest range of the subspecies, being limited to C Washington along the interface of the Middle Cascade Mountains with the western edge of the Columbia Plateau.

Environment

Dissected plateaus, hilly piedmont, and mountains at low to middle elevations, on upper slopes of rolling terrain, and ridges, 250–8000 ft. (75–2400 m); *MAP* 15–40 in. (38–102 cm); *habitat* dry to dry-mesic; *soils* typically zonal, not excessively stony; *parent material* basalt on the Columbia Plateau, granite and serpentine in the Klamath Mountains and Sierra Nevada.

Subspecies *lignosa* is strongly associated with the Yakima District of the Columbia Plateau, a distinctive region where normally flat-lying basalt lava flows have been buckled steeply into long folds termed anticlines (Fenneman 1931). These folds appear in the landscape as high ridges that extend eastward from the Cascades onto the Columbia Plateau. This distinctive subspecies occurs on several of these anticlines, including Horse Heaven Hills, Manastash Ridge, Rattlesnake Hills, and Toppenish Ridge. The upper portions of these ridges are characterized by shallow, stony soils (lithosols) and patterned ground. At the northern limits of its range, subsp. *lignosa* is associated with the rocky basalt outcrops of the Channeled Scablands. Subspecies *lignosa* occurs in the driest portion of the Columbia Plateau where the rain-shadow effect of the Cascade Range is most pronounced, compounded by the xeric nature of the rocky habitat.

Associations

Conifer woodland/savanna, oak-conifer woodland, chaparral, shrubland, shrub-steppe, and steppe/grassland.

Showy phlox is associated with a variety of forested systems. In the northern portion of its range in E Washington and N Idaho, showy phlox (subsp. *speciosa*) is associated with ponderosa pine (*Pinus ponderosa*) woodland, typically in a zone between shrub-steppe and conifer forest. Antelope bitterbrush (*Purshia tridentata*) is often an understory shrub in these open woodlands. In the southern portion of its range, from the Klamath Mountains southward through the North Coast Ranges and the Sierra Nevada, showy phlox (subsp. *occidentalis*) occurs in openings in forest and woodland dominated by conifers or conifers mixed with oaks. At higher elevations these forests are dominated by California red fir (*Abies magnifica*). At middle to lower elevations they are comprised of Douglas-fir (*Pseudotsuga menziesii*), Jeffrey pine (*Pinus jeffreyi*), or ponderosa pine, with oaks such as Oregon white oak (*Quercus garryana*) and California black oak (*Q. kelloggii*) becoming more important at lower elevations.

In the Siskiyou-Klamath Mountain region of N California and S Oregon, showy phlox (subsp. *nitida*) occurs in unique conifer woodlands that develop at relatively low elevations on exposures of serpentine rock in the mountains. This vegetation consists of mixed conifers in open growth above a scrub-oak stratum with grassy openings. The more mesic of these communities have Jeffrey pine, western white pine (*Pinus monticola*), sugar pine (*P. lambertiana*), incense cedar (*Calocedrus decurrens*), and Douglas-fir in open stands with patches of shrubs dominated by huckleberry oak (*Quercus vacciniifolia*). On drier slopes and ridges these woodlands take on a savanna-like aspect, with Jeffrey pine and occasionally incense cedar the only trees, and with the shrub layer replaced by an herbaceous layer of perennial grasses including Lemmon's needlegrass (*Achnatherum lemmonii*), California oatgrass (*Danthonia californica*), and prairie Junegrass (*Koeleria macrantha*). Showy phlox grows in shade and duff of these dry woodlands, in a sparse layer of herbaceous plants. On even more xeric sites, it occurs in open woodlands dominated by knobcone pine (*P. attenuata*).

Showy phlox (subsp. *lignosa*) occurs in association with a narrow belt of Oregon white oak woodland in the Horse Heaven Hills of SC Washington, in an ecotone and mosaic with ponderosa pine forest, shrub-steppe, and steppe. This critically imperiled plant community extends into adjacent Oregon along the eastern Cascades near the Columbia River Gorge. Bluebunch wheatgrass (*Pseudoroegneria spicata*) and Idaho fescue (*Festuca idahoensis*) dominate the herbaceous understory of these open woodlands, with antelope bitterbrush occasionally associated.

At the northern edges of the Columbia Basin in E and C Washington, showy phlox occurs in shrub-steppe comprised of antelope bitterbrush with Idaho fescue as codominant and, to a lesser extent, big sagebrush (*Artemisia tridentata*) with bluebunch wheatgrass as codominant. Showy phlox tends to occur in stonier soils and more mesic habitat which favors antelope bitterbrush over big sagebrush. Showy phlox also occurs with rabbitbrush (*Chrysothamnus* spp.) on rocky slopes. Showy phlox sometimes grows into and over shrubs, and it can be difficult to determine which stems and branches belong to which plant. Such plants are more lax and have longer stems than those growing in the open.

At the western edge of the Columbia Plateau in C Washington, showy phlox (subsp. *lignosa*) occurs in lithosol or scabland habitat associated with exposures of basalt bedrock and patterned ground. These shallow, rocky soils support unique, sparsely vegetated communities dominated by one of several dwarf shrubs-scabland sagebrush (*Artemisia rigida*), snow buckwheat (*Eriogonum niveum*), Douglas' wild buckwheat (*E. douglasii*), or thyme-leaf wild buckwheat (*E. thymoides*). Showy phlox tends to occur at the interface between the dwarf shrub community and shrub-steppe dominated by antelope bitterbrush. It can be locally abundant in lithosol habitat.

Showy phlox (subsp. *speciosa*) occurs in association with bunchgrass steppe communities in SE Washington and adjacent Idaho and Oregon, in a distinctive vegetation region of the Pacific Northwest known as Palouse prairie or grassland. Showy phlox occurs most frequently in relatively mesic grassland characterized by Idaho fescue with a dwarf shrub element, primarily common snowberry (*Symphoricarpos albus*) and Nootka rose (*Rosa nutkana*). This community, for which plant ecologist Rex Daubenmire (1992) used the European term "meadow-steppe," has a profuse and conspicuous forb element including arrow-leaf balsam-root (*Balsamorhiza sagittata*), red coral-drops (*Besseya rubra*), elegant mariposa lily (*Calochortus elegans*), sticky geranium (*Geranium viscosissimum*), prairie smoke (*Geum triflorum*), Pursh's silky lupine (*Lupinus sericeus*), and fanleaf cinquefoil (*Potentilla gracilis*). The grasslands of the Palouse region have been almost totally converted to cultivated agriculture, with only about one-tenth of one percent intact, and are considered one of the most endangered ecosystems in the United States. Piper and Beattie described showy phlox as "common in moist ground" in their 1901 *Flora of the Palouse Region*. Showy phlox occurs with two rare Palouse endemics in remnants of this community in Idaho, Palouse goldenweed (*Pyrrocoma liatriformis*) and the endangered Spalding's catchfly (*Silene spaldingii*). A mosaic of ponderosa pine woodland and Palouse grassland characterizes the area where Meriwether Lewis collected the type of showy phlox in present-day Nez Perce County, Idaho. In NW Montana, showy phlox occurs in montane grassland dominated by rough fescue (*Festuca campestris*) accompanied by Idaho fescue and bluebunch wheatgrass.

Cultivation
Showy phlox has been cultivated to a limited extent, but the exciting horticultural potential of this species complex remains essentially untapped. The taller expressions (subsp. *speciosa* and subsp. *occidentalis*) that occur in Palouse grasslands, moun-

tain meadows, and other mesic habitat would likely prove adaptable to garden culture in many parts of the United States and Canada. The low growing, flower-smothered expressions (subsp. *lignosa*) associated with xeric rocky habitat have potential as drought-tolerant landscape plants for drier parts of the Pacific Northwest, and would be striking subjects for the rock garden. Whatever the application, the beautiful, starry flowers of showy phlox would make any garden more glorious. Grow in full sun to light shade in dry to slightly moist, well-drained soil.

Phlox stansburyi
Mojave phlox

Coastal redwoods have drawn their fill, and Sierran forests have been to the trough, by the time Pacific air reaches California's White Mountains. What's left is just enough moisture to support a thin grizzled society of pinyon pine, juniper, and sagebrush. But these desert slopes are not as drab you might expect. Summertime finds penstemons, eriogonums, daisies, and other vivid wildflowers brightening breaches in the gray and olive overtones. Mojave phlox is one of them, proclaiming its grit with uplifted trumpets of pink and white.

Mojave phlox has a scattered distribution across the southern Great Basin, from eastern California into New Mexico, most often occurring on the slopes of desert-bound mountain ranges. Relative of the wide-ranging longleaf phlox, Mojave phlox is a more compact plant with elongate, sky-offered flowers. This distinctive flower structure and orientation favors pollination by long-tongued hawkmoths, aloft at dusk and at work through the night, sipping nectar and shuffling genes by the light of the Milky Way.

Phlox stansburyi (Torrey) A. Heller, Bull. Torrey Bot. Club 24: 478. 1897. Basionym: *P. speciosa* Pursh var.? *stansburyi* Torrey, Rep. U.S. Mex. Bound., Bot. (2) 1: 145. 1858 [after Howard Stansbury (1806–1863)] Upstanding suffrutescent perennial with erect-ascending flowering shoots 10–45 cm tall with ca. 5 leaf nodes below the inflorescence; *leaves* linear to oblong-lanceolate, short-accuminate, finely ciliate and pilose, max. length 40–80 mm (20–60 mm for variant treated as *P. superba*) and width 2–6 mm; *inflorescence* 3- to 21-flowered, its herbage pubescent with partly gland-tipped hairs, max. pedicel length 5–25(–40) mm; *calyx* (7–)9–14 mm long, united ⅜–⅝ its length, lobes linear-subulate with obscure midrib and with apex sharp-cuspidate to subaristate, membranes plicate; *corolla* tube 19–33 mm long, lobes narrowly obovate (dimensions varying from 9 × 5 mm to 10 × 6 mm) with apex obtuse and entire or erose to emarginate, hue purple to pink or white, often with a pale eye bearing deep-hued striae; *stamens* borne on upper corolla tube, with some anthers placed at orifice; *style* 15–28 mm long, free for 1–2 mm, with stigma placed among the anthers near corolla orifice; *flowering season* spring, April–June.

∽∾∾

Taxonomic notes

Based on examination of the type material, Wherry (1942) determined the chief diagnostic character distinguishing *Phlox stansburyi* to be corolla tube length of 22–23 mm, with an overall range of 20–25 mm for the typical expression of the species. In his 1955 monograph, Wherry placed *P. stansburyi* in a series of long-flowered species running from *P. dolichantha* (corolla tube length 35–45 mm) through *P. superba* (26–33 mm) and *P. stansburyi* (19–25 mm) to *P. grayi* (13–16 [–18] mm). While *P. dolichantha* is consistently recognized as distinct, most subsequent workers (Cronquist 1984, Peabody 1979, Welsh 2003, Wilken and Porter 2005) have treated *P. superba*, *P. stansburyi*, and *P. grayi* as variants of *P. longifolia*. Kartesz (1988) considered *P. stansburyi* a species distinct from *P. longifolia* in the flora of Nevada, comprised of the nominative *P. stansburyi* subsp. *stansburyi* (corolla tube 20–25 mm) plus a longer-flowered expression, *P. stansburyi* subsp. *superba* (26–33 mm). A recent study of the pollination biology of southwestern phloxes found differences in corolla tube length may be under the influence of pollinator-mediated selection, suggesting corolla tube length may have ecological and evolutionary significance and that such differences appear to warrant some level of taxonomic recognition (Strakosh 2004; Strakosh and Ferguson 2005). Yet, an ecological study of the flora of SC Nevada found corolla tube length to be "highly variable" between and within populations of Mojave phlox (Beatley 1976).

Pending further study, *Phlox stansburyi* is interpreted here as a complex of relatively long-flowered (corolla tube 19–33 mm) entities separate from both *P. dolichantha* (corolla tube > 35 mm) and from the *P. longifolia* complex (average corolla tube length of 13–17 mm). This interpretation incorporates *P. superba* into *P. stansburyi* and *P. grayi* into *P. longifolia*.

Geography

Western interior of the United States, concentrated in the Great Basin and western Colorado Plateau, ranging from the eastern slope of Sierra Nevada in California, eastward across Nevada and Utah, and south into N Arizona. Apart from this roughly contiguous core range, Mojave phlox occurs remotely in S New Mexico, and (possibly) W (Trans-Pecos) Texas. Mojave phlox has been described as "a mountain-hopper across the desert southwest" (Krantz 1994). It is reported from the White/Inyo, Panamint, and San Bernardino Mountains of E and S California; the Monitor Range, Wassuk, Spring (Charleston), and Pahranagat mountains of Nevada; the White Mountains of E Arizona; and the Organ Mountains of S New Mexico. Mojave phlox is the only phlox occurring within the geographic boundaries of the Mojave Desert, hence the common name.

Environment
Dissected plateaus and isolated mountain ranges mostly at low to middle elevations, occasionally higher, associated with canyons, mesas, bajadas (coalesced alluvial fans), slopes, and ridges, 3000–10,000 ft. (900–3000 m); *MAP* 8–12 in. (20–30 cmm); *habitat* xeric to dry; *soils* medium- to coarse-textured, stony, shallow; *parent material* volcanic, carbonate, sandstone.

Associations
Primarily conifer woodland. In the core of the range of Mojave phlox in the Great Basin these woodlands are codominated by singleleaf pinyon (*Pinus monophylla*) and Utah juniper (*Juniperus osteosperma*). At the eastern limits of its range in the mountains of S New Mexico (where the type was collected), Mojave phlox occurs in woodland dominated by singleseed juniper (*J. monosperma*) and enriched by Chihuahuan succulent shrubs including Sacahuista bar-grass (*Nolina microcarpa*) and smooth sotol (*Dasylirion leiophyllum*), with black grama (*Bouteloua eriopoda*) and other grasses typical of desert grasslands comprising the herbaceous layer.

In the White Mountains of E California, Mojave phlox extends from the pinyon-juniper (*Pinus monophylla–Juniperus osteosperma*) zone up to the lower edge of subalpine woodland dominated by Intermountain bristlecone pine (*P. longaeva*). Here at an elevation of ca. 9000 ft (2700 m) it is a component of a rather diverse herbaceous layer dominated by graminoids including prairie Junegrass (*Koeleria macrantha*), Indian mountain-ricegrass (*Achnatherum hymenoides*), bottlebrush squirrel-tail (*Elymus elymoides*), and needle-and-thread (*Hesperostipa comata*), with Mojave phlox often growing within or at the base of grass clumps. Frequently associated forbs include silvery fleabane (*Erigeron* argentatus), matted wild buckwheat (*Eriogonum caespitosum*), Westgard beardtongue (*Penstemon scapoides*), and two members of the Phlox family (Polemoniaceae)—granite prickly-phlox (*Leptodactylon pungens* subsp. *hallii*) and Nuttall's linanthastrum (*Linanthastrum nuttallii*).

In the Mojave Desert of SE California and S Nevada, Mojave phlox occurs in mixed desert scrub in the transition zone above creosotebush (*Larrea tridentata*) desert scrub but below pinyon-juniper woodland. Shrubs characterize these communities, with blackbrush (*Coleogyne ramosissima*), spiny hop-sage (*Grayia spinosa*), and redberry desert-thorn (*Lycium andersonii*) being major species. On the eastern slope of the Sierra Nevada, and on the lower slopes of Great Basin mountains in C Nevada, Mojave phlox occurs above Mojave desert scrub communities dominated by saltbush (*Atriplex* spp.) and in shrubland dominated by Wyoming big sagebrush (*Artemisia tridentata* subsp. *wyomingensis*). On the Wassuk Range of W Nevada, Mojave phlox occurs in subalpine shrubland dominated by dwarf sagebrush (*A. arbuscula*).

Cultivation

The name *Phlox stansburyi* appears in horticultural literature, but given the confused and complicated nomenclature of the long-flowered phloxes of the American Southwest it would be hard to say which species is actually in view. Whatever the botanists finally determine Mojave phlox to be, this variable but always beautiful plant would be a welcome addition to the West's palette of drought-tolerant landscape perennials. Short-statured expressions would also make fine rock garden subjects. Grow in full sun in dry, well-drained soil.

Phlox stolonifera
Cherokee phlox

There is a grandfatherly composure to the Great Smoky Mountains that sooths like a kiss to the brow. Compared to the hard-at-it Rockies or the edgy Sierras, the Smokies, glories of the Appalachian chain, seem warmer, deeper, more knowable, more at ease—they have time for you. A "hazy blue moodiness and dreamy ancient beauty" is what poet Merrill Gilfillan feels in the presence of these old timers, with their "thousand secret nooks and hollows and countless half-hidden sweet-smelling things."

Cherokee phlox is among those sweet-smelling things, dwelling in nooks and hollows known to ecologists as cove forests. These magnificent, primeval forests, one of the world's most diverse plant communities, are named for their occurrence in protected habor-like mountain folds. More species of trees occur here than in any other forest type in North America. The wildflower component is even more lavish, particularly in spring when Cherokee phlox is at its prime, spreading fragrant drifts of violet across the forest floor.

Phlox stolonifera Sims, Curtis's Bot. Mag. 16: plate 563. 1802 [producing stolons; "having trailing stalks which take root at every joint, much in the manner of *Ajuga reptans* and *Viola odorata*, whence our trivial name"] Stoloniferous perennial herb with flowering shoots arising from elongate modified stems (stolons) that spread horizontally along the surface of the ground, these producing terminal rosettes of persistent leaves, sending up from rooted nodes ascending flowering shoots 15–25 cm tall with ca. 4 well-spaced leaf nodes below the inflorescence; *leaves* on stolons partly persistent, obovate to oblanceolate, with long coarse-ciliate petiole, the blade up to 45 mm long and 18 mm wide, on flowering shoots deciduous, about half as large, oblong to ovate, sessile, sparsely ciliate and pilose; *inflorescence* of ca. 6 flowers, rather lax, its herbage glandular-pubescent, max. pedicel length 5–30 mm; *calyx* 9–11 mm long, united ⅜–⅝ its length, lobes linear-subulate with apex cuspidate, membranes somewhat plicate; *corolla* tube 21–25 mm long, pilose with partly gland-tipped hairs, lobes obovate (average dimensions 14 × 9 mm) with apex usually entire, hue predominantly violet in more southern colonies, varying from violet to lavender and from purple to lilac at mid-range, and uniformly purple at northern portion of range, the eye only exceptionally paled or striate; *stamens* with 1 or more anthers often exserted; *style* 20–24 mm, united to tip which is free for 1 mm, with stigma placed among the anthers at or near corolla orifice (sometimes exserted); *flowering season* spring, late April through early June.

Geography

Appalachian Highlands, chiefly in the mountains of the Blue Ridge and Appalachian Plateaus physiographic provinces but extending locally onto the Appalachian Piedmont. Cherokee phlox occurs as an escape from cultivation in parts of the northeastern United States and adjacent Canada. The common name commemorates the Cherokee people, whose ancestral homeland in the southern Appalachian Mountains is the core of the range of this species and the provenance of its original collection.

Environment

Mountains at middle elevations, also hilly piedmont, on slopes, in coves, gorges, ravines, and other protected landscape features, and along streams and alluvial terraces, 1000–4500 ft. (300–1400 m); *habitat* primarily mesic, sometimes dry-mesic; *soils* humus-rich loam, occasionally stony; *parent material* sandstone, metavolcanic mafic rock (rich in magnesium). Within the Appalachian Mountains, Cherokee phlox occurs in the gorges of the southeastern Blue Ridge Escarpment of North Carolina, South Carolina, and Georgia and in the sheltered, mid-elevation coves of the Great Smoky Mountains of North Carolina and Tennessee. Wherry (1932b) described its habitat in S Ohio as "cool ravines."

Associations

Primarily high-diversity mesophytic forest, usually in somewhat protected landscape positions. These predominantly deciduous forests are dominated by sugar maple (*Acer saccharum*), American beech (*Fagus grandifolia*), tuliptree (*Liriodendron tulipifera*), American basswood (*Tilia americana*), and cucumber magnolia (*Magnolia acuminata*). In the Great Smoky Mountains (W North Carolina and E Tennessee), Cherokee phlox occurs in distinctive communities known as cove forests, the composition of which varies based on slope position. Cherokee phlox occurs most often in mid-elevation (2000–4600 ft. [600–1400 m]) cove forests of the buckeye-basswood association dominated by yellow buckeye (*Aesculus flava*) and Appalachian basswood (*Tilia americana* var. *heterophylla*). Old growth stands of the buckeye-basswood association have a high canopy with a dense and luxurious herbaceous layer with high species diversity and little intervening woody undergrowth. Cherokee phlox occurs to a lesser extent in cove forests dominated by eastern hemlock (*Tsuga canadensis*), sugar maple, and beech. This association typically has a denser understory dominated by heaths including mountain laurel (*Kalmia latifolia*) and great laurel (*Rhododendron maximum*).

Within these various forest types, Cherokee phlox is a component of a rich stratum of herbs including American hog-peanut (*Amphicarpaea bracteata*), wood anemone (*Anemone quinquefolia*), twoleaf toothwort (*Cardamine diphylla* [*Dentaria diphylla*]), narrowleaf springbeauty (*Claytonia virginica*), squirrel-corn (*Dicentra canadensis*), yellow trout-lily (*Erythronium americanum*), white wood aster (*Eurybia divaricata* [*Aster divaricatus*]), wild crane's-bill (*Geranium maculatum*), blunt-leaf waterleaf (*Hydrophyllum canadense*), partridge-berry (*Mitchella repens*), mayapple (*Podophyllum peltatum*), Tennessee chickweed (*Stellaria corei* [*Alsine tennesseensis*]), purple meadow-parsnip (*Thaspium trifoliatum*), heartleaf foamflower (*Tiarella cordifolia*), trilliums (*Trillium* spp.), and violets (especially *Viola sororia*).

In the Blue Ridge Mountains of North Carolina and Georgia, and at a few sites on the Piedmont of South Carolina, Cherokee phlox occurs in plant communities associated with ultramafic rock substrates. Where the substrate is more weathered and the site more sheltered these communities are relatively mesic forests comprised of red maple (*Acer rubrum*), American beech, and sweet birch (*Betula lenta*). Where rock is exposed or close to the surface and the unusual chemistry more strongly influences soil properties, these communities are characterized by drier woodlands with a more open canopy structure, with white oak (*Quercus alba*) and pitch pine (*Pinus rigida*) typical dominants. A shrub layer may be present, with mountain laurel being typical, but rockier sites take on the aspect of a barren, with stunted trees and a graminoid-dominated herbaceous layer.

Cultivation

Cherokee phlox was the first "Plant-of-the-Year" of the Perennial Plant Association, selected for the honor in 1990. But this Appalachian beauty was a familiar
garden plant in Europe long before it became the subject of an American marketing campaign, with Reginald Farrer referring to it as "this favourite joy" in *The
English Rock-Garden*, seven decades prior. Cherokee phlox has been quietly appreciated by American rock gardeners for years, particularly as a ground-covering
plant for shade and woodland gardens. Grow in shade to partial shade in evenly
moist, well-drained, humus-rich acid to neutral soil. Cultivars are available, some
of which have a reputation for being aggressive, an admirable attribute when a
groundcover is desired but problematic in a rock garden. Horticultural hybrids are
documented (see Appendix B).

Phlox subulata
creeping phlox
PLATES 67, 68

Spring retakes the Allegheny Mountains with a sensuous grace. Pent-up, young-
of-the-year oak leaves spread a pinkish green haze across ridge and hollow as they
crack open winter buds. Dogwoods float rafts of ivory saucers between canopy and
forest floor. Ruffed grouse drum up the sun and hens from atop hollow logs. And
mats of creeping phlox foliage disappear beneath a fresh washing of pink, white,
and blue-gray stars.

While creeping phlox ranges throughout much of the northeastern United
States, it reigns in the Alleghenies of Virginia and West Virginia. Here it is tied to
shale barrens, a severe but fragile habitat that develops where shale formations
have been cut and laid open by erosion. The dry crumbling slopes resist the root of
oak and pine, presenting sun-drenched openings favored by a specialized flora
with a welter of endemic species. Creeping phlox is not among the endemics, but
its constancy and colorful presence make it the signature plant of the shale barrens.

∽∾

Phlox subulata Linnaeus, Sp. Pl. 1: 152. 1753 [awl-shaped, of the leaves; "foliis subulatis"] Caespitose suffrutescent perennial forming carpetlike mats or cushions from
a dense network of prostrate nonflowering stems with persistent leaves that send up
from rooted nodes numerous erect flowering shoots 4–10(–20) cm tall with leaf
nodes numerous and crowded, 3–6 nodes below the inflorescence, fascicles of
smaller, nonflowering shoots present in leaf axils; *leaves* linear to subulate, fine-

ciliate and the upper pilose, max. length (8–)10–20(–25) mm and width 1–2(–3) mm; *inflorescence* 3- to 6- (12-) flowered, lax, its herbage pubescent with hairs mostly glandless for nominate expression (subsp. *subulata*), mostly glandular for subsp. *brittonii* and subsp. *setacea*, max. pedicel length 5–20(–30) mm; *calyx* 5–9 mm long, united ⅜–⅝ its length, lobes linear-subulate to narrow-triangular with apex cuspidate to subaristate, membranes flat to somewhat plicate; *corolla* tube 8–16 mm long, lobes obovate (dimensions ranging from 6 × 3 mm to 12 × 8 mm) with apex varying from entire to notched 1–1.5(–3) mm deep, hue predominantly purple to lavender, varying to somewhat reddish or less often bluish, exceptionally to pink and to white, often with deep-hued eye striae which may coalesce to a star or ring; *stamens* with 1 or 2 anthers exserted; *style* 7–12 mm long, united to tip which is free for 1 mm, with stigma placed among the anthers at or near corolla orifice (sometimes exserted); *flowering season* spring, through much of April and May, into June in cooler localities, with additional flowers sometimes developed in autumn and sometimes even into winter.

Taxonomic notes

What is recognized here as *Phlox subulata* subsp. *setacea*, Wherry (1955) treated as *P. subulata* subsp. *australis*. Reveal et al. (1982) demonstrated *setacea* has priority and reinstated *P. subulata* var. *setacea* (Brand 1907), which Locklear (2009) elevated to subspecies rank.

Key to the Subspecies of *Phlox subulata*

1a. Inflorescence-herbage pubescent with glandless (rarely gland-tipped) hairs, or glandular only in rare variants; corolla tube length averaging 10.5 mm, corolla lobe dimensions 8 × 5.5 mm, and notch-depth 1 mm; hue predominantly purple, but markedly variable; northern portion of species distribution ....
.... **subsp. *subulata*** | creeping phlox
1b. Inflorescence-herbage pubescent with gland-tipped (rarely glandless) hairs; corolla lobe notch ca. 1.5 mm deep 2
2a. Corolla mostly purple, its tube averaging 12 mm long with lobe dimensions 8.5 × 6 mm; southern portion of species distribution ....
.... **subsp. *setacea*** (Linnaeus) Locklear, J. Bot. Res. Inst. Texas 3: 649. 2009 |
Blue Ridge creeping phlox
2b. Corolla mostly pale lavender, its tube averaging 10.5 mm long with lobe dimensions 7.5 × 5 mm; typical of shale barrens region of Virginia and West Virginia ....
... **subsp. *brittonii*** (Small) Wherry, Castanea 16: 98. 1951 | shalebarren creeping phlox

Geography

Eastern and central United States and extending into southeastern Canada, concentrated in the Appalachian Highlands, extending into the eastern Interior Plains. Creeping phlox has escaped cultivation and has become naturalized in many areas outside of its historic range, particularly in the northeastern United States. It has commonly been planted over graves in cemeteries, where it has acquired the name "graveyard plant." Subspecies *subulata* and *setacea* essentially occupy the northern and southern portions of this overall range, respectively, with an area of overlap for all three subspecies in the central Appalachian Highlands. The nominate expression of creeping phlox (subsp. *subulata*) ranges from New Jersey through the northern Appalachian Highlands and eastern Great Lakes region to S Michigan. Subspecies *brittonii* has the most limited distribution, being most common in the shale barrens region of the central Appalachian Highlands (W Virginia and E West Virginia). Subspecies *setacea* has a somewhat bi-centric distribution, with one area of occurrence in the Blue Ridge Mountains of the Appalachian Highlands and the other at the interface of the western Appalachian Highlands with the Interior Low Plateaus (NE Kentucky and S Ohio).

Environment

Hilly piedmont and mountains, on slopes, escarpments, glades, rock ledges, and cliffs, 300–4000 ft. (90–1200 m); *habitat* xeric to dry; *soils* minimally developed, sandy to stony, with little organic matter, shallow with bedrock at or near the surface; *parent material* gneiss, serpentine, limestone, sandstone, shale. Creeping phlox occurs in association with rock formations that support some of the most fragile and interesting ecological systems in eastern North America—serpentine barrens and shale barrens.

The nominate expression of creeping phlox (subsp. *subulata*) occurs in association with a chain of serpentine rock outcroppings in SE Pennsylvania and N Maryland. The open nature of serpentine vegetation is influenced by the unique attributes of serpentine bedrock, and is sustained by the action of fire in this habitat, which prevents the encroachment of woody vegetation by directly killing woody competitors and by removing organic matter that insulates plants from the mineral conditions of the bedrock.

Subspecies *brittonii* is strongly associated with exposed shale formations in the Valley and Ridge province of the Appalachian Highlands of western Virginia and adjacent West Virginia, with isolated occurrences on shale northward into Maryland and S Pennsylvania. This habitat has been termed "shale barrens" due to the paucity of vegetation in association with shale formations. Shale layers capable of developing barrens habitat occur in association with Ordovician, Silurian, and Devonian rocks. Barrens often develop where a stream abuts a shale slope, undercutting and exposing the shale formation. Continual undercutting of the relatively weak shale formation by the stream maintains the barren. Shale barrens are steeply sloped (more than 20 percent), with slope aspect typically south facing but ranging from west to southeast facing. Steep slopes, the friable nature of shale, low moisture availability, and high soil surface temperatures (up to 150 degrees F) result in xeric habitat that does not favor the establishment of the surrounding dominant forest vegetation. The trailing, matlike growth habit of creeping phlox is well-suited to the steep, shifting slopes of crumbling shale.

Associations

Mixed conifer/hardwood woodland, oak woodland/savanna, grassland, riverwash, and rock outcrop communities.

At the northwestern limits of its range (Ohio and S Ontario), creeping phlox occurs in a mosaic of oak savanna/barrens and tallgrass prairie communities, typically in the driest portions of these communities. Black oak (*Quercus velutina*) and white oak (*Q. alba*) are the dominant trees of the savannas, with the grasses big bluestem (*Andropogon gerardii*), little bluestem (*Schizachyrium scoparium*), and yellow Indiangrass (*Sorghastrum nutans*) dominating the herbaceous flora of the savannas and associated tallgrass prairie.

Creeping phlox is often an important component of the herbaceous flora of serpentine barrens. Historically, serpentine habitat was characterized by woodland, savanna, and grassland. In serpentine woodland and savanna, stunted blackjack oak (*Quercus marilandica*) and post oak (*Q. stellata*) grow in open stands with herbaceous vegetation typical of serpentine grassland. Serpentine grassland is characterized by prairielike vegetation dominated by little bluestem, with arrowfeather three-awn (*Aristida purpurascens*), yellow Indiangrass, and northern dropseed (*Sporobolis heterolepis*) also important grasses. These plant communities were historically maintained by periodic fires, and have succeeded to forests dominated by Virginia pine (*Pinus virginiana*) or eastern red-cedar (*Juniperus virginiana*) in the absence of such fires. Creeping phlox occurs on the driest sites within serpentine communities, where bedrock is close to the surface. It is frequently associated with the serpentine specialists and indicator species such as barrens chickweed (*Cerastium velutinum* var. *villosissimum*), serpentine aster (*Symphyotrichum depauperatum*), and roundleaf fameflower (*Talinum teretifolium*).

The often dramatic association of creeping phlox with serpentine barrens was noted as early as 1745 by Pennsylvania botanist John Bartram in a letter to Johannes

Gronovius of Holland (Darlington 1849), and has inspired local place names like "Pink Hill." William Darlington wrote of creeping phlox in 1853: "This species is chiefly confined to the Serpentine rock, with us [in Chester County, Pennsylvania]; and when it is in full bloom, the hills at a distance, appear as if covered with a sheet of flame." Stout (1917) noted creeping phlox "covers about one hundred contiguous acres with a dense mat growth" on serpentine barrens near Unionville, Pennsylvania. Harshberger (1909) recognized a "Phlox Association" on Pennsylvania serpentine barrens where creeping phlox and barrens chickweed codominated.

The mid-Appalachian shale barrens are characterized by herbaceous vegetation in an area otherwise dominated by oak and oak-pine forest. Shale barrens typically occur within the context of woodland dominated by Virginia pine and rock chestnut oak (*Quercus prinus*). Shale barrens vegetation is sparse, with considerable areas of exposed rock and soil. A unique flora of herbaceous species is associated with shale barrens, with a high number of endemic species. While not restricted to shale barrens, creeping phlox (subsp. *brittonii*) is a relatively constant aspect of the flora, and occurs with a number of shale barrens endemics or specialists including shalebarren rockcress (*Arabis serotina*), white-hair leatherflower (*Clematis albicoma*), Millboro leatherflower (*Clematis viticaulis*), shalebarren wild buckwheat (*Eriogonum allenii*), shalebarren evening primrose (*Oenothera argillicola*), cat's-paw ragwort (*Packera antennariifolia*), and Kate's Mountain clover (*Trifolium virginicum*). A number of these endemics are of conservation concern. Brooks (1965) noted the close association of creeping phlox with shale barrens, observing "[c]ertain Allegheny slopes are so thickly overgrown as to appear snow-covered when creeping phlox is in bloom."

Creeping phlox occurs in association with exposed rock outcrops within the flood zone of major Piedmont and Appalachian mountain range rivers, particularly the gorges of the Potomac, Shenandoah, and James rivers in Maryland, Virginia, and the District of Columbia. These riverside outcrop barrens are subject to periodic flood scouring that keeps the habitat relatively free of shading tree canopy. In these same river systems, creeping phlox also occurs in basalt outcrop scour prairie dominated by big bluestem, little bluestem, and other herbaceous species.

In the Blue Ridge Mountains of North Carolina and E Tennessee, creeping phlox (subsp. *setacea*) occurs in high elevation rock outcrop and glade communities and, to a lesser extent, in grassy balds adjacent to rock outcrops. Glades occur on gently sloping or flat exposures of rock, and have more vegetation than more steeply sloped outcrops and cliffs. Grassy balds, so called for the dominance of herbaceous plants in otherwise forested mountains, are famous for their rare and endemic species. At lower elevations, creeping phlox occurs in association with sparsely vegetated barrens, glades, cliffs, ledges, and other exposed rock outcrop habitat derived from a variety of bedrock types including limestone in Kentucky, Virginia, and West Virginia, mafic rock in the Blue Ridge Mountains of North Carolina, and sandstone along New River in West Virginia. It occurs with several rare plants in limestone glade and cliff habitat, including Canby's mountain-lover (*Paxistima canbyi*) in Kentucky and West Virginia, and mountain parsley (*Taenidia montana*) in West Virginia. It is sometimes a dominant aspect of the flora of these communities.

Cultivation

Creeping phlox has a long history in horticulture (see chapter 2), and is one of the most widely grown herbaceous perennials in the world. Of all the horticultural praise lavished upon this species over the years, none is as odd as that of English plantsman B. H. B Symons-Jeune who wrote, "it has as many 'little' names as ever a Zulu chanted in praise of a conquering chief. But when a plant has many 'little' names it is a sure sign of affection." Yet, because it is a plant of the masses, creeping phlox has been disdained by more sophisticated gardeners with tags like "plebeian," "outhouse plant," and "a garish commonplace." Creeping phlox does end up in some of our most mundane and pedestrian landscapes, but, no matter the setting, its flower-smothered mats are still a refreshing sight after a long, hard winter. Rock gardeners shouldn't look past creeping phlox, selecting instead cultivars of more restrained growth habit and more subtle flower coloration.

Many European cushion phlox cultivars of longstanding use are held to be selections of *Phlox subulata*, but may actually be hybrids. Nomenclatural formulations are available for hybrids of known parentage, or the formulation *Phlox* Scotia Alpines Group can be used for cushion phlox cultivars of unknown parentage (see Appendix B).

Creeping phlox can be used as a subject for rock gardens or wall gardens, or massed and used as a groundcover. Grow in full sun to light shade in dry well-drained soil. Colonies of creeping phlox can form by the spread and rooting of prostrate stems, an admirable attribute when a groundcover is desired but problematic in a smaller rock garden. Selections of subsp. *brittonii* are more refined and restrained in growth habit. Many cultivars are available. Horticultural hybrids are documented (see Appendix B).

Phlox tenuifolia
Apache phlox
PLATES 69, 70

The term *desert* does disservice to the fantastic saguaro forests of Arizona. Defined by the presence of the giant saguaro cactus, there is an exuberance to this Sonoran ecosystem that feels tropical, with cartoonish cacti and prickly shrubs sheltering Gila monsters, elf owls, and more kinds of hummingbirds than you can shake a stick at. Apache phlox leads a quiet life amidst this biological bedlam, a straggling, somewhat woody plant growing among, often within, acacia, jojoba, bush buckwheat, and other shrubs, comprised more of twig than of leaf.

The range of Apache phlox is concentrated in southeast Arizona, where once the Chiricahua and other Apache bands held sway. Here it dwells in the *sky islands*—an archipelago of mountain ranges surrounded by desert basins. Saguaro-thronged desert scrub dominates the lower slopes, while chaparral and Madrean evergreen woodland succeeds it farther up. Apache phlox works its way up into the understory of these xeric woodlands of oak, pine, and juniper, where its creamy white flowers impart a whiff of vanilla to the crackly shade.

☙ ❧

Phlox tenuifolia A. Gray ex E. E. Nelson, Revis. W. N. Amer. Phlox. 27. 1899 [thin-leaved; "leaves linear-attenuate"] Upstanding suffrutescent perennial 25–75 cm tall with slender ascending flowering shoots 15–50 cm long with ca. 6–12 well-spaced leaf nodes below the inflorescence, the habit more lax and sprawling when growing in among shrubs, with stems up to 125 cm long; *leaves* linear to narrowly elliptic or lanceolate, glabrous to sparsely pilose, max. length 25–45 mm and width 1–3 mm; *inflorescence* 6- to 18-flowered, its herbage copiously pubescent with rather coarse hairs, a few sometimes gland-tipped, max. pedicel length 7.5–40 mm; *calyx* 7–12 mm long, united ⅜–⅝ its length, the lobes linear-subulate with apex subaristate, membranes flat to somewhat plicate; *corolla* tube 7–11 mm long, often shorter than the calyx, distinctly expanded upward, lobes oblanceolate (average dimensions 9 × 5 mm) with apex obtuse and erose or emarginate, hue white, creamy, or rarely somewhat lavender-shaded; *stamens* inserted on the upper tube; *style* 4–8 mm long, united ca. ½ its length; *flowering season* spring, February–May, again in autumn in years of abundant precipitation.

☙ ❧

Geography
Basin and Range Physiographic Province, concentrated in the Mexican Highland

section (SC and SE Arizona), with outlying occurrence in the Sonoran Desert section (Ajo Mountains in Organ Pipe National Monument) in SW Arizona, possibly disjunct to the state of Chihuahua, Mexico. The distribution of Apache phlox coincides with the northeastern portion of the Arizona Upland subdivision of the Sonoran Floristic Province, as well as the Apachian District of the Madrean Floristic Province. Most occurrences of Apache phlox are associated with the island-like mountain ranges scattered across SE Arizona, including the Galiuro, Mescal, Mule, Pinaleno, Rincon, Santa Catalina, Sierra Ancha, Superstition, and Tucson mountains. These mountains, isolated by basins vegetated by desert scrub and desert grassland, have been referred to as the "sky islands" and link the Southern Rocky Mountains with the Sierra Madre Occidental. The common name reflects close association with the Apachian floristic area (McLaughlin 2007) centered in SE Arizona, a name derived from the historical influence of the Apache people on this region.

Environment

Mountains at low to middle elevations, on slopes and in canyons and ravines, 1400–5000 ft. (400–1500 m); *MAP* 12–14 in. (30–35 cm); *habitat* xeric to dry; *soils* medium- to coarse-textured, stony, shallow. The flowering season of Apache phlox is influenced by the bi-seasonal precipitation regime of SE Arizona, with gentle rains in winter through spring (late November to mid-April), and monsoon-like storms in summer (July to October), interrupted by spring and fall aridity.

Associations

Primarily Sonoran desert scrub characterized by a large variety of cacti species in a mosaic with drought-deciduous shrub-trees, shrubs, and grasses. Dominants in the overstory include the iconic saguaro (*Carnegiea gigantea*) cactus, along with little-leaf paloverde (*Parkinsonia microphylla*) and velvet mesquite (*Prosopis velutina*). Important shrubs include catclaw acacia (*Acacia greggii*), mescat acacia (*A. constricta*), triangle bursage (*Ambrosia deltoidea*), California wild buckwheat (*Eriogonum fasciculatum*), ocotillo (*Fouquieria splendens*), sangre de cristo (*Jatropha cardiophylla*), creosotebush (*Larrea tridentata*), and jojoba (*Simmondsia chinensis*). Various cholla cacti (*Opuntia* subgenus *Cylindropuntia*) and prickly-pear cacti (*Opuntia* subgenus *Platyopuntia*) also are typical.

At its upper elevation limits, Apache phlox occurs in mixed woodland comprised of evergreen oaks, juniper, and pine. Arizona oak (*Quercus arizonica*), Emory's oak (*Q. emoryi*), and Mexican blue oak (*Q. oblongifolia*) are dominants of these woodlands, with alligator juniper (*Juniperus deppeana*) or Mexican pinyon

(*Pinus cembroides*) also typically associated. These open woodlands grade into desert grassland at their lower, more xeric limits. In Organ Pipe National Monument, Apache phlox occurs with Ajo Mountain oak (*Q. ajoensis*). At elevations below those that support mixed woodland, Apache phlox occurs in chaparral dominated by shrub live oak (*Q. turbinella*) with other shrubs including common sotol (*Dasylirion wheeleri*), Fremont's mahonia (*Mahonia fremontii*), Sacahuista beargrass (*Nolina microcarpa*), and redberry buckthorn (*Rhamnus crocea*).

Cultivation
Apache phlox has been cultivated in desert gardens in Arizona. It is an attractive, fragrant species with potential as a drought-tolerant landscape plant for the southwestern United States. Grow in full sun to light shade in dry, well-drained soil.

Phlox villosissima
Comanche phlox

The Devils River wrestles a living from the limestone rock of the Edwards Plateau. Quarrying its way to the Rio Grande, this pristine stream runs on springs recharged by infrequent but now and again monstrous thunderstorms. Comanche phlox benefits from these Texas toad stranglers, running roots deep into the flood-cleansed cobble bars of the Devils, the Nueces, and other plateau-born streams, scoured of thornscrub and soaked in sunshine.

The distribution of Comanche phlox is centered on the Edwards Plateau of Texas, where it frequents oak-and-juniper-wooded canyons along the Balcones Escarpment. It also occurs in rocky remnants of the Fort Worth Prairie, and on Trans-Pecos mesas spiked with Chihuahuan desert succulents. Fine settings all, but there is something about its cobble bar habitat that makes your knees buckle, an awful atmosphere-on-lithosphere simplicity, a terrible purity, like the sight of an osprey shaking the Devils River from its wings.

∽∾

Phlox villosissima (A. Gray) Small, Fl. S.E. U.S. 977. 1903. Basionym: *P. drummondii* Hooker var. *villosissima* A. Gray, Proc. Amer. Acad. Arts 8: 257. 1870 [hairiest, of the foliage; "pilis viscosis longis crebris"] Upstanding perennial herb with erect-ascending flowering shoots, those of the nominate expression (subsp. *villosissima*) tending to be branched, 10–30 cm tall with leaf nodes (7–15) rather crowded, those of subsp. *latisepala* tending to be simple and stout, 22–45 cm tall with leaf nodes (8–12) more remote, shoots of subsp. *villosissima* arising from well-developed woody rhizome; *vesture* copious and glandular-pubescent throughout for subsp. *villosissima*, glandularity moderate for subsp. *latisepala*; *leaves* linear below, becoming lanceolate upward, the largest (at 2 or 3 nodes below the inflorescence) 30–60 mm long and 3–8 mm wide, the upper alternate; *inflorescence* 12- to 24-flowered (18- to 36-flowered for subsp. *latisepala*), compact, max. pedicel length 3–15 mm; *calyx* 8–15 mm long, united ⅜–½ its length, lobes linear-subulate with obscure midrib and with apex aristate with awn 1–2 mm long, membranes plicate; *corolla* tube 12–15 mm long, pubescent, lobes obovate (average dimensions ranging from 11 × 7 mm for subsp. *latisepala* to 12 × 9 mm for subsp. *villosissima*) with apex obtusish to apiculate, hue purple to pink, sometimes with a pale eye bearing one or two purple striae; *stamens* with anthers included; *style* 1.5–3 mm long, united ⅜–⅝ its length, with stigma included; *flowering season* spring, April–May; tetraploid, $2n = 28$ (Levin 1966; Smith and Levin 1967).

☙❧

Taxonomic notes

What is recognized here as *Phlox villosissima* subsp. *villosissima* and *P. villosissima* subsp. *latisepala*, Wherry (1955, 1966) treated as *P. pilosa* subsp. *riparia* and *P. pilosa* subsp. *latisepala* and Levin (1968) treated as *P. villosissima* and *P. aspera*. Turner (1998a) combined them under the first published *P. villosissima*. Locklear (2009) recognized them as subspecies under *P. villosissima*. See also Ferguson (1998), Ferguson et al. (1999), Ferguson and Jansen (2002), Levin (1966), and Levin and Schaal (1970).

Key to the Subspecies of *Phlox villosissima*

1a. Plant with well-developed slender woody rhizomes producing multiple erect-ascending stems; stems tending to be branched with nodes rather crowded, 12–25 cm tall; inflorescence-herbage densely glandular-pubescent; xeric habitat in western portion of species range, notably riverwash habitat and talus slopes . . . . **subsp. *villosissima*** | Comanche phlox
1b. Plant not rhizomatous; stems tending to be simple, moderately stout, 22–45 cm tall; pubescent throughout with pointed hairs, with inflorescence-herbage moderately glandular; occurs in more mesic habitat in eastern portion of species range . . . . **subsp. *latisepala*** (Wherry) Locklear, J. Bot. Res. Inst. Texas 3: 649. 2009 | Guadalupe Comanche phlox

Geography

Chiefly Great Plains, nearly endemic to Texas, concentrated on the Edwards Plateau and Central Texas sections, extending north into the Osage Plains section of the Central Lowlands in NC Texas, with outlying occurrence (subsp. *latisepala*) in northern Coahuila, Mexico. The common name reflects close association with the Comanchian floristic area (McLaughlin 2007) centered on the Edwards Plateau, a name derived from the historical influence of the Comanche people on this region. The nominate expression of Comanche phlox (subsp. *villosissima*) is most common along the southern and western edge of the Edwards Plateau. Its westernmost occurrence is on the Stockton Plateau, which is separated from the main body of the Edwards Plateau by the Pecos River. Subspecies *latisepala* occupies the northern portion of the species range, notably on the Comanche Plateau of NC Texas, where it occurs on the Lampasas Cut Plain and Fort Worth Prairie regions.

Environment

Cuestas and dissected plateaus, on escarpments, mesas, bluffs, talus slopes, and in canyons and on gravel beds of streams, 500–1500 ft. (150–450 m); *MAP* 25–35 in. (64–89 cm), to as low as 12 in. (30 cm) at the western edge of its range; *habitat* primarily dry-mesic, xeric at western limits of its range; *soils* stony, shallow with bedrock sometimes at or near the surface; *parent material* limestone.

In the southwestern portion of its range, Comanche phlox is associated with gravel and cobble bars in draws, creeks, and rivers, particularly as these emerge from canyons cut into the escarpments of the Edwards and Stockton plateaus. These streams, notably the Devils, Frio, Nueces, Sabinal and their tributaries, are dry throughout most of the year or may occupy only a small part of the streambed except during periods of flooding. Scouring by periodic catastrophic floods keep the gravel bars relatively free of the dense thornscrub vegetation that characterizes the region. Whitehouse (1935) observed the woody roots of Comanche phlox individuals growing on gravel bars extending three feet or more in length, possibly an adaptation to the coarse, deep substrate of the streambeds. In one occurrence on the upper reaches of the South Llano River, Comanche phlox occurred in gravel-filled pockets in the exposed limestone bedrock. This unique association with river (riparian) habitat, which Whitehouse (1935) called "a strange environment for a phlox," inspired Wherry's (1955) name for this plant, *Phlox pilosa* subsp. *riparia* (of river banks).

Heller (1895) described the habitat of his specimen *1641*, the type of subsp. *latisepala*, as "rich soil about Kerrville, especially in damp places," which would seem to indicate association with relatively mesic floodplain woodland of the Guadalupe River near Kerrville, as opposed to the dry rocky uplands typical of this region of the Edwards Plateau.

Associations

Juniper woodland and savanna, oak woodland, shrubland, grassland, and river-wash communities. On the dry western edge of the Edwards Plateau and adjacent Stockton Plateau, Comanche phlox (subsp. *villosissima*) occurs in savanna dominated by Ashe's juniper (*Juniperus ashei*) or Pinchot's juniper (*J. pinchotii*), intermixed with shrubby, succulent species characteristic of Chihuahuan desert scrub, including Texas sotol (*Dasylirion texanum*), lechuguilla agave (*Agave lechuguilla*), and ocotillo (*Fouquieria splendens*).

Along the eastern edge of the Edwards Plateau and northward onto the Lampasas Cut Plain, Comanche phlox (subsp. *latisepala*) occurs in relatively mesic woodlands dominated by Buckley oak (*Quercus buckleyi*), Ashe's juniper, and Texas ash (*Fraxinus texensis*). It also occurs in floodplain woodland dominated by open stands of pecan (*Carya illinoinensis*), sugarberry (*Celtis laevigata* var. *laevigata*), sycamore (*Platanus occidentalis*), and American elm (*Ulmus americana*). The herbaceous layer of these woodlands is dominated by little bluestem (*Schizachyrium scoparium*), sideoats grama (*Bouteloua curtipendula*), and seep muhly (*Muhlenbergia reverchonii*).

On the southern edge of the Edwards Plateau, Comanche phlox (subsp. *latisepala*) occurs in woodlands dominated by Lacey oak (*Quercus laceyi*), Buckley oak, and Ashe's juniper. This association occurs in north- and northeast-facing canyon slopes that are relatively mesic compared to the ridges and upland slopes where woodlands of Ashe's juniper and plateau live oak (*Q. fusiformis*) are dominant. The herbaceous layer of these woodlands is characterized by grasses typical of mixed-grass prairie, including Texas grama (*Bouteloua rigidiseta*), common curly-mesquite (*Hilaria belangeri*), Texas needlegrass (*Nassella leucotricha*), and buffalograss (*Buchloe dactyloides*).

In riverwash habitat, Comanche phlox (subsp. *villosissima*) occurs in association with scattered to dense stands of shrubs, most notably little walnut (*Juglans microcarpa*) but also including split-leaf brickell-bush (*Brickellia laciniata*), common buttonbush (*Cephalanthus occidentalis*), desert-willow (*Chilopsis linearis*), and Lindheimer's indigo (*Indigofera lindheimeriana*). This shrubby vegetation type is characteristic of flood-scoured dry streambeds of escarpment rivers on the Edwards Plateau, and is more open than the surrounding Tamaulipan thornscrub vegetation dominated by honey mesquite (*Prosopis glandulosa*) and acacia (*Acacia* spp.). In this habitat, Comanche phlox occurs in the zone of grasses and other herbaceous vegetation that occurs just above the main creek bed, sometimes at the base of grass clumps. Whitehouse (1935) noted Comanche phlox taking advantage of the shade provided by scattered trees and shrubs on gravel bars.

The northern portion of the range of Comanche phlox (subsp. *latisepala*) corresponds to the Grand Prairie region of Texas, which encompasses the Fort Worth Prairie and the Lampasas Cut Plain. Historically, this area was characterized by grassland with woodland inclusions. The floristic composition of the Fort Worth Prairie is distinct from that of the tallgrass prairie to the northeast and the Blackland prairies to the southeast, primarily due to the influence of limestone bedrock on the soils and surface topography of this region, with little bluestem the dominant grass of high quality relicts, followed by sideoats grama, yellow Indiangrass (*Sorghastrum nutans*), tall dropseed (*Sporobolus compositus*), and hairy grama (*Bouteloua hirsuta*). Grasslands of the Lampasas Cut Plain are characterized by similar species, with sideoats grama becoming more important. The forb component of these grasslands and related savannas is very diverse, sharing species with the Edwards Plateau to the south and the tallgrass prairie to the north, as well as the xeric limestone prairies of the Ozark Plateau of Arkansas and Missouri.

Cultivation

Comanche phlox has not been reported in cultivation, at least not under the name of *Phlox villosissima*. Given its natural habitat, this species would likely be tolerant of challenging dry shade situations in the landscape. Its colonizing habit could make Comanche phlox an effective groundcover for the southern Great Plains and southwestern United States. Grow in full sun to light shade in dry to slightly moist, well-drained soil.

Phlox viscida
basalt phlox
PLATES 71, 72

Can power of place kindle greatness of character? Consider the Joseph-Imnaha Plateau. Set in Oregon's northeast corner, the cloud-wreathed Wallowas on the west, staggering Hells Canyon to the east, a more exquisitely framed landscape is beyond imagining. The plateau itself is mantled by the magnificent Zumwalt Prairie, the largest sweep of bunchgrass steppe left on the continent. Raptors of all sorts know of this country, and battle for the privilege of fledging their young here, hopeful for heroes like another born on the plateau—Chief Joseph of the Nez Perce.

Founded upon colossal floods of basalt, heaved up from where rock is made by the collision of tectonic plates, the plateau is a vast rock garden in April and May and a stronghold of basalt phlox. This beautiful, loosely tufted plant is a specialist of scabland and lithosol habitat on the Columbia Plateau. Abundant on the Zumwalt, basalt phlox graces other heady settings, including Washington's Horse Heaven Hills and Idaho's Seven Devils Mountains, melding with other showy forbs to salve these rocklands with a fleeting balm of color in spring.

∽᠗᠙᠘

Phlox viscida E. E. Nelson, Revis. W. N. Amer. Phlox. 24. 1899 [sticky, of the foliage and inflorescence due to dense covering of glandular hairs; "viscid pubescent throughout, the pedicels and calyx densely so"] Upstanding suffrutescent perennial with erect-ascending flowering shoots 10–20 cm tall with 4 or 5 leaf nodes below the inflorescence; *leaves* linear to lanceolate, thinnish, short- to long-accuminate, ciliate and pilose with many hairs land-tipped, max. length 25–45 mm and width 2–4 mm; *inflorescence* 3- to 15-flowered, compact, its herbage copiously glandular-pubescent, max. pedicel length 10–60 mm but most often ca. 30 mm; *calyx* 11–16 mm long, united ⅜–⅝ its length, lobes linear with weak midrib and with apex cuspidate, membranes flat to inconspicuously plicate; *corolla* tube 12–18 mm long, lobes obovate (average dimensions 10 × 6 mm) with apex entire to emarginate, hue purple, pink, or white, sometimes with a pale eye bearing deep-hued striae; *stamens* borne on upper corolla tube, with some anthers exserted; *style* 7–14 mm long, united to tip which is free for 1 mm, with stigma placed among the anthers near corolla orifice; *flowering season* spring and early summer, May–June.

∽᠗᠙᠘

Taxonomic notes
Wherry (1955) described *Phlox mollis*, an entity known from only a few collections in the Snake River region of the Columbia Plateau taken within the larger distribution of basalt phlox. Cronquist (1959) reduced *P. mollis* to synonymy under *P. viscida*, noting it could warrant recognition as a loosely woolly, nonglandular local variety. Wherry (1962) took exception, noting *P. viscida* has "multiflorous inflorescences" while *P. mollis* has flowers solitary or in threes. Pending further study, this entity is incorporated here into *P. viscida*, but clarification of the ecological relationships and actual distribution may prove it worthy of some level of taxonomic recognition.

Geography
Columbia Plateau, concentrated along the eastern edge at its interface with the Northern Rocky Mountains, occurring most abundantly in the Hells Canyon

region in the Blue Mountains (NE Oregon and SE Washington), the Wallowa and Ochoco Mountains (NE Oregon), and Craig Mountain and the Seven Devils Mountains (W Idaho). From the Hells Canyon region the distribution follows the arc of the northern edge of the Columbia Plateau north into C Washington then south through the Horse Heaven Hills region (SC Washington) into the watersheds of the Deschutes and John Day rivers (NC Oregon).

Environment

Dissected plateaus and mountains at low to middle elevations, on upper slopes of rolling terrain and divides, bluffs, scablands, and ridges, 800–4500 ft. (250–1400 m); *MAP* 13–18 in. (33–45 cm); *habitat* xeric to dry; *soils* minimally developed to medium-textured, stony, shallow with bedrock at or near the surface; *parent material* primarily basalt.

Basalt phlox is most often associated with lithosols (rock soils) that develop in areas underlain by Columbia River basalt. Lithosols consist of a thin layer of frost-jumbled basalt blocks resting on the jointed bedrock from which they were weathered (Daubenmire 1970). Lithosols are also referred to as "scablands" because rock and bare ground are conspicuous and soil deposition is insignificant and primarily restricted to cracks in the basalt. In this habitat, surface soil temperatures can reach 150 degrees F by June, and extremely active winter frost heaving is frequent because the basalt bedrock inhibits water drainage resulting in saturated soils. Lithosol habitat is apparently too rigorous to support the steppe vegetation that dominates much of the Columbia Plateau. Basalt phlox plants growing in lithosol habitat often form large, rounded cushions. The very sticky foliage of basalt phlox may be an adaptation to this harsh environment.

While basalt phlox is most strongly associated with basalt outcrops and basalt-derived soils, it occurs in deeper soils in NC Oregon, where a large colony was observed to be associated with soil disturbed by pocket gopher activity. In Klickitat County, Washington, it occurs with some frequency along grassy roadsides, and was observed in an overgrazed pasture and among native grasses in a cemetery. Plants growing in habitat with greater soil development are smaller and more lax. Basalt phlox occurs to a minor extent on outwash derived from the volcanic deposits in the John Day River basin in NC Oregon.

Associations

Oak woodland, shrubland, shrub-steppe, grassland/steppe, and rock outcrop communities. In the Horse Heaven Hills region, basalt phlox occurs in association with a narrow belt of Oregon white oak (*Quercus garryana*) woodland that forms an ecotone and mosaic between ponderosa pine forest (*Pinus ponderosa*), shrub-steppe, and steppe. This critically imperiled plant community extends into adjacent Oregon along the east slope of the Middle Cascade Mountains near the Columbia River Gorge. Basalt phlox is more common in shrub-steppe and steppe communities adjacent to the oak woodlands, but in several places occurs in the herbaceous layer of these open woodlands with bluebunch wheatgrass (*Pseudoroegneria spicata*) or Idaho fescue (*Festuca idahoensis*) the dominant graminoids. Antelope bitterbrush (*Purshia tridentata*) is occasionally associated with these woodlands.

On very rocky lithosols in the Hells Canyon region, basalt phlox occurs in dwarf shrub-steppe characterized by a carpet of curly bluegrass (*Poa secunda*) in association with scabland sagebrush (*Artemisia rigida*) or Douglas' wild buckwheat (*Eriogonum douglasii*). These communities occur as small patch openings within conifer forest, from the ponderosa pine and Douglas-fir (*Pseudotsuga menziesii*) zones up into the subalpine fir (*Abies lasiocarpa*) zone. In the Horse Heaven Hills region, basalt phlox occurs in shrub-steppe comprised of dwarf sagebrush (*Artemisia arbuscula*) in association with Idaho fescue, bluebunch wheatgrass, or curly bluegrass, in a complex mosaic of openings within woodland of ponderosa pine, Oregon white oak, or western juniper (*Juniperus occidentalis* var. *occidentalis*). Basalt phlox occurs to a lesser extent in shrubland and shrub-steppe dominated by big sagebrush (*A. tridentata*). Basalt phlox occurs in sparsely vegetated badlands associated with the volcanic John Day and Clarno formations (NC Oregon), in the context of open western juniper woodland and big sagebrush shrubland.

Basalt phlox is particularly abundant on the Zumwalt Prairie on the Joseph-Imnaha Plateau (NE Oregon), the largest and highest quality expanse of fescue bunchgrass steppe remaining in North America, where it occurs on gently sloping ridge-crests, ridge-brows, and plateaus at elevations of 4500–5500 ft. (1400–1700 m) in an Idaho fescue / prairie junegrass (*Koeleria macrantha*) association. On more xeric sites with thin, rocky soils, basalt phlox occurs in a lithosolic phase of a bluebunch wheatgrass / curly bluegrass association. These bunchgrass-dominated associations have a large and diverse forb component and are very showy in spring. Basalt phlox also occurs in grasslands associated with the Horse Heaven Hills region of SC Washington, and the divide between the Deschutes and John Day Rivers in NC Oregon.

A number of forbs are relatively constant associates of basalt phlox including alliums (*Allium* spp.), Hooker's balsam-root (*Balsamorhiza hookeri*), red coral-drops (*Besseya rubra*), prairie smoke (*Geum triflorum*), dwarf waterleaf (*Hydrophyllum capitatum*), Oregon bitterroot (*Lewisia rediviva*), woodland-stars (*Lithophragma* spp.), desert-parsleys (*Lomatium* spp.), Gairdner's beardtongue (*Penstemon gaird-*

neri), lanceleaf stonecrop (*Sedum lanceolatum*), narrow-petal stonecrop (*S. stenopeta-lum*), and large-head clover (*Trifolium macrocephalum*). Basalt phlox occasionally occurs with broad-fruited mariposa lily (*Calochortus nitidis*) and lovely beardtongue (*P. elegantulus*), rare endemics of the Hells Canyon region. The forbs of lithosol communities flower mostly in the spring, and are dormant by June.

Cultivation
Basalt phlox has not been reported in cultivation, but it is a showy species with potential as a drought-tolerant landscape plant for drier portions of the Pacific Northwest. It would also be a striking subject for the rock garden. Grow in full sun in dry, well-drained soil.

Phlox woodhousei
Mogollon phlox
PLATE 73

The Mogollon Rim is the most profound biological boundary in the American Southwest. Marking the southern edge of the Colorado Plateau, this command-ing escarpment cuts northwest-to-southeast across the heart of Arizona. Below its brink, where the Rim falls away into cliffs and canyons, are the northern fringes of the Sonoran Desert, replete with saguaro, agave, ocotillo, and a host of other succulent prickly things. On top of the plateau is an evergreen forest cooled by pine, oak, and juniper. Mogollon phlox takes to the latter.

Mogollon phlox belongs to the great ponderosa pine forests of central Arizona, where it prospers in shade and needle cast. From the Rim northward to the Grand Canyon, its bright pink, delicately cleft flowers are sign of spring's arrival. It is particularly abundant in Oak Creek Canyon, which climbs the Rim between Sedona and Flagstaff. I saw it here one early April morning, in the company of bitterroot and mertensia, all gently frosted by a strange snowy fog, and sparkling where light had won passage through the piney rafters.

❧❧❧

Phlox woodhousei (Torrey ex A. Gray) E. E. Nelson, Revis. W. N. Amer. Phlox. 31. 1899. BASIONYM: *P. speciosa* Pursh var. *woodhousii* Torrey ex A. Gray, Proc. Amer. Acad. Arts 8: 256. 1870 [after Samuel Washington Woodhouse (1821–1903), col-lector of the type] Upstanding suffrutescent perennial (5–)10–15 cm tall, with erect-ascending flowering shoots 2.5–12.5 cm long; *leaves* oblong to narrowly elliptic or lanceolate, thick-textured, acutish or obtusish, the lower glabrous and upper pubescent with gland-tipped or rarely glandless hairs, max. length 20–45 mm and width 2.5–7.5 mm, the leaves successively widen upward along the shoot; *inflorescence* 3- to 12-flowered, its herbage glandular-pubescent, max. pedicel length 5–25 mm; *calyx* 7–12 mm long, united ½–⅝ its length, lobes linear with obscure midrib and with apex cuspidate, membranes flat or nearly so; *corolla* tube 8–15 mm long, lobes obovate (dimensions variable but averaging 9 × 6 mm) with apex notched 1–2 mm deep, hue purple to pink, often with prominent eye striae; *stamens* borne on mid corolla tube, with anthers included; *style* 1.5–3.5 mm long, united ¼–½ its length, with stigma deeply included and placed below the anthers; *flowering season* spring, April–June, again in autumn following seasonal rains.

Geography

Colorado Plateau, concentrated along the southern edge in N Arizona, extending into a limited area of WC New Mexico. The distribution of Mogollon phlox roughly parallels the northwest-to-southeast trend of the southern escarpment of the Colorado Plateau, known as the Mogollon (pronounced *MUGgy-own*) Rim, hence the common name. Mogollon phlox occurs from the Mogollon Rim northward to the southern rim of the Grand Canyon, with outlying occurrences on the northern rim of the Grand Canyon. In the southern portion of its range it occurs in the Sierra Ancha (C Arizona), the White Mountains (E Arizona), and the Mogollon Mountains (WC New Mexico).

Environment

Dissected plateaus and mountains at low to middle elevations, on upper slopes of rolling terrain and divides, escarpments, and in upper reaches of canyons and ravines, 3500–8000 ft. (1000–2400 m); *MAP* 20–30 in. (51–76 cm); *habitat* dry-mesic; *soils* medium-textured, stony, shallow with bedrock at or near the surface. Mogollon phlox is most abundant in the upper portions of the canyons descending from the southern escarpment of the Colorado Plateau, such as Oak Creek Canyon between Flagstaff and Sedona.

Associations

Primarily ponderosa pine (*Pinus ponderosa*) woodland, with alligator juniper (*Juniperus deppeana*) and Gambel oak (*Quercus gambelii*) associated. Mature stands of ponderosa pine woodland have an open structure, with an herbaceous layer dominated by graminoids including Arizona fescue (*Festuca arizonica*), mountain muhly (*Muhlenbergia montana*), screwleaf muhly (*M. virescens*), hairy dropseed (*Blepharo-*

neuron tricholepis), muttongrass (*Poa fendleriana*), and bottlebrush squirrel-tail (*Elymus elymoides*). At higher elevations receiving more moisture, Mogollon phlox occurs in mixed conifer forest where white fir (*Abies concolor*), Douglas-fir (*Pseudotsuga menziesii*), and mountain maple (*Acer glabrum*) are important species. Frequently associated forbs include Nuttall's pussytoes (*Antennaria parvifolia*), purple cluster crane's-bill (*Geranium caespitosum*), many-flower viguiera (*Heliomeris multiflora*), Cooper's bitterweed (*Hymenoxys cooperi*), Oregon bitterroot (*Lewisia rediviva*), Franciscan bluebells (*Mertensia franciscana*), New Mexico groundsel (*Packera neomexicana*), dwarf lousewort (*Pedicularis centranthera*), toadflax beardtongue (*Penstemon linarioides*), and white prairie aster (*Symphyotrichum falcatum* subsp. *commutatum*).

Cultivation

Mogollon phlox has been cultivated to a limited extent as a rock garden plant. This showy species with its attractive cleft flowers also has potential as a drought-tolerant landscape plant for the southwestern United States. Grow in full sun to light shade in dry to slightly moist, well-drained, humus-rich soil

Appendix A

State and Provincial *Phlox* Lists

The following lists are based on regional and state or provincial floristic references, web-based floras and atlases, and the distribution maps of Wherry (1955) in the *Genus Phlox*. Identification is to subspecies where known or discernable. Listings are annotated to indicate taxa endemic (E) to or of conservation concern (CC) within the state or province.

United States of America

Alabama
P. *amplifolia*
P. *carolina*
P. *divaricata* subsp. *divaricata*; subsp. *laphamii*
P. *floridana*
P. *glaberrima*
P. *nivalis*
P. *pilosa* subsp. *ozarkana*; subsp. *pilosa*
P. *pulchra* (E, CC)

Alaska
P. *hoodii* subsp. *hoodii* (CC)
P. *richardsonii* subsp. *alaskensis*; subsp. *borealis*; subsp. *richardsonii*

Arizona
P. *amabilis* (E, CC)
P. *austromontana* subsp. *austromontana*; subsp. *lutescens* (CC)
P. *cluteana* (CC)
P. *griseola* subsp. *griseola*
P. *hoodii* subsp. *canescens*
P. *longifolia*
P. *nana*

P. *tenuifolia*
P. *woodhousei*

Arkansas
P. *amplifolia*
P. *bifida* subsp. *arkansana* (E); subsp. *bifida* (CC); subsp. *stellaria* (CC)
P. *carolina* subsp. *angusta*
P. *divaricata* subsp. *laphamii*
P. *glaberrima* subsp. *interior*
P. *paniculata*
P. *pilosa* subsp. *ozarkana*; subsp. *pilosa*

California
P. *adsurgens*
P. *austromontana* subsp. *austromontana*
P. *condensata* subsp. *covillei*
P. *diffusa*
P. *dispersa* (E, CC)
P. *dolichantha* (E, CC)
P. *douglasii*
P. *hirsuta* (E, CC; U.S.F.W.S. endangered species)
P. *hoodii* subsp. *canescens*)
P. *longifolia*
P. *muscoides* (CC)
P. *pulvinata*

P. speciosa subsp. *nitida*; subsp. *occidentalis*
P. stansburyi

Colorado
P. andicola (cc)
P. austromontana subsp. *austromontana*
P. caryophylla (cc)
P. condensata subsp. *condensata*
P. hoodii subsp. *canescens*; subsp. *hoodii*
P. kelseyi (cc)
P. longifolia
P. multiflora
P. muscoides
P. pulvinata

Connecticut
P. maculata
P. pilosa subsp. *pilosa*

Delaware
P. maculata (cc)

District of Columbia
P. divaricata subsp. *divaricata*
P. subulata subsp. *subulata*

Florida
P. carolina
P. divaricata subsp. *laphamii*
P. floridana (cc)
P. nivalis
P. pilosa subsp. *pilosa*

Georgia
P. amoena
P. amplifolia (cc)
P. carolina
P. divaricata subsp. *divaricata*; subsp.
  laphamii
P. floridana (cc)
P. glaberrima subsp. *glaberrima*
P. maculata
P. nivalis subsp. *hentzii*; subsp. *nivalis*
P. ovata (cc)

P. paniculata
P. pilosa subsp. *ozarkana*; subsp. *pilosa* (cc)
P. stolonifera (cc)

Hawaii—None

Idaho
P. aculeata
P. albomarginata
P. austromontana subsp. *austromontana*
P. caespitosa
P. colubrina
P. diffusa
P. douglasii
P. hoodii
P. idahonis (E, cc)
P. kelseyi (cc)
P. longifolia
P. multiflora
P. muscoides
P. speciosa subsp. *speciosa*
P. viscida

Illinois
P. bifida subsp. *bifida*; subsp. *stellaria* (cc)
P. carolina subsp. *angusta*
P. divaricata subsp. *divaricata*; subsp.
  laphamii
P. glaberrima subsp. *interior*
P. maculata
P. paniculata
P. pilosa subsp. *fulgida*; subsp. *pilosa*;
  subsp. *sangamonensis* (E, cc)

Indiana
P. amplifolia (cc)
P. bifida subsp. *bifida*; subsp. *stellaria* (cc)
P. carolina
P. divaricata subsp. *divaricata*
P. glaberrima subsp. *interior*
P. maculata
P. ovata (cc)
P. paniculata

P. pilosa subsp. *deamii*; subsp. *fulgida*;
  subsp. *pilosa*
P. subula (introduced?)

Iowa
P. bifida subsp. *bifida* (cc)
P. divaricata subsp. *laphamii*
P. maculata
P. pilosa subsp. *fulgida*; subsp. *pilosa*

Kansas
P. andicola (cc)
P. divaricata subsp. *laphamii*
P. oklahomensis (cc)
P. pilosa subsp. *fulgida*; subsp. *ozarkana*;
  subsp. *pilosa*

Kentucky
P. amoena (cc)
P. amplifolia
P. bifida subsp. *bifida* (cc); subsp. *stellaria*
  (cc)
P. carolina
P. divaricata subsp. *divaricata*; subsp.
  laphamii
P. glaberrima
P. maculata
P. paniculata
P. pilosa subsp. *deamii* (cc); subsp. *pilosa*
P. stolonifera (cc)
P. subulata subsp. *setacea*

Louisiana
P. carolina
P. cuspidata
P. divaricata subsp. *laphamii*
P. pilosa subsp. *ozarkana* (cc); subsp. *pilosa*

Maine—None

Maryland
P. divaricata subsp. *divaricata*
P. maculata
P. paniculata

P. pilosa subsp. *pilosa*
P. subulata subsp. *brittonii*; subsp. *subulata*

Massachusetts
P. macula (introduced?)

Michigan
P. bifida subsp. *bifida* (cc)
P. divaricata subsp. *divaricata*
P. maculata
P. pilosa subsp. *pilosa*
P. subulata subsp. *subulata*

Minnesota
P. divaricata subsp. *laphamii*
P. maculata
P. pilosa subsp. *fulgida*

Mississippi
P. carolina
P. divaricata subsp. *laphamii*
P. paniculata
P. pilosa subsp. *pilosa*

Missouri
P. amplifolia (cc)
P. bifida subsp. *bifida* (cc); subsp. *stellaria*
  (cc)
P. carolina (cc)
P. divaricata subsp. *laphamii*
P. glaberrima subsp. *interior*
P. maculata (cc)
P. paniculata
P. pilosa subsp. *fulgida*; subsp. *ozarkana*;
  subsp. *pilosa*

Montana
P. albomarginata (cc)
P. alyssifolia
P. andicola (cc)
P. caespitosa
P. diffusa
P. hoodii
P. kelseyi

P. longifolia
P. missoulensis (E, CC)
P. multiflora
P. muscoides
P. speciosa subsp. *occidentalis*

Nebraska
P. alyssifolia
P. andicola
P. divaricata subsp. *laphamii*
P. hoodii subsp. *hoodii*
P. muscoides (CC)
P. pilosa subsp. *fulgida*

Nevada
P. austromontana subsp. *austromontana*
P. condensata subsp. *covillei*
P. diffusa
P. gladiformis (CC)
P. griseola subsp. *griseola*; subsp. *tumulosa*
P. hoodii subsp. *canescens*
P. kelseyi
P. longifolia
P. multiflora
P. muscoides
P. pulvinata
P. speciosa
P. stansburyi

New Hampshire—None

New Jersey
P. divaricata subsp. *divaricata* (CC)
P. maculata (CC)
P. pilosa subsp. *pilosa*
P. subulata subsp. *subulata*

New Mexico
P. austromontana subsp. *austromontana*
  (CC)
P. caryophylla (CC)
P. cluteana (CC)
P. condensata subsp. *condensata*
P. hoodii subsp. *canescens*

P. longifolia
P. nana
P. stansburyi
P. woodhousei

New York
P. divaricata subsp. *divaricata*
P. maculata (CC)
P. pilosa subsp. *pilosa*
P. subulata subsp. *subulata*

North Carolina
P. amoena (CC)
P. amplifolia (CC)
P. carolina (CC)
P. divaricata subsp. *divaricata* (CC)
P. glaberrima subsp. *glaberrima*
P. maculata (CC)
P. nivalis subsp. *hentzii*; subsp. *nivalis*
P. ovata
P. paniculata
P. pilosa subsp. *pilosa* (CC)
P. stolonifera (CC)
P. subulata subsp. *setacea* (CC)

North Dakota
P. alyssifolia (CC)
P. andicola
P. hoodii subsp. *hoodii*
P. pilosa subsp. *fulgida*

Ohio
P. divaricata subsp. *divaricata*
P. glaberrima
P. maculata
P. ovata (CC)
P. paniculata
P. pilosa subsp. *pilosa*
P. stolonifera (CC)
P. subulata subsp. *setacea*; subsp. *subulata*

Oklahoma
P. cuspidata
P. divaricata subsp. *laphamii*

P. glaberrima subsp. *interior*
P. oklahomensis (cc)
P. pilosa subsp. *longipilosa* (e, cc); subsp.
   ozarkana; subsp. *pilosa*

Oregon
P. aculeata
P. adsurgens
P. austromontana subsp. *austromontana*
P. colubrina (cc)
P. diffusa
P. douglasii
P. hendersonii (cc)
P. hoodii subsp. *canescens*
P. longifolia
P. multiflora (cc)
P. muscoides (cc)
P. pulvinata
P. speciosa subsp. *lignosa*; subsp. *nitida*;
   subsp. *occidentalis*
P. viscida

Pennsylvania
P. divaricata subsp. *divaricata*
P. maculata
P. ovata
P. paniculata
P. pilosa subsp. *pilosa*
P. stolonifera
P. subulata subsp. *brittonii*; subsp. *subulata*

Rhode Island—None

South Carolina
P. amoena
P. carolina
P. divaricata subsp. *divaricata*
P. glaberrima subsp. *glaberrima*
P. maculata
P. nivalis subsp. *hentzii*; subsp. *nivalis*
P. ovata
P. paniculata
P. pilosa subsp. *pilosa*
P. stolonifera

South Dakota
P. alyssifolia
P. andicola
P. divaricata subsp. *laphamii*
P. hoodii subsp. *hoodii*
P. pilosa subsp. *fulgida*

Tennessee
P. amoena
P. amplifolia
P. bifida subsp. *stellaria* (cc)
P. carolina
P. divaricata subsp. *divaricata*; subsp.
   laphamii
P. glaberrima
P. maculata
P. ovata
P. paniculata
P. pilosa subsp. *deamii*; subsp. *ozarkana*
   (cc); subsp. *pilosa*
P. stolonifera
P. subulata subsp. *setacea*

Texas
P. bifida subsp. *bifida*
P. carolina subsp. *angusta*
P. cuspidata
P. divaricata subsp. *laphamii*
P. drummondii subsp. *drummondii* (e);
   subsp. *glabriflora*; subsp. *johnstonii*
   (e, cc); subsp. *mcallesteri* (e); subsp.
   tharpii (e)
P. nana
P. nivalis subsp. *texensis* (e, cc;
   U.S.F.W.S. endangered species)
P. pilosa subsp. *pilosa*
P. pulcherrima (e)
P. roemeriana (e)
P. villosissima subsp. *latisepala*; subsp.
   villosissima (e)

Utah
P. albomarginata (cc)

P. austromontana subsp. *austromontana*;
subsp. *jonesii* (E); subsp. *lutescens* (CC);
subsp. *prostrata* (E)
P. cluteana (CC)
P. gladiformis (CC)
P. griseola subsp. *griseola* (CC); subsp.
tumulosa (CC)
P. hoodii subsp. *canescens*
P. longifolia
P. multiflora
P. muscoides
P. opalensis (CC)
P. pulvinata

Vermont
P. divaricata subsp. *divaricata*

Virginia
P. amplifolia (CC)
P. buckleyi (CC)
P. carolina
P. divaricata subsp. *divaricata*
P. glaberrima subsp. *glaberrima*
P. maculata
P. nivalis subsp. *hentzii*
P. ovata
P. paniculata
P. pilosa subsp. *pilosa* (CC)
P. stolonifera
P. subulata subsp. *brittonii*; subsp. *setacea*;
subsp. *subulata*

Washington
P. aculeata
P. colubrina
P. diffusa subsp. *longistylis*
P. douglasii
P. hendersonii
P. hoodii subsp. *canescens*
P. longifolia
P. speciosa subsp. *lignosa*; subsp. *nitida*;
subsp. *speciosa*
P. viscida

West Virginia
P. buckleyi (CC)
P. carolina (CC)
P. divaricata subsp. *divaricata*
P. maculata
P. ovata
P. paniculata
P. stolonifera
P. subulata subsp. *brittonii*

Wisconsin
P. bifida subsp. *bifida* (CC)
P. divaricata subsp. *laphamii*
P. glaberrima subsp. *interior*
P. pilosa subsp. *fulgida*; subsp. *pilosa*

Wyoming
P. albomarginata (CC)
P. alyssifolia (CC)
P. andicola (CC)
P. diffusa
P. hoodii
P. kelseyi (CC)
P. longifolia
P. multiflora
P. muscoides
P. opalensis (CC)
P. pulvinata
P. pungens (E, CC)

Canada

Alberta
P. alyssifolia (CC)
P. hoodii subsp. *hoodii*

British Columbia
P. alyssifolia (CC)
P. caespitosa
P. diffusa subsp. *longistylis*
P. hoodii subsp. *hoodii*
P. longifolia
P. speciosa subsp. *occidentalis* (CC); subsp.
speciosa (CC)

Manitoba

P. hoodii subsp. *hoodii*
P. pilosa subsp. *fulgida*

Northwest Territories

P. richardsonii subsp. *richardsonii*

Nunavut

P. richardsonii subsp. *alaskensis*; subsp. *richardsonii*

Ontario

P. divaricata subsp. *divaricata*
P. pilosa subsp. *pilosa* (presumed extirpated)
P. subulata subsp. *subulata* (cc)

Quebec

P. divaricata subsp. *divaricata*

Saskatchewan

P. alyssifolia (cc)
P. hoodii subsp. *hoodii*

Yukon Territory

P. hoodii subsp. *hoodii*
P. richardsonii subsp. *alaskensis*; subsp. *richardsonii*

Mexico

Baja California

P. austromontana subsp. *austromontana*

Chihuahua

P. nana
P. tenuifolia

Coahuila

P. pattersonii
P. villosissima subsp. *latisepala*

Durango

P. nana

Nuevo Leon

P. pattersonii

Sonora

P. nana

Tamaulipas

P. drummondii subsp. *glabriflora*

Appendix B

Phlox Horticultural Nomenclature

Recognized *Phlox* Horticultural Hybrids

Phlox ×arendsii (Arends) Wherry, Natl. Hort. Mag. 14: 229. 1935
P. arendsii Arends, J. Roy. Hort. Soc. 38: 151, fig. 114. 1912
Parent species: *P. divaricata* × *P. paniculata*

Phlox ×frondosa Hort. ex Wherry, Natl. Hort. Mag. 14: 213. 1935
P. frondosa Hort. ex. Vilmorin, Fl. Pleine Terre (ed. 2) 683. 1866
Parent species: *P. nivalis* × *P. subulata*

Phlox ×henryae Wherry, Natl. Hort. Mag. 25: 151. 1941
Parent species: *P. bifida* × *P. nivalis*

Phlox ×lilacina Hort. ex Wherry, Natl. Hort. Mag. 14: 213. 1935
Parent species: *P. bifida* × *P. subulata*

Phlox ×procumbens (Lehmann) Wherry, Natl. Hort. Mag. 14: 219. 1935
P. procumbens Lehmann, Sem. Bot. Hamburg. 17. 1828
Parent species: *P. stolonifera* × *P. subulata*

Innovations in *Phlox* Horticultural Nomenclature

The following nomenclatural innovations are **designated here** for the purpose of standardizing horticultural nomenclature in the genus *Phlox* Linnaeus.

Phlox Chattahoochee Group
The phlox cultivar 'Chattahoochee' was introduced in the mid-1940s, and has since been a popular garden subject in North America and Europe. The original material of 'Chattahoochee' was collected from the wild by Mary Gibson ("Mrs. J. Norman") Henry in the watershed of the Chattahoochee River in northwestern

Florida. Henry grew the original transplants in her garden in Gladwyne, Pennsylvania, and later made stock available for distribution through Mayfair Nurseries of Bergenfield, New Jersey. Listed under the name *Phlox divaricata* 'Chattahoochee', it was described in the Mayfair catalog as "[o]ne of the most handsome and impressive of Phlox, 10 to 12 inches tall, with ample heads of large pale lavender blue flowers with crimson centers. Almost prostrate mats of rough evergreen leaves that turn a deep purple in winter."

Wherry (1946, 1955) considered Henry's 'Chattahoochee' to be a strain of *Phlox divaricata* subsp. *laphamii*, but others have speculated the cultivar to be of hybrid origin, involving crosses between *P. divaricata* subsp. *laphamii* and one of two other eastern short-styled phloxes—*P. amoena* (Goodwin 1990) or *P. pilosa* (Foster and Foster 1990). Wherry later (1970) noted other cultivars ('Marietta' and 'Opelousas') with Chattahoochee-like traits circulating in horticulture. Georgia nurseryman Don Jacobs (1990) is of the opinion that two different entities were circulating under the name 'Chattahoochee' and, while attractive, neither of the "imposters" represented the original material, one being a hybrid of *P. amoena* with either *P. pilosa* or *P. subulata*, the other a selection of *P. pilosa* subsp. *fulgida*, for which he proposed the cultivar name 'Moody Blue'. Given the confusion, Jacobs suggested the term "'Chattahoochee' type" for existing forms associated with this name and assigning new, precise cultivar names for red-eyed forms of *P. divaricata* such as 'Eco Regal', a selection Jacobs made from the vicinity of the Chattahoochee River in western Georgia that he felt epitomized Henry's original type.

Whatever its true identity, the cultivar 'Chattahoochee' is highly regarded, having received the Royal Horticultural Society Award of Merit and rating highest of all phlox species and cultivars in trials of the Chicago Botanic Garden (Hawke 1999). Cultivars of *Phlox divaricata* subsp. *laphamii*, or its hybrids with *P. amoena* or *P. pilosa*, that exhibit the characters of the original 'Chattahoochee' (unnotched corolla lobes, lavender or purple corolla hue, and red-purple corolla eye) are collectively designated the *Phlox* Chattahoochee Group.

Phlox Madrean Group

While the botanical nomenclature of the *Phlox nana* complex is far from settled (see species account), named selections from wild populations have circulated in horticulture as cultivars, as have crosses made between these cultivars. Referred to as the "Mexican phlox" or the "Chihuahuan Hybrids," these are mostly listed as cultivars of *P. mesoleuca* or *P. triovulata*, which are names applied in the past to taller, more southern expressions of the first-described *P. nana*. The name *P. mesaleuca* (sic) is particularly well-known in England, since a plant shown under this name received the Royal Horticultural Society Award of Merit in 1930. Paul Maslin (1979), who was responsible for the discovery and initial introduction of many of the exciting color forms from Chihuahua in the late 1970s, proposed recognition of two additional species within the complex—*P. lutea* and *P. purpurea*.

Given the unsettled nomenclature of this complex, and the uncertain parentage of some of the cultivars selected from cultivated material, it seems best at this time to treat these as a cultivar group rather than assign cultivar names to a particular species. The formulation *Phlox* Madrean Group is suggested to indicate the association of this complex with the Madrean floristic region of the southwestern United States and northern Mexico. Cultivars of unknown parentage that exhibit the characters of the *P. nana* complex (see species account) are collectively designated the *Phlox* Madrean Group. A cultivar of this group is thus rendered *Phlox* (Madrean Group) 'Arroyo'.

Phlox Scotia Alpines Group

A great many cushion or "alpine" phlox cultivars circulating in horticulture are of unknown parentage. These have typically been assigned to either *Phlox subulata* or to *P. douglasii*, the latter a western cushion-forming species discovered by Scottish plant explorer David Douglas in the Pacific Northwest. Placement of cultivars under the name *P. douglasii* occurred as early as 1931 in an article on "alpine phlox" in cultivation at the Royal Horticultural Society's garden at Wisley. Some of these cultivars, also called "Douglasii Hybrids," are among the most popular and most easily grown cushion phlox in Europe, and several have received the Award of Garden Merit form the Royal Horticultural Society, including 'Boothman's Variety', 'Crackerjack', 'Iceberg', and 'Red Admiral'.

Whatever the original source of this material, it was not Douglas himself and most likely was not true *Phlox douglasii* (see species account). Of the many cultivars circulating under the name "Douglasii Hybrids," at least some are suspected to represent hybrids involving spontaneous garden crosses between *P. bifida*, *P. nivalis*, or *P. subulata*, or between one of these eastern mat-forming phloxes and one of the western cushion-forming phloxes. *Phlox diffusa* is the suspected westerner because, like the cultivars in question but unlike most of the western cushion-forming phloxes, it is relatively easy to propagate.

The formulation *Phlox* ×*douglasii* has been used to designate these cushion phlox cultivars, but is unacceptable according to the International Code of Botanical Nomenclature since at least one of the parental species involved in a hybrid entity must be known with certainty for this formulation to be employed. The formulation *Phlox* Douglasii Group also is not acceptable, since *P. douglasii* is in current use as a species name. The formulation *Phlox* Scotia Alpines Group is suggested, to honor the important role played by Scottish explorer-naturalists David Douglas, Thomas Drummond, and John Richardson in bringing the first of the western cushion-forming phloxes to light, as well as that of Scottish nurseryman John Fraser in introducing *P. nivalis* and *P. subulata* to horticulture. While not true ecological alpines, these species represent "alpines" in the old horticultural sense of the term.

Cultivars of unknown parentage that exhibit the characters of the eastern mat-forming phloxes and/or the western cushion-forming phloxes are collectively designated the *Phlox* Scotia Alpines Group. A cultivar of this group is thus rendered *Phlox* (Scotia Alpines Group) 'Crackerjack'.

Phlox Suffruticosa Group

Early summer (early June into July) flowering border phlox cultivars have been informally classified under the name "Suffruticosa," derived from *Phlox suffruticosa* (Ventenat 1804; Willdenow 1809), an early synonym of *P. carolina*. Members of this group represent either: (1) horticultural selections of *P. carolina* or *P. maculata*; (2) crosses between different strains or cultivars of each species; (3) crosses between the two species. The name "Maculata Hybrids" (Jelitto and Schacht 1990) also has been used for this group. Some border phlox cultivars such as 'Omega' exhibit the elongate, columnar inflorescence so characteristic of *P. maculata*, and these can safely be associated with that name. Others, such as the popular 'Miss Lingard' show characteristics of both *P. carolina* and *P. maculata*. 'Miss Lingard' has been listed as a cultivar of either *P. carolina* or *P. maculata*, but is in fact a triploid (Meyer 1944) that produces no viable seed (Wherry 1932a), indicating hybrid origin. Cultivars of unknown parentage that exhibit the characters of *P. carolina* or *P. maculata* or their hybrids are collectively designated the *Phlox* Suffruticosa Group. A cultivar of this group is thus rendered *Phlox* (Suffruticosa Group) 'Miss Lingard'.

Horticultural Hybrids

Phlox ×*oliveri* Locklear, hybr. nov.

Parent species: *Phlox adsurgens* Torrey ex A. Gray × *Phlox nivalis* Loddiges val. Sweet

Etymology: Commemorates Charles and Martha Oliver, proprietors of Primrose Path Nursery, Scottdale, Pennsylvania, who crossed *P. adsurgens* 'Wagon Wheel' with *P. nivalis* 'Eco Flirty Eyes', the resultant plant with needlelike foliage and peach-pink flowers to which they gave the cultivar name 'Sunshine'.

Documentation: Primrose Path 1990 catalog

Phlox ×whittakeri Locklear, hybr. nov.

Parent species: *Phlox adsurgens* Torrey ex A. Gray × *Phlox stolonifera* Sims

Etymology: Commemorates Robert Harding Whittaker (1920–1980), distinguished American plant ecologist who wrote important papers on the floristic connection between the Siskiyou Mountains (home of *P. adsurgens*) and the Appalachian Mountains (home of *P. stolonifera*) (Whittaker 1960, 1966).

Documentation: Foster and Foster 1990

Glossary

acerose needle-shaped

acicular needle-shaped, very slender and elongate

acuminate tapering gradually along slightly concave sides to a slender sharp tip

acute tapering evenly along straight sides to a sharp tip

alternate one after the other along an axis; in reference to leaves along a stem

annual a plant which completes its life cycle in one year or less. The annual phloxes, *Phlox cuspidata*, *P. drummondii*, and *P. roemeriana*, are winter annuals, germinating in autumn, overwintering in a vegetative state (usually as a rosette aboveground), reproducing sexually from early spring through summer, and dying promptly thereafter

anther the pollen-bearing portion of the stamen

apex the tip or end of a structure

apiculate ending in an abrupt short-pointed tip

appressed lying flat against, in reference to leaves against a stem

arachnoid beset with fairly sparse, fine, white, loosely tangled hairs; cobwebby

areolate divided into small angular spaces; in reference to the spaces between veins of a leaf

aristate awned; tipped by a stiff slender bristle

ascending rising in an upward direction

attenuate tapering gradually to a slender tip

bract a reduced leaf, typically subtending an inflorescence

caespitose low-growing with numerous stems arising in discrete tuftlike clusters

calyx the outer whorl of floral parts (sepals); the collective name for all of the sepals of a flower. The *Phlox* calyx is comprised of five sepals with their lower portions fused into a tube for ⅜–¾ their length and terminating in five spreading lobes

calyx lobe the free terminal portion of the calyx extending beyond the united calyx tube

calyx membrane thin, dry, translucent tissue occurring at the junction of the sepals in the calyx tube; also termed the intercostal or intersepaline membrane

calyx tube the fused lower portion of the calyx

canescent beset with dense vesture of fine, relatively short, gray hairs; hoary

carinate raised lengthwise into a keel-like ridge; in reference to calyx membrane

caudex the woody base of an otherwise herbaceous perennial; the short, persistent, thickened, more or less erect, main stem of a perennial plant that otherwise has annual stems. The more compact and elaborately branched the caudex, the more hemispherical the profile of the plant, ranging from cushionlike (pulvinate) to extremely mounded (tumulose) forms

chartaceous like paper; very thin, flexible, and readily torn

cilia hairs confined to the margin of a leaf

ciliate fringed with cilia

ciliolate fringed with minute cilia

connate fused or united to a similar part; in reference to the bases of leaves

cordate heart-shaped

corolla the inner whorl of floral parts (petals); the collective name for all of the petals of a flower. The *Phlox* corolla is comprised of five petals with their lower portions fused into a tube for ½–¾ its length and terminating in five spreading lobes

corolla lobe the free terminal portion of the corolla extending beyond the united corolla tube

corolla tube the fused lower portion of the corolla

corymb an inflorescence with the flower stalks arising at different levels, the flowers reaching one level, the outer flowers having longer pedicels than the inner ones and opening before them

costate possessing a longitudinal rib (a costa or midrib)

cuneate wedge-shaped

cushion or cushion-forming caespitose growth habit with short flowering shoots arising from a compact branching caudex, the stems closely bunched together in a hemispherical shape resembling a cushion (syn. pulvinate)

cuspidate bearing an abrupt, small, sharp point

cyme an inflorescence arranged as an open flat-topped cluster, the inner flowers of which open first

cymule a diminutive cyme, with few flowers, few or no branches, and short axes

deciduous having leaves which are shed after one year's growth

decumbent reclining on the ground with tip ascending or erect

deltoid a broadly triangular shape

diffuse widely or loosely spreading

eglandular without glands

elliptic a circular shape which has been laterally compressed, widest about the middle and narrower at the ends

emarginate with a shallow notch at the tip; in reference to the corolla lobe apex

entire uninterrupted, without serration, incision, or notch along the margin of the structure; lacking convexities or concavities

erose ragged or unevenly cut or incised; in reference to the corolla lobe apex

exserted protruding beyond the rim of an enveloping or confining structure; in reference to the position of the anthers or the stigma in relationship to the corolla tube orifice

eye descriptive of the corolla tube orifice when of a different hue than the corolla lobes

falcate sickle-shaped; slenderly curved and tapered to a sharp tip

fascicle a cluster or bunch

filament the anther-bearing stalk of the stamen

foliaceous resembling a leaf

forb a broad-leaved herbaceous plant, not a grass or sedge

funnelform gradually expanding outward in the shape of a funnel; in reference to the corolla

glabrate nearly glabrous, with only a very sparse covering of hairs; becoming smooth

glabrous lacking hairs; smooth in the sense of lacking hairs

glandular bearing a globular waxy tip

graminoid a grass or grasslike plant such as a sedge or rush

habit the general appearance of a plant

helicoid cyme a cyme in which each axial segment branches to only one side, and all branches to the same side, the whole thus appearing to have an elongate main axis that curves or coils toward its unbranched side

herb an annual or perennial plant with no woody (lignified) tissue in any part of the shoot; when persisting over more than one growing season, the parts of the shoot die back seasonally

herbaceous not woody, having no significant degree of secondary growth in any part of the shoot, which thus does not develop woody (lignified) tissue

hirsute beset with coarse, rough, elongate, more or less straight and erect hairs

included contained within; in reference to anthers or the stigma which do not exceed the corolla tube in length

inflorescence an assemblage of flowers that forms a recognizable group

inflorescence-herbage collectively the bracts, inflorescence stalks (peduncles), and flower stalks (pedicels)

internode the space along the axis of a shoot or stem between leaf nodes

involucre a whorl of bracts subtending an inflorescence

lanceolate lance-shaped, widest below the middle and longer than wide, tapering to a sharp point

lax open or loose growth habit

leaf axil the region between the adaxial (upper surface) side of the leaf and the stem

leaf node place on the stem at which a leaf arises

limb the expanded portion of the corolla above the tube

linear narrow with sides parallel over most of length

lobe the free terminal part of a calyx or corolla

locule a discriminate cavity or chamber of space within an ovary or capsule

maculate spotted

MAP mean annual precipitation

mat or mat-forming caespitose growth habit with short flowering shoots arising from stems and branches spreading along the ground, the branches often rooting in places

morphology form and structure

mound or mound-forming caespitose growth habit with short flowering shoots arising from a compact, elaborately branched root crown, with shoots so densely packed or interwoven together as to resemble a domed mound in profile (syn. tumulose)

mucronate having an abrupt, sharp point at the tip

nominate expression a subspecies that represents the first-described expression of a species; *Phlox divaricata* subsp. *divaricata* is the nominate expression of the species *P. divaricata*

oblanceolate lance-shaped, with wider end toward the tip

oblong two or more times longer than wide and having parallel sides

obovate egg-shaped, with wider end toward the tip

obtuse blunt

open growth habit with stems not greatly condensed or densely crowded in relation to other stems; internodes apparent along the stems (syn. diffuse)

opposite arranged in pairs along an axis; in reference to leaves along a stem

orbicular rounded or nearly so in outline

ovary the portion of the pistil that contains the ovules

ovate egg-shaped in outline, widest below the middle and attached toward the wide end

ovule the body which, after fertilization, becomes the seed

panicle a compound inflorescence

papillate having small cone-shaped protrusions overall; appearing pimply

pedicel the stalk supporting a single flower in a multiflowered inflorescence

pedicelate having a pedicel (stalk)

peduncle the stalk supporting an inflorescence, applied even when the inflorescence is reduced to a solitary flower

perennial a plant which normally persists more than two years, with no definite limit to its life span

persistent leaves that remain green and functional through more than one growing season; evergreen

petiolate having a petiole

petiole the stalk supporting the leaf blade

pilose beset with relatively sparse, soft, slender, more or less erect hairs

pistil organ comprised of ovary, style, and stigma

plicate raised lengthwise into a flattened fold; in reference to the calyx membrane

procumbent prostrate or trailing on the ground but not rooting at the nodes

prostrate lying flat upon the substrate

puberulent beset with minute, fine, short, loose, curled hairs

pubescent bearing hairs; hairy

pulvinate see cushion or cushion-forming

pungent very sharp

recurved curving outward, downward, or backward

reticulate forming a network of interconnecting veins

retuse notched slightly at a usually rounded apex

rhizome modified stem that spreads laterally below ground, usually rooting at the nodes and becoming upturned at the apex

rhombic four-sided

rosette leaves clustered and crowned around a common basal point of attachment

rosulate growing in rosettes

salverform corolla with slender tube abruptly spreading into a flat limb

serrulate minutely toothed along the margin, the teeth sharp and forward-pointing

sessile lacking a stalk, attached directly to a stem

shoot the part of an herbaceous plant that is distinct from the root, comprised of stem, leaves, and, in flowering shoots, the inflorescence

simple having a single blade, not divided or compound

spinescent becoming spinulose

spinulose terminating in a rigid, tapering, sharp continuation of the primary vein

stamen pollen-producing structure comprised of the anther and the filament

stigma the part of the pistil receptive to pollen

stolon modified elongate stem that spreads horizontally along the surface of the ground, rooting at its nodes or apex and giving rise to new plants

stoloniferous growing from stolons

striae a fine, long, longitudinal line; *Phlox* flowers may have striae that occur as deeper-hued lines at the base of the corolla lobe toward the corolla orifice

striate marked with fine, long, longitudinal lines

style slender stalk connecting the stigma to the ovary

subshrub shrublike growth habit, branched from the base like a shrub and annually producing herbaceous flowering growth that dies back to woody tissue at the end of the growing season

subtend a structure situated at the base of another structure

subulate shaped like an awl; narrowly triangular

succulent fleshy and juicy

suffrutescent stems woody at the base

surficial occurring on the surface of a structure

taproot the primary, central, downward-growing root. Most caespitose phloxes arise from a taproot which is surmounted by a caudex

taxa plural of taxon

taxon a taxonomic entity of any rank, such as species or subspecies

tomentose beset with dense vesture of short, stiff hairs

truncate ending abruptly, as if cut straight across

tuft caespitose growth habit with short flowering shoots bunched together basally to form a tussock, the internodes typically apparent

tumulose see mound or mound-forming

turbinate top-shaped; inversely conical

upstanding growth habit with essentially erect or ascending flowering shoots

villous beset with sparse, long, slender, soft hairs that are not intertwined; shaggy

woolly beset with relatively long, moderately stout, intertwined, somewhat matted hairs

Bibliography

Literature Cited

Abbott, C. 1979. *How to Know and Grow Texas Wildflowers*. Kerrville, Texas: Green Horizons Press.

Aiton, W. 1789. *Hortus Kewensis*. London.

Allison, J. R., and T. E. Stevens. 2001. Vascular flora of Ketona dolomite outcrops in Bibb County, Alabama. *Castanea* 66 (1–2): 154–205.

Ambrose, S. E. 1996. *Undaunted Courage: Meriwether Lewis, Thomas Jefferson, and the Opening of the American West*. New York: Simon & Schuster.

Bailey, L. H. 1924. *Manual of Cultivated Plants*. New York: The Macmillan Company.

Baker, R. H. 1956. Mammals of Coahuila, Mexico. *University of Kansas Publications Museum of Natural History* 9 (7): 125–335.

Balandin, S. A., and V. Y. Razzhivin. 1980. The influence of snow cover on the distribution of vegetation in the south-east of Chukotka Peninsula. *Botanicheskii zhurnal* 71: 1088–1097. [In Russian, with summary in English]

Balogh, G. J. 1987. New localities for *Schinia indiana* (Smith) (Noctuidae). *Ohio Lepidopterist* 9 (2): 15–16.

Barneby, R. C. 1947. Distributional notes and minor novelties. *Leaflets of Western Botany* 5 (4): 61–66.

Barnett, J. K., and J. A. Crawford. 1994. Pre-laying nutrition of sage grouse hens in Oregon. *Journal of Range Management* 47: 114–118.

Barr. C. A. 1962. The gardener's pocketbook: *Phlox alyssifolia*. *American Horticultural Magazine* 41 (3): 170–171.

Barr, C. A. 1983. *Jewels of the Plains*. Minneapolis, Minnesota: University of Minnesota Press.

Barrell, J. 1969. *Flora of the Gunnison Basin: Gunnison, Saguache, and Hinsdale Counties, Colorado*. Rockford, Illinois: Natural Land Institute.

Bass, R. 1994. On Willow Creek. In *Heart of the Land: Essays on Last Great Places*. Ed. J. Barbato and L. Weinerman. New York: Pantheon Books. 7–24.

Beatley, J. C. 1976. *Vascular Plants of the Nevada Test Site and Central-Southern Nevada: Ecologic and Geographic Distributions*. Oak Ridge, Tennessee: Technical Information Center, Office of Technical Information, Energy Research and Development Administration.

Beck, L. C. 1826. Contributions towards the botany of the states of Illinois and Missouri. *American Journal of Science* 11 (1): 167–182.

Benedict, N. B. 1983. Plant associations of subalpine meadows, Sequoia National Park, California. *Arctic and Alpine Research* 15 (3): 383–396.

Benedict, N. B., and J. Major. 1982. A physiographic classification of subalpine meadows of the Sierra Nevada, California. *Madrono* 29 (1): 1–12.

Bentham, G. 1845. Polemoniaceae. A. P. de Candolle, *Prodomus systematis naturalis regni vegetabilis* 9: 302–322.

Bentham, G. 1849. *Plantae Hartwegianae*. London.

Berry, W. 2000. *Jayber Crow*. Washington, D.C.: Counterpoint.

Bir, D. 2003. Phlox under fire. *Nursery Management & Production* 19 (2): 45–46.

Bjork, C. R., and M. Fishbein. 2006. *Astragalus asotinensis* (Fabaceae), a newly discovered species from Washington and Idaho, United States. *Novon* 16: 299–303.

Boyd, R. S., and C. D. Hilton. 1994. Ecological studies of the endangered species *Clematis socialis* Kral. *Castanea* 59 (1): 31–40.

Brand, A. 1907. Polemoniaceae. *Das Pflanzenreich*, 4, 250: 1–203. Leipzig: Wilhelm Engelmann.

Breck, J. 1846. A chapter on phloxes. *The Horticulturist and Journal of Rural Art and Rural Taste* 1 (3): 122–127.

Breck, J. 1851. *The Flower Garden or Breck's Book of Flowers*. Boston: John P. Jewett.

Britton, N. L. 1892. New and noteworthy North American phanerograms–VI. *Bulletin of the Torrey Botanical Club* 19 (7): 219–226.

Britton, N. L. 1901. *Manual of the Flora of the Northern United States and Canada*. New York: Henry Holt and Company.

Brooks, M. 1965. *The Appalachians*. Boston, Massachusetts: Houghton Mifflin.

Buist, R. 1832. *American Flower-Garden Directory*. Philadelphia: Hibbert and Buist.

Butler, A. C. 1979. *Mima Mound Grasslands of the Upper Coastal Prairie of Texas*. MS Thesis, Texas A & M University.

Callaway, E. E. 1966. *In the Beginning: Creation, Evolution, Garden of Eden, and Noah's Ark*. New York: Carlton Press.

Campbell, L. M. 1992. Biosystematics of *Phlox kelseyi* (Polemoniaceae). M.S. thesis, University of Montana, Missoula.

Chapell, C. B. 2006. Plant associations of balds and bluffs of western Washington. Washington Department of Natural Resources Natural Heritage Report 2006–02.

Cholewa, A. F., and D. M. Henderson. 1984. *Primula alcalina* (Primulaceae): a new species from Idaho. *Brittonia* 36 (1): 59–62.

Clute, W. N. 1911a. The flora of the Chicago Plain. *American Botanist* 17 (3): 65–70.

Clute, W. N. 1911b. The smooth or meadow phlox. *American Botanist* 17 (4): 97–98.

Clute, W. N. 1919. A trip to Navajo Mountain. *American Botanist* 25 (3): 81–87.

Clute, W. N., and J. H. Ferriss. 1911. A new species of *Phlox*. *American Botanist* 17 (3): 74–76.

Correvon, H. 1930. *Rock Garden and Alpine Plants*. Ed. L. Barron. New York: Macmillan.

Coville, F. V. 1893. Botany of the Death Valley Expedition: A report on the botany of the expedition sent out in 1891 by the U.S. Department of Agriculture to make a biological survey of the region of Death Valley, California. *Contributions of the U.S. National Herbarium* 4: 1–363.

Cronquist, A. 1959. *Phlox*. In *Vascular Plants of the Pacific Northwest*. Ed. C. L. Hitchcock, A. Cronquist, M. Ownbey, and J. W. Thompson. Seattle, Washington: University of Washington Press. 124–137.

Cronquist, A. 1984. *Phlox*. In *Intermountain Flora*, volume 4. Ed. A. Cronquist, A. H. Holmgren, N. H. Holmgren, J. L. Reveal, and P. K. Holmgren. Bronx, New York: The New York Botanical Garden. 95–107.

Curtis, W. 1791. Phlox divaricata. Early-flowering lychnidea. *Botanical Magazine* 5: plate 163

Curtis, W. 1798a. Phlox subulata. Awl-leaved phlox, or lychnidea. *Botanical Magazine* 12: plate 411.

Curtis, W. 1798b. Phlox setacea. Fine-leaved phlox. *Botanical Magazine* 12: plate 415.

Cypher, E. A. 1993. Taxonomy, site characteristics, and rarity of the *Phlox bifida* Beck complex. PhD Thesis, Southern Illinois Univerity.

Darlington, W. 1849. *Memorials of John Bartram and Humphrey Marshall, with Notices of their Botanical Contemporaries*. Philadelphia, Pennsylvania: Lindsay and Blakison.

Daubenmire, R. 1970. Steppe vegetation of Washington. *Washington Agricultural Experiment Station Technical Bulletin* 62: 1–131.

Daubenmire, R. 1992. Palouse Prairie. In *Ecosystems of the World 8A: Natural Grasslands – Introduction and Western Hemisphere*. Ed. R. T. Coupland. Amsterdam, The Netherlands: Elsevier Science Publishers. 297–312.

Deam, C. C. 1940. *Flora of Indiana*. Indianapolis, Indiana: Department of Conservation, Division of Forestry.

del Moral, R. 1979a. High-elevation species clusters in the Enchantment Lakes Basin, Washington. *Madrono* 26 (4): 164–172.

del Moral, R. 1979b. High elevation vegetation of the Enchanted Lakes Basin, Washington. *Canadian Journal of Botany* 57: 1111–1130.

del Moral, R. 1983. Initial recovery of subalpine vegetation on Mount St. Helens, Washington. *American Midland Naturalist* 109 (1): 72–80.

Dillard, A. 1974. *Pilgrim at Tinker Creek*. New York: Harper's Magazine Press.

Dillenius, J. J. 1732. *Hortus Elthamensis, seu Plantarum rariorum quas in horto suo Elthami in Cantio coluit*. London.

Dillwyn, L. W. 1843. *Hortus Collinsonianus*. Swansea (Wales): W. C. Murray and D. Rees.

Dorn, R. D. 1988. *Vascular Plants of Wyoming*. Cheyenne, Wyoming: Mountain West Publishing.

Dorn, R. D. 1992. *Vascular Plants of Wyoming*, edition 2. Cheyenne, Wyoming: Mountain West Publishing.

Dorn, R. D. 2001. *Vascular Plants of Wyoming*, edition 3. Cheyenne, Wyoming: Mountain West Publishing.

Elam, D. 1998. Endangered and threatened wildlife and plants: proposed endangered status for the plant *Phlox hirsuta* (Yreka phlox) from northern California. Federal Register 63 (62): 15820–15825.

Erbe, L., and B. L. Turner. 1962. A biosystematic study of the *Phlox cuspidata-Phlox drummondii* complex. *American Midland Naturalist* 67 (2): 257–281.

Erlicht, G. 1985. *The Solace of Open Places*. New York: Penguin Books.

Ermakov, N., M. Cherosov, and P. Gogoleva. 2002. Classification of ultracontinental boreal forests in central Yakutia. *Folia Geobotanica* 37: 419–440.

Ermakov, N., B. Chytry, and M. Valachovic. 2006. Vegetation of the rock outcrops and screes in the forest-steppe and steppe belts of the Altai and Western Sayan Mts., southern Siberia. *Phytocoenologia* 36 (4): 509–545.

Eyster-Smith, N. M. 1984. Tallgrass prairies: an ecological analysis of 77 remnants. PhD Thesis, University of Arkansas.

Farrer, R. J. 1919. *The English Rock-Garden*. London, England: T. C. & E. C. Jack.

Fenneman, N. M. 1931. *Physiography of Western United States*. New York: McGraw-Hill Book Company.

Ferguson, C. J. 1998. Molecular systematics of eastern *Phlox* L. (Polemoniaceae). PhD Thesis, University of Texas at Austin.

Ferguson, C. J., and R. K. Jansen. 2002. A chloroplast DNA phylogeny of eastern *Phlox* (Polemoniaceae): implications of congruence and incongruence with the ITS phylogeny. *American Journal of Botany* 89 (8): 1324–1335.

Ferguson, C. J., F. Kramer, and R. K. Jansen. 1999. Relationships of eastern North American *Phlox* (Polemoniaceae) based on ITS sequence data. *Systematic Botany* 24 (4): 616–631.

Fertig, W. 1996. Status report on *Phlox opalensis* in southwestern Wyoming and northeastern Utah. Wyoming Natural Diversity Database, Laramie, Wyoming.

Flory, W. S., Jr. 1931. Chromosome numbers in *Phlox. American Naturalist* 65 (700): 473–476.

Foster, H. L. 1966. Waterworks Hill. *Quarterly Bulletin of the Alpine Garden Society* 34: 222–227.

Foster, H. L. 1968. *Rock Gardening: A Guide to Growing Alpines and Other Wildflowers in the American Garden*. Boston: Houghton Mifflin Company.

Foster, H. L., and L. L. Foster. 1990. *Cuttings from a Rock Garden*. Portland, Oregon: Timber Press.

Fraser, J. 1789. *A short history of the Agrostis cornucopiae; or, the new American grass: and a botanical description of the plant. To which are added, experiments tending to point out the proper mode of cultivating this plant: and also, some account of a journey to the Cherokee Nation, in search of new plants*. London.

Frazier, C. 1997. *Cold Mountain*. New York: Atlantic Monthly Press.

Gabrielson, I. N. 1932. *Western American Alpines*. New York: The MacMillan Company.

Gattinger, A. 1901. *Flora of Tennessee and a Philosophy of Botany Respectfully Dedicated to the Citizens of Tennessee*. Nashville, Tennessee: Press of Gospel Advocate Publishing Company.

Gentry, H. S. 1957. *Los Pastizales de Durango, Estudio Ecologico, Fisiografico y Floristico*. Ediciones del Instituto Mexicano de Recursos Naturales Renovables, Mexico City.

Gibson, A. C., P. W. Rundel, and M. R. Sharifi. 2008. Ecology and ecophysiology of a subalpine fellfield community on Mount Pinos, Southern California. *Madrono* 55 (1): 41–51.

Gilfillan, M. 1997. *From Burnt House to Paw Paw: Appalachian Notes*. West Stockbridge, Massachusetts: Hard Press, Inc.

Gleason, H. A., and A. Cronquist. 1991. *Manual of Vascular Plants of Northeastern United States and Adjacent Canada*. Bronx, New York: New York Botanical Garden Press.

Gmelin, J. G. 1747–1769. *Flora Sibirica sive Historia Plantarum Sibiriae*. 4 volumes. St. Petersburg.

Gold, T. S. 1850. Flowers for the million. *The Horticulturist and Journal of Rural Art and Rural Taste* 4 (9): 419–421.

Gold, W. G., K. A. Glew, and L. G. Dickson. 2001. Functional influences of cryptobiotic surface crusts in an alpine tundra basin of the Olympic Mountains, Washington, U.S.A. *Northwest Science* 75 (3): 315–326.

Goodwin, N. 1990. *Phlox*. In *A Rock Garden in the South*. Ed. N. Goodwin and A. Lacy. Durham, North Carolina: Duke University Press. 98–100.

Grant, V. 1998. Primary classification and phylogeny of the Polemoniaceae, with comments on molecular cladistics. *American Journal of Botany* 85:741–752.

Gray, A. 1870. Revision of the North American Polemoniaceae. *Proceedings of the American Academy of Arts and Sciences* 8: 247–282.

Gray, A. 1887. Contributions to North American botany. *Proceedings of the American Academy of Arts and Sciences* 22: 270–314.

Greene, E. L. 1896. New or noteworthy species–XV. *Pittonia* 3: 13–28.

Greene, E. L. 1905. New plants from southwestern mountains. *Leaflets of Botanical Observation and Criticism*. 145–160.

Hansen-Bristow, K. J., and L. W. Price. 1985, Turf-banked terraces in the Olympic Mountain, Washington, U.S.A. *Arctic and Alpine Research* 17 (3): 261–270.

Hardwick, D. F. 1958. Taxonomy, life history, and habits of the elliptoid-eyed species of *Schinia*, with notes on the Heliothinidinae. *Canadian Entomologist* 90: Suppl. 6: 1–116.

Harmer, J., and J. Elliott. 2001. *Phlox*. North East Lincolnshire [United Kingdom]: The Hardy Plant Society.

Harper, F. 1958. *The Travels of William Bartram: Naturalist's Edition. Edited with commentary and annotated index*. New Haven, Conneticutt: Yale University Press.

Harper, R. M. 1906. A phytogeographical sketch of the Altamaha Grit region of the Coastal Plain of Georgia. *Annals of the New York Academy of Science* 17: 1–415.

Harper, R. M. 1920. A week in eastern Texas. *Bulletin of the Torrey Botanical Club* 47 (7): 289–317.

Harrison, C. S. 1906. *A Manual on the Phlox*. York, Nebraska: Published by author.

Harshberger, J. W. 1909. The plant formations of the Nockamixon Rocks, Pennsylvania. *Bulletin of the Torrey Botanical Club* 36 (12): 651–673.

Hawke, R. G. 1999. An evaluation report of selected *Phlox* species and hybrids. *Chicago Botanic Garden Plant Evaluation Notes* 13: 1–4.

Heller, A. A. 1895. Botanical explorations in Southern Texas during the season of 1894. *Contributions from the Herbarium of Franklin and Marshall College* No. 1: 1–116.

Herre, A. W. 1940. An early Illinois prairie. *American Botanist* 46: 39–44.

Homoya, M. A., and D. B. Abrell. 2005. A natural occurrence of the federally endangered Short's goldenrod (*Solidago shortii* T. & G.) [Asteraceae] in Indiana: Its discovery, habitat, and associated flora. *Castanea* 70 (4): 255–262.

Hooker, W. J. 1835b. Phlox Drummondii. Mr. Drummond's Phlox. *Curtis's Botantical Magazine* 62. N.S. 9: pl. 3441.

Hooker, W. J. [1829–]1833–1840. *Flora Boreali-Americana; or, the Botany of the Northern Parts of British North America. . . .* 2 volumes, in 12 parts. London, Paris, and Strasbourg.

Hosack, D. 1811. *Hortus Elginensis: or a Catalogue of Plants, Indigenous and Exotic, Cultivated in the Elgin Botanic Garden, in the Vicinity of the City of New-York*, ed. 2. New York: T. & J. Swords.

Hovey, C. M. 1838. On the cultivation of Phlox Drummondii, Nicotiana longiflora, Petunias, and other annuals. *Magazine of Horticulture, Botany, and all Useful Discoveries and Improvements in Rural Affairs* 4: 97–103.

Hovey, C. M. 1846. Some remarks on the cultivation of Phlox, with descriptions of twenty-four new and beautiful varities. *Magazine of Horticulture, Botany, and all Useful Discoveries and Improvements in Rural Affairs* 4: 322–329.

Hovey, C. M. 1849. New Phloxes. *Magazine of Horticulture, Botany, and all Useful Discoveries and Improvements in Rural Affairs* 4: 220.

Hreha, A. M. 1993. The significance of edaphic and light factors to calciphile endemics restricted to the Claron Limestone in Red Canyon, Utah. PhD Thesis, University of Utah.

Hreha, A. M. 1995. Red Canyon, Utah: geology and plants. *Rock Garden Quarterly* 53: 259–264.

Hutchinson, I. W. 1934. *North to the Rime-Ringed Sun: Being the Record of an Alaskan-Canadian Journey Made in 1933–34.* London: Blackie & Son Limited.

Ingwersen, W. 1978. *Manual of Alpine Plants.* London: Cassell Publishers.

Jacobs, D. L. 1990. Tracking the timber phlox. *Bulletin of the American Rock Garden Society* 48 (3): 162–167.

Jelitto, L., and W. Schacht. 1990. *Hardy Herbaceous Perennials*, volume. 2, L–Z. Portland, Oregon: Timber Press.

Johansen, F. 1924. General observations on the vegetation. *Report of the Canadian Arctic Expedition 1913–18*, 5 (C): 1–85. Ottawa, Canada: F. A. Acland.

Johnson, C. G., Jr., and S. A. Simon. 1987. Plant Associations of the Wallowa-Snake Province. USDA Forest Service Publication R6-ECOL-TP-255A-86. USDA Forest Service, Pacific Northwest Region, Portland, Oregon.

Jones, M. E. 1894. Contributions to western botany VI. *Zoe* 4: 366–369.

Jones, M. E. 1895. Contributions to western botany. No. VII. *Proceedings of the California Academy of Science* II, 5: 611–732.

Jones, W. A. 1874. Report upon the reconnaissance of northwestern Wyoming, made in the summer of 1873. 43rd Congress, 1st Session, House Ex. Doc. No. 285. Serial 470.

Jordal, L. H. 1952. Some new entities in the flora of the Brooks Range, Alaska. Rhodora 54: 35–39.

Kartesz, J. T. 1988. A flora of Nevada. PhD Thesis, University of Nevada.

Kelaidis, P. 1984. Fiery phloxes of Chihuahua. *Pacific Horticulture* 45 (4): 38–40.

Kelaidis, P. 1986. Dryland bunneries: Persian carpets of the West. *Bulletin of the American Rock Garden Society* 44: 70–76.

Kelly, J. P. 1915. Cultivated varieties of *Phlox drummondii*. *Journal of the New York Botanical Garden* 16: 179–191.

Kempis, T. à. 1973. *Of the Imitation of Christ*. Grand Rapids, Michigan: Baker Book House.

Kiener, W. B. 1967. Sociological studies of the alpine vegetation on Longs Peak. *University of Nebraska Studies* No. 34.

Klebenow, D. A. and G. M. Gray. 1968. Food habits of juvenile sage grouse. *Journal of Range Management* 21 (2): 80–83.

Knight, P. J., B. Johnston, and S. L. O'Kane Jr. 1986. Status report on *Phlox caryophylla*. Colorado Natural Areas Program, Denver, Colorado.

Knyazev, M. S., S. A. Mamaev, and V. E. Vlasenko. 2007. Relict communities and populations of petrophilous plant species in northern regions of Sverdlovsk Oblast and problems in their conservation. *Russian Journal of Ecology* 38 (5): 317–322.

Krantz, T. P. 1983. *Phlox dolichantha*, the Sugarloaf phlox: a botanical survey of the species throughout its range. Report to the San Bernardino National Forest, San Bernardino National Forest, San Bernardino, California.

Krantz, T. 1990. *A Guide to the Rare and Unusual Wildflowers of the Big Bear Valley Preserve*. Big Bear Lake, California: Friends of the Big Bear Valley Preserve.

Krantz, T. P. 1994. *A Phytogeography of the San Bernardino Mountains, San Bernadino County, California*. PhD Thesis, University of California at Berkeley.

Kuc, M. 1996. Plant formations and their bio-products from the western Banks Island (N.W.T., Canada). *Polarforschung* 64 (3): 115–122.

Kwiat, A. 1908. One day's collecting, with a description of a new Noctuid. *Entomological News* 19: 420–424.

Leopold, A. 1949. *Sand County Almanac*. New York: Oxford University Press.

Lesica, P., and P. B. Kannowski. 1998. Ants create hummocks and alter structure and vegetation of a Montana fen. *American Midland Naturalist* 139 (1): 58–68.

Leverich, W. J. 1983. Pocket gophers, tanks, and plant community composition. *Southwestern Naturalist* 28 (3): 378.

Levin, D. A. 1966. The *Phlox pilosa* complex: crossing and chromosome relationships. *Brittonia* 18 (April–June): 142–162.

Levin, D. A. 1968. The genome constituents of eastern North American *Phlox* amphiploids. *Evolution* 22 (September): 612–632.

Levin, D. A., and B. A. Schaal. 1970. Reticulate evolution in *Phlox* as seen through protein electrophoresis. *American Journal of Botany* 57 (8): 977–987.

Levin, D. A., and D. M. Smith. 1965. An enigmatic *Phlox* from Illinois. *Brittonia* 17 (July): 254–266.

Levin, D. A., and D. M. Smith. 1966. Hybridization and evolution in the *Phlox pilosa* complex. *American Naturalist* 100 (913): 289–302.

Lichthardt, J., and K. Gray. 2001. Population monitoring and conservation assessment of Idaho phlox (*Phlox idahonis*). Boise, Idaho: Idaho Department of Fish and Game Conservation Data Center.

Lindley, J. 1843. Van Houtte's phlox. *Edwards's Botanical Register* (series the second) 16: plate 5.

Linnaeus, C. 1737. *Genera Plantarum*. Leiden.

Linnaeus, C. 1738. *Hortus Cliffortianus*. Amsterdam.

Linnaeus, C. 1748. *Hortus Upsaliensis*. Stockholm.

Linnaeus, C. 1750. *Plantae Rariores Camschatcenses* [dissertation of J. P. Halenius]. Uppsala.

Linnaeus, C. 1753. *Species Plantarum*. Stockholm.

Linnaeus, C. 1762. *Species Plantarum*, second ed., vol. 1. Uppsala.

Locklear, J. H. 2009. Nomenclatural innovations in *Phlox* (Polemoniaceae), with updated circumscription of *P. caespitosa*, *P. douglasii*, *P. missoulensis*, and *P. richardsonii*. *Journal of the Botanical Research Institute of Texas* 3 (2): 645–658.

Locklear, J. H. 2010 (in press). *Phlox ovata* L. (Polemoniaceae): clarification of the nomenclature of the Allegheny phlox. *Castanea* 75.

Loddiges, C. 1823. Phlox nivalis. *Botanical Cabinet* 8: plate 780

Lundell, C. L. 1942. Studies of American spermatophytes–III. *Contributions from the University of Michigan Herbarium* 8: 75–88.

Lundell, C. L. 1945. New American spermatophytes. *Field & Laboratory* 13 (1): 1–23.

Lupp, R. 2003. The north side of Mt. Adams, an alpine paradise. *Northwestern Chapter Newsletter* (North American Rock Garden Society). April 2003: 3–4.

Marsh, D. L. 1960. Relationship of *Phlox oklahomensis* to *Phlox bifida* complex: Including a new subspecies of *Phlox bifida*. *Transactions of the Kansas Academy of Science* 63 (1): 12–19.

Martyn, J. 1728. *Historiae plantarum rariorum centuriae primae decas I–V*. London.

Maslin, T. P. 1979. The rediscovery of *Phlox lutea* and *Phlox purpurea*. *Bulletin of the American Rock Garden Society* 37 (2): 62–69.

McBryde, J. B. 1933. The vegetation and habitat factors of the Carrizo Sands. *Ecological Monographs* 3 (2): 247–297.

McLaughlin, S. P. 2007. Tundra to tropics: the floristic plant geography of North America. *Sida, Botanical Miscellany*, No. 30. 1–58.

McPherson, S. 1996. *Edge Effect: Trails and Portrayals*. Middletown, Connecticut: Wesleyan University Press.

Meyer, J. R. 1944. Chromosome studies of *Phlox*. *Genetics* 29: 19–216.

Michaux, A. 1803. *Flora Boreali-Americana*. . . . 2 vols. Paris and Strasbourg.

Miller, P. 1731. *Gardeners Dictionary*. London

Miller, P. 1733. *Gardeners Dictionary*, second edition. London.

Miller, P. 1752. *Gardeners Dictionary*, sixth edition. London.

Miller, P. 1759. *Gardeners Dictionary*, seventh edition. London.

Mirkin, B. M., P. A. Gogoleva, and K. E. Kononov. 1985. The vegetation of central Yacutian Alases. *Folia Geobotanica & Phytotaxonomica* 20 (4): 345–395.

Moench, C. 1794. *Methodus plantas horti botanici et agri Marburgensis*. . . . Marburg.

Mohlenbrock, R. H. 1988. This land: Red Canyon, Utah. *Natural History* 97 (12): 18–21.

Mohr, C. 1901. Plant life of Alabama. *Contributions from the United States National Herbarium* 6: 1–921.

Moseley, R. K. 1990. Report on the conservation status of *Douglasia idahoensis*, in Idaho. Boise, Idaho: Idaho Department of Fish and Game.

Moseley, R. K., and R. C. Crawford. 1995. Fifteen-year population and habitat changes in a narrow Idaho endemic, *Phlox idahonis*. *Bulletin of the Torrey Botanical Club* 122 (2): 109–114.

Muir, J. 1914. *The Yosemite*. Boston, Massachusetts: Houghton Mifflin.

Muller, C. H. 1939. Relations of the vegetation and climatic types in Nuevo Leon, Mexico. *American Midland Naturalist* 21 (3): 687–729.

Muller, C. H. 1947. Vegetation and climate of Coahuila, Mexico. *Madrono* 9: 33–57.

Nelson, A. 1898. New plants from Wyoming–II. *Bulletin of the Torrey Botanical Club* 25 (5): 275–284.

Nelson, A. 1911. New plants from Idaho. *Botanical Gazette.* 52 (4): 261–274.

Nelson, A. 1922. Flora of the Navajo Reservation, II. *American Botanist* 28: 20–25.

Nelson, E. E. 1899. Revision of the western North American phloxes. *Contributions of the Department of Botany and Rocky Mountain Herbarium* 27: 1–35.

Nuttall, T. 1834. A catalogue of a collection of plants made chiefly in the valleys of the Rocky Mountains or Northern Andes, towards the sources of the Columbia River, by Mr. Nathaniel B. Wyeth, and described by T. Nuttall. *Journal of the Academy of Natural Sciences of Philadelphia* 7: 5–60.

Nuttall, T. 1848. Descriptions of plants collected by Mr. William Gambel in the Rocky Mountains and Upper California. *Proceedings of the Academy of Natural Sciences of Philadelphia* 4: 7–26.

O'Kane, S. L. 1988. Colorado's rare flora. *Great Basin Naturalist* 48 (4): 434–484.

Ott, J. R., J. A. Nelson, and T. Caillouet. 1998. The effect of spider-mediated flower alteration on seed production in golden-eye phlox. *Southwestern Naturalist* 43 (4): 430–436.

Parry, C. C. 1862. Physiographic sketch of that portion of the Rocky Mountain range, at the head waters of south Clear Creek, and east of Middle Park: with an enumeration of the plants collected in this district, in the summer months of 1861. *American Journal of Science* 33: 231–243.

Parry, C. C. 1874. Botanical observations in western Wyoming. *American Naturalist* 8: 9–14, 102–108, 175–180, 211–215.

Paxton, J. 1836. Phlox drummondii (Mr. Drummond's Lichnidea). *Paxton's Magazine of Botany* 2: 221.

Peabody, F. J. 1979. *Phlox longifolia* Nutt. (Polemoniaceae) complex of North America. *Great Basin Naturalist* 39 (1): 1–14.

Phillips, A. M, III, B. G. Phillips, and N. Brian. 1982. Status report: *Phlox cluteana* A. Nels. U.S. Fish and Wildlife Service, Albuquerque, New Mexico.

Piper, C. V. 1902. New and noteworthy northwestern plants–VII. *Bulletin of the Torrey Botanical Club* 29 (11): 642–646.

Piper, C. V. 1906. Flora of the State of Washington. *Contributions from the United States National Herbarium* 11: 1–637.

Piper, C. V., and R. K. Beattie. 1901. *The Flora of the Palouse Region*. Pullman, Washington: Washington Agricultural College and School of Science.

Pittillo, J. D. 2000a. Blue Ridge Parkway tour: Asheville to Cherokee. In *Exploring North Carolina's Natural Areas: Parks, Nature Preserves, and Hiking Trails*. Ed. D.

Frankenberg. Chapel Hill, North Carolina: University of North Carolina Press. 335–347.

Pittillo. J. D. 2000b. Cherohala Skyway to Joyce Kilmer Forest. In *Exploring North Carolina's Natural Areas: Parks, Nature Preserves, and Hiking Trails*. Ed. D. Frankenberg. Chapel Hill, North Carolina: University of North Carolina Press. 359–366.

Plukenet, L. 1691. *Phytographia*. London.

Plukenet, L. 1696. *Almagestum botanicum sive Phytographiae Pluknetianae onomsaticon*. London.

Plukenet, L. 1700. *Almagesti botanici mantissa . . . cum indice totius operis*. London.

Porter, J. M., and L. A. Johnson. 2000. A phylogenetic classification of Polemoniaceae. *Aliso* 19: 55–91.

Prather, A. L. 1994. A new species of *Phlox* (Polemoniaceae) from northern Mexico with an expanded circumscription of subsection *Divaricatae*. *Plant Systematics and Evolution* 192: 61–66.

Pursh, F. [1814]1813. *Flora Americae Septentrionalis; or, a Systematic Arrangement and Description of the Plants of North America*. 2 vols. London.

Radford, A. E., H. E. Ahles, and C. R. Bell. 1964. *Manual of the Vascular Flora of the Carolinas*. Chapel Hill, North Carolina: University of North Carolina Press.

Rasputin, V. G. 1996. *Siberia, Siberia*. Evanston, Illinois: Northwestern University Press. Translated, and with an introduction by Margaret Winchell and Gerald Mikkelson.

Ray, J. 1999. *Ecology of a Cracker Childhood*. Minneapolis, Minnesota: Milkweed Editions.

Ray, J. 1704. *Historia Plantarum . . . supplementum*. London: Smith & Walford.

Razzhivin, V. Y. 1986. The effect of soil reaction on the plant distribution of snowbed plant communities of the southeasternmost Chukotka peninsula. *Botanicheskii zhurnal* 71: 1088–1097. [in Russian]

Razzhivin, V. Y. 1994. Snowbed vegetation of far northeastern Asia. *Journal of Vegetation Science* 5 (6): 829–842.

Reveal, J. L., C. R. Broome, M. L. Brown, and G. F. Frick. 1982. Comments on the typification of two Linnaean species of *Phlox* (Polemoniaceae). *Taxon* 31: 733–736.

Richardson, J. 1823. Botanical appendix. In. J. Franklin. *Narrative of a Journey to the Shores of the Polar Sea, in the Years 1819, 20, 21, and 22*. London: John Murray. 729–762.

Richardson, J. 1828. Narrative of the proceedings of the Eastern Detachment of the Expedition. In. J. Franklin. *Narrative of a Second Expedition to the Shores of the Polar Sea, in the Years 1825, 1826, and 1827*. London: John Murray. 187–285.

Ripley, D. D. 1944. More western Americans. *Quarterly Bulletin of the Alpine Garden Society* 12 (2): 65–74.

Ripley, D. D. 1945. Nevada, 1944. *Bulletin of the Alpine Garden Society* 13 (1): 28–37.

Rogers, L. E., and K. A. Gano. 1980. Townsend ground squirrel (*Spermophilus townsendii*) diets in the shrub steppe of south central Washington. *Journal of Range Management* 33 (6): 463–465.

Roosevelt, T. 1893. *Hunting the Grisly and Other Sketches: An Account of the Big Game of the United States and its Chase with Horse, Hound, and Rifle*. New York: G. P. Putnam's Sons.

Royal Horticultural Society (London), Published under the direction of. 1914. *Journal kept by David Douglas during his travels in North America, 1823–1827, together with a particular description of thirty-three species of American oaks and eighteen species of* Pinus, *with appendices containing a list of plants introduced by Douglas and an account of his death in 1834.* London.

Sandoz, M. 1962. *Old Jules.* Lincoln, Nebraska: University of Nebraska Press.

Schaal, B. A., and W. J. Leverich. 1982. Survivorship patterns in an annual plant community. *Oecologia* 54: 149–151.

Schafale, M. P., and P. A. Harcombe. 1983. Presettlement vegetation of Hardin County, Texas. *American Midland Naturalist* 109 (2): 355–366.

Scheele, G. 1848. Beitrage zur flor von Texas. *Linnaea* 21: 747–768.

Scheele, G. 1850. Beitrage zur flora von Texas. *Linnaea* 23: 139–146.

Schwaegerle, K. E., and F. A. Bazzaz. 1987. Differentiation among nine populations of *Phlox*: response to environmental gradients. *Ecology* 68 (1): 54–64.

Schwaegerle, K. E., and D. A. Levin. 1990. Environmental effects on growth and fruit production in *Phlox drummondii. Journal of Ecology* 78 (1): 15–26.

Sharsmith, C. W. 1958. A new species of *Luzula* and of *Phlox* from the High Sierra Nevada of California. *Aliso* 4 (1): 125–129.

Shea, A. B., and T. H. Roulston. 1996. Cumberland Rosemary Recovery Plan. U.S. Fish and Wildlife Service, Atlanta, Georgia.

Shinners, L. H. 1951. Botanical notes. *Field and Laboratory* 19: 126–127.

Shinners, L. H. 1961. *Phlox bifida* var. *induta* (Polemoniaceae) a new endemic in north central Texas. *Southwestern Naturalist* 6 (1): 50–51.

Shreve, F. 1939. Observations on the vegetation of Chihuahua. *Madrono* 5: 1–13.

Shreve, F. 1942. Grasslands and related vegetation in northern Mexico. *Madrono* 6: 190–198.

Sims, J. 1802. Phlox stolonifera. Creeping Phlox. *Curtis's Botanical Magazine* 16: plate 563.

Sims, J. 1810. Phlox amoena. Fraser's Hairy Phlox. *Curtis's Botanical Magazine* 32: plate 1308.

Small, J. K. 1903. *Flora of the Southeastern United States.* New York. Published by the author.

Smith, D. M., and D. A. Levin. 1967. Karyotypes of eastern North American *Phlox. American Journal of Botany* 54: 324–334.

Sommers, P. 1994. Recovery Plan for Tennessee yellow-eyed grass (*Xyris tennesseensis* Kral). U.S. Fish and Wildlife Service, Jackson, Mississippi.

Springer, T. L. 1983. Distribution, habitat and reproductive biology of *Phlox oklahomensis* (Polemoniaceae). MS Thesis, Oklahoma State University.

Springer, T. L., and R. J. Tyrl. 1989. Distribution, habitat and reproductive biology of *Phlox oklahmensis* Wherry (Polemoniaceae). *Proceedings of the Oklahoma Academy of Science* 69: 15–22.

Springer, T. L. and R. J. Tyrl. 2003. Status of *Phlox oklahomensis* (Polemoniaceae) in northwestern Oklahoma and adjacent Kansas: assessment 20 years later. *Proceedings of the Oklahoma Academy of Science* 83: 89–92.

Steyermark, J. A. 1963. *Flora of Missouri.* Ames, Iowa: Iowa State University Press.

Strakosh, S. C. 2004. Systematic studies in *Phlox* (Polemoniaceae) with a focus on *P. dolichantha*, *P. superba*, *P. stansburyi* and *P. grayi*. MS Thesis, Kansas State University.

Strakosh, S. C., and C. J. Ferguson. 2005. Pollination biology of four southwestern species of *Phlox* (Polemoniaceae): insect visitation in relation to corolla tube length. *Southwestern Naturalist* 50 (3): 291–301.

Stout, A. B. 1917. Variation in the moss pink, *Phlox subulata*. *Journal of the New York Botanical Garden* 18 (208): 75–83.

Sweet, R. 1827. Phlox nivalis. *British Flower Garden* 2: plate 185.

Swengel, A. B., and S. R. Swengel. 199b. Observations on *Schinia indiana* and *Schinia lucens* in the Midwestern United States. *Holarctic Lepidoptera* 6 (1): 11–21.

Symons-Jeune, B. H. B. 1953. *Phlox: A Flower Monograph*. London, England: Collins.

Thayer, C. L. 1919. Hardy phlox for present planting. *The Garden Magazine* 29: 125–126.

Thomas, G. S. 1989. *The Rock Garden and Its Plants: From Grotto to Alpine House*. Portland, Oregon: Sagapress.

Tisdale, E. W. 1986. Canyon grasslands and associated shrublands of west-central Idaho and adjacent areas. *University of Idaho Forest, Wildlife and Range Experiment Station Bulletin* 40: 1–42.

Tolkien, J. R. R. 1983. *The Book of Lost Tales, Part 1*. London. Ed. C. Tolkien. Boston: Allen & Unwin.

Torrey, J. 1859. Botany. In *Report on the United States and Mexican Boundary Survey, Made Under the Direction of the Secretary of the Interior*, volume 2, part 1. Ed. W. H. Emory. Washington, D.C.

Tucker, G. E. 1972. The vascular flora of Bluff Mountain, Ashe County, North Carolina. *Castanea* 37: 2–26.

Turner, B. L. 1998a. Atlas of the Texas species of *Phlox* (Polemoniaceae). *Phytologia* 85 (5): 309–326.

Turner, B. L. 1998b. *Phlox drummondii* (Polemoniaceae) revisited. *Phytologia* 85 (4): 280–287.

Uttal, L. J. 1971. Five amendments to the flora of southwest Virginia. *Castanea* 36 (2): 79–81.

Van Horne, B., R., L. Schooley, and P. B. Sharpe. 1998. Influence of habitat, sex, age, and drought on the diet of Townsend's ground squirrels. *Journal of Mammology* 79 (2): 521–537.

Ventenat, E. P. 1803. *Jardin de la Malmaison*. Paris.

Wallace, D. R. 2003. *The Klamath Knot*. Berkeley, California: University of California Press.

Warnock, M. J. 1995. Texas Trailing Phlox (*Phlox nivalis* ssp. *texensis*) Recovery Plan. U.S. Fish and Wildlife Service, Albuquerque, New Mexico.

Warrick, R. B. 1987. A floristic survey of the Pine Valley Mountains, Utah. MS Thesis, Brigham Young University.

Weber, W. A. 1973. Additions to the flora of Colorado, V, with nomenclatural revisions. *Southwestern Naturalist* 18: 319–329.

Welsh, S. L. 2003. Polemoniaceae. In *A Utah Flora*, Third Edition, Revised. Eds. S. L. Welsh, N. D. Atwood, S. Goodrich, and L. C. Higgins. Provo, Utah: Jones Endowment Fund, Monte L. Bean Life Science Museum, Brigham Young University. 479–496.

Wenk, E. H., and T. E. Dawson. 2007. Interspecific differences in seed germination, establishment, and early growth in relation to preferred soil type in an alpine community. *Arctic, Antarctic and Alpine Research* 39 (1): 165–176.

Wherry, E. T. 1928. Selecting a national flower. *Journal of the New York Botanical Garden* 29: 209–210.

Wherry, E. T. 1929a. Picking out the Polemoniaceae. *Bartonia* 11: 1–4.

Wherry, E. T. 1929b. The eastern subulate-leaved phloxes. *Bartonia* 11: 5–35, plates 1–4.

Wherry, E. T. 1930. A long lost *Phlox. Journal of the Washington Academy of Sciences* 20 (2): 25–28.

Wherry, E. T. 1931a. Save our Wild Flowers. *Scientific Monthly* 33 (6): 535–538.

Wherry, E. T. 1931b. The eastern short-styled phloxes. *Bartonia* 12: 24–53.

Wherry, E. T. 1932a. The eastern long-styled phloxes, part I. *Bartonia* 13: 18–37.

Wherry, E. T. 1932b. The eastern long-styled phloxes, part 2. *Bartonia* 14: 14–26.

Wherry, E. T. 1933. The eastern veiny-leaved phloxes. *Bartonia* 15: 14–26.

Wherry, E. T. 1934. Supplemental notes of eastern phloxes. *Bartonia* 16: 38–44.

Wherry, E. T. 1935a. A new variety of *Phlox ovata* from the Alabama Mountains. *Bartonia* 16: 37–39.

Wherry, E. T. 1935b. Our native Phloxes and their horticultural derivatives. *National Horticultural Magazine* 14: 209–231.

Wherry, E. T. 1935c. Fifteen notable shale-barren plants. *Claytonia* 2: 19–22.

Wherry, E. T. 1937. Synonyms and variety names for *Phlox nivalis. Gardeners Chronicle of America* 41: 212–213.

Wherry, E. T. 1938. The Phloxes of Oregon. *Proceedings of the Academy of Natural Sciences of Philadelphia* 90: 133–140.

Wherry, E. T. 1939. Four southwestern subspecies of *Phlox. Journal of the Washington Academy of Sciences* 29: 518–519.

Wherry, E. T. 1941a. A new hybrid *Phlox* (*P. Henryae*). *Nat. Hort. Mag.* 20: 151.

Wherry, E. T. 1941b. Remarks on the name *Phlox nivalis. Rhodora* 43: 71

Wherry, E. T. 1941c. The Phloxes of Idaho. *Notulae Naturae of the Academy of Natural Sciences of Philadelphia* 87: 1–15.

Wherry, E. T. 1942. The phloxes of Nevada. *Notulae Naturae of the Academy of Sciences of Philadelphia* 113: 1–11.

Wherry, E. T. 1943a. *Microsteris, Phlox*, and an Intermediate. *Brittonia* 5:60–63.

Wherry, E. T. 1943b. Variation in *Phlox floridana. Bartonia* 22: 1–2.

Wherry, E. T. 1944. New phloxes from the Rocky Mountains and neighboring regions. *Notulae Naturae of the Academy of Natural Sciences of Philadelphia* 146: 1–11.

Wherry, E. T. 1945. The *Phlox carolina* complex. *Bartonia* 23: 1–9.

Wherry, E. T. 1946. Rock garden phloxes. *Bulletin of the American Rock Garden Society* 4 (2): 17–27.

Wherry, E. T. 1951. Subspecies of three eastern phloxes. *Castanea* 16: 97–100.

Wherry, E. T. 1953. Shale-barren plants on other geological formations. *Castanea* 18: 64–65.

Wherry, E. T. 1955. *The Genus Phlox*. Philadelphia, Pennsylvania: The Morris Arboretum of the University of Pennsylvania.

Wherry, E. T. 1956. Validation of new combinations in *Phlox*. *Baileya* 4: 5–7.

Wherry, E. T. 1961. A new annual *Phlox*. *Wrightia* 2 (4): 198–199.

Wherry, E. T. 1962. A controversial treatment of the Polemoniaceae. *Madrono* 16 (8): 270–271.

Wherry, E. T. 1964. New combinations in Texas Polemoniaceae. *Sida* 1 (4): 250–251.

Wherry, E. T. 1965a. Relationships of *Phlox caespitosa*, *pulvinata*, and *douglasii* (Polemoniaceae). *Sida* 2 (2): 154–156.

Wherry, E. T. 1965b. *The Genus Phlox*, ten years after. *Bartonia* 35: 13–16.

Wherry, E. T. 1966. Polemoniaceae. In *Flora of Texas*, volume 1. Ed. C. L. Lundell. Renner, Texas: Texas Research Foundation. 283–321.

Wherry, E. T. 1969. *Phlox* and *Polemonium* (Polemoniaceae) in the Intermountain Region. *Sida* 3 (6): 441–444.

Wherry, E. T. 1970. Notes on phloxes in the Gulf Coast states. *Castanea* 35: 198–199.

Wherry, E. T., and L. Constance. 1938. A new phlox from the Snake River Canyon, Idaho-Oregon. *American Midland Naturalist* 19: 433–435.

Whitehouse, E. 1935. Notes on Texas phloxes. *Bulletin of the Torrey Botanical Club* 62: 381–401.

Whitehouse, E. 1936. *Texas Flowers in Natural Colors*. Austin, Texas: Published by author

Whitehouse, E. 1945. Annual *Phlox* species. *American Midland Naturalist* 34 (2): 388–401.

Whittaker, R. H. 1960. Vegetation of the Siskiyou Mountains, Oregon and California. *Ecological Monographs* 30 (3): 279–338.

Whittaker, R. H. 1966. Forest dimensions and productions in the Great Smoky Mountains. *Ecology* 47 (1): 103–121.

Wilken, D. H. 1986. Polemoniaceae. In *Flora of the Great Plains*. Great Plains Flora Association. Lawrence, Kansas: University Press of Kansas. 666–677.

Wilken, D. H., and J. M. Porter. 2005. Vascular plants of Arizona: Polemoniaceae. *Canotia* 1: 1–37.

Willdenow, C. L. 1809. *Enumeratio Plantarum Horti Regii Botanici Berolinensis.* . . . Berlin.

Williams, L. R., G. N. Cameron, S. R. Spencer, B. D. Eshelman, and M. J. Gregory. 1986. Experimental analysis of the effects of pocket gopher mounds on Texas coastal prairie. *Journal of Mammology* 67 (4): 672–679.

Witherspoon, J. T. 1971. Plant succession on gravel bars along the Jacks Fork and Current Rivers in the Southcentral Missouri Ozarks. MS Thesis, Southwest Missouri State University.

Wolfe-Bellin, K. S., and K. A. Moloney. 2001. Successional vegetation dynamics on pocket gopher mounds in an Iowa tallgrass prairie. In *Seeds for the Future; Roots of the Past: Proceedings of the Seventeenth North American Prairie Conference*. Ed. N. P. Bernstein and L. J. Ostrander. Mason City, Iowa: Northern Iowa Community College. 155–163.

Young, J. A., R. A. Evans, and J. Major. 1977. Sagebrush steppe. In *Terrestrial Vegetation of California*. Ed. M. G. Barbour and J. Major. New York: Wiley-Interscience. 763–796.

Yurtsev, B. A. 1972. Phytogeography of northeastern Asia and the problem of Transberingian floristic interrelations. In *Floristics and Paleofloristics of Asia and Eastern North America*. Ed. A. Graham. Amsterdam, Netherlands: Elsevier Publishing Company. 19–54.

Works Consulted: Botanical History

Berkeley E., and D. S. Berkeley. 1963. *John Clayton: Pioneer of American Botany*. Chapel Hill, North Carolina: University of North Carolina Press.

Elisens, W. J. 1985. The Montana collections of Francis Duncan Kelsey. *Brittonia* 37 (4): 382–391.

Ewan, J. 1950. *Rocky Mountain Naturalists*. Denver, Colorado: University of Denver Press.

Ewan, J. 1952. Fredrick Pursh, 1774–1820, and his botanical associates. *Proceedings of the American Philosophical Society* 96 (5): 599–628.

Ewan, J., and N. Ewan. 1970. *John Banister and His Natural History of Virginia, 1678–1692*. Urbana, Illinois: University of Illinois Press.

Fogg, J. M., Jr. 1983. Edgar Theodore Wherry (1885–1982). *Bartonia* 49: 1–5.

Frick, G. F., J. L. Reveal, C. R. Broome, and M. L. Brown. 1987. Botanical explorations and discoveries in colonial Maryland, 1688 to 1753. *Huntia* 7: 5–59.

Frick, G. F., and R. P. Stearns. 1961. *Mark Catesby: The Colonial Audubon*. Urbana, Illinois: University of Illinois Press.

Gieser, S. W. 1948. *Naturalists of the Frontier*, second edition. Dallas, Texas: Southern Methodist University Press.

Graham, A. 1966. *Plantae Rariores Camschatcenses*: a translation of the dissertation of Jonas P. Halenius, 1750. *Brittonia* 18: 131–139

Graunstein, J. E. 1967. *Thomas Nuttall, Naturalist; Explorations in America, 1808–1841*. Cambridge, Massachusetts: Harvard University Press.

Harvey, A. G. 1947. *Douglas of the Fir: a Biography of David Douglas, Botanist*. Cambridge, Massachusetts: Harvard University Press.

Heller, J. L. 1968. Linnaeus's *Hortus Cliffortianus*. *Taxon* 17 (6): 663–719.

Hooker, W. J. 1834. Notice concerning Mr. Drummond's collections, made in the southern and western parts of the United States. *Journal of Botany* 1: 50–60.

Hooker, W. J. 1835a. Notice concerning the late Mr. Drummond's journeys and his collections, made chiefly in the southern and western parts of the United States. *Companion to the Botanical Magazine* 1: 39–49

Hooker, W. J. 1836a. A brief memoir of the life of Mr. David Douglas, with extracts from his letters. *Companion to the Botanical Magazine* 2: 79–182.

Houston, C. S. 1983. Robert Hood (1797–1821). *Arctic* 36 (2): 210–211.

Jarvis, C. 2007. *Order Out of Chaos: Linnaean Plant Names and their Types*. London: The

Linnaean Society of London in association with the Natural History Museum, London.

Jarvis, C. E., S. R. Majorov, D. D. Sokoloff, S. A. Balandin, I. A. Gubanov, and S. S. Simonov. 2002 [2001]. The typification of *Astragalus physodes* L., based on material in the Herbarium of Moscow (MW), and the discovery of an isolectotype of *Phlox sibirica* L. *Taxon* 50: 1129–1135.

Johnson, R. E. 1976. *Sir John Richardson: Arctic Explorer, Natural Historian, Naval Surgeon*. London: Taylor & Francis LTD.

Juel, H. O., and J. W. Harshberger. 1929. New light on the collections of North American plants made by Peter Kalm. *Proceedings of the Academy of Natural Sciences of Philadelphia* 81: 297–303.

Lenz, L. W. 1986. *Marcus E. Jones: Western Geologist, Mining Engineer and Botanist*. Claremont, California: Rancho Santa Ana Botanic Garden.

Lewis, J. 1860. Biographical notice of the late Thomas Nuttall. *Proceedings of the American Philosophical Society* 7 (63): 297–315.

McIntosh, R. P. 1983. Edward Lee Greene: The Man. In *Landmarks of Botanical History / Edward Lee Green*, Part I. Stanford, California: Stanford University Press. 16–53.

McKelvey, S. D. 1955. *Botanical Exploration of the Trans-Mississippi West, 1790–1850*. Jamaica Plain, Massachusetts: Arnold Arboretum of Harvard University.

Moulton, G. E. (ed.). 1999. *The Journals of the Lewis and Clark Expedition. Herbarium of the Lewis and Clark Expedition*. Volume 12. Lincoln, Nebraska: University of Nebraska Press.

Pennell, F. W. 1936. Travels and scientific collections of Thomas Nuttall. *Bartonia* 18: 1–51.

Rembert, D. H., Jr. 2004. Andre Michaux's travels and plant discoveries in the Carolinas. *Castanea Occasional Papers in Eastern Botany* 2: 107–118.

Reveal, J. L. 2004. No man is an island: the life and times of Andre Michaux. *Castanea Occasional Papers in Eastern Botany* 2: 22–68

Reveal, J. L., C. R. Broome, M. L. Brown, and G. F. Frick. 1987a. The identification of pre-1753 polynomials and collections of vascular plants from the British colony of Maryland. *Huntia* 7: 91–208.

Reveal, J. L., C. R. Broome, M. L. Brown, and G. F. Frick. 1987b. On the identities of Maryland plants mentioned in the first two editions of Linnaeus's *Species plantarum*. *Huntia* 7: 209–245.

Reveal, J. L., G. E. Moulton, and A. E. Schuyler. 1999. The Lewis and Clark collections of vascular plant types: names, types, and comments. *Proceedings of the Academy of Natural Sciences of Philadelphia* 149: 1–64.

Rowell, M. 1980. Linnaeus and botanists in eighteenth-century Russia. *Taxon* 29 (1): 15–26.

Sargent, C. S. 1889. Portions of the journal of Andre Michaux, botanist, written during his travels in the United States and Canada, 1785–1796. *Proceedings of the American Philosophical Society* 26 (129): 1–145.

Savage, H., Jr., and E. J. Savage. 1986. *Andre and Francois Andre Michaux*. Charlottesville, Virginia: University of Virginia Press.

Savage, S. 1945. *A Catalogue of the Linnaen Herbarium*. London: Taylor & Francis, LTD.

Sokoloff, D. D., S. A. Balandin, I. A. Gubanov, C. E. Jarvis, S. R. Majorov, and S. S. Simonov. 2002. The history of botany in Moscow and Russia in the 18th and early 19th centuries in the context of the Linnaean Collection and Moscow University (MW). *Huntia* 11 (2): 129–191.

Simpson, M. B., S. Moran, and S. W. Simpson. 1997. Biographical notes on John Fraser (1750–1811): plant nurseryman, explorer, and royal botanical collector to the Czar of Russia. *Archives of Natural History* 24: 1–18.

Stearn, W. T. 1957. An introduction to the Species Plantarum and cognate botanical works of Carl Linnaeus. Prefixed to Ray Society facsimile of Linnaeus *Species Plantarum*, ed. 1, 1753. Vol. 1. London: Ray Society. 1–176.

Stearn, W. T. 1972. Philip Miller and the plants from the Chelsea Physic Garden presented to the Royal Society of London, 1723–1796. *Transactions of the Botanical Society of Edinburgh* 41: 293–307.

Stearns, R. P. 1970. *Science in the British Colonies of America*. Urbana, Illinois: University of Illinois Press.

Tharp, B. C. 1939. The vegetation of Texas. *Texas Academy Publications in Natural History* 1: 1–74.

Thomas, J. H. 1979. Botanical explorations in Washington, Oregon, California and adjacent regions. *Huntia* 3: 1–66.

Tiehm, A. 1996. Nevada vascular plant types and their collectors. *Memoirs of the New York Botanical Garden* 77: 1–104.

Uttal, L. J. 1984. The type localities of the *Flora Boreali-Americana* of Andre Michaux. *Rhodora* 86 (845): 1–66.

Wagner, W. H., Jr. 1982. Edgar T. Wherry 1885–1982. *Bulletin of the Torrey Botanical Club* 109 (4): 545–548.

Wagner, W. H., Jr. 1983. Bibliography of Edgar T. Wherry. *Bartonia* 49: 6–14.

Weber, W. A. 1997. *King of Colorado Botany: Charles Christopher Parry, 1823–1890*. Niwot, Colorado: University Press of Colorado.

Williams, R. L. 1984. *Aven Nelson of Wyoming*. Boulder, Colorado: Colorado Associated University Press.

Williams, R. L. 2003. *"A Region of Astonishing Beauty": The Botanical Exploration of the Rocky Mountains*. Lanham, Maryland. Roberts Rinehart Publishers.

Young, F. G. 1899. The correspondence and journals of Captain Nathaniel J. Wyeth, 1831–6. *Sources of the History of Oregon* 1 (3–6): 1–262.

Works Consulted: Horticultural History

Britten, J. 1899. Bibliographical notes. XXI.–Frasers' Catalogues. *Journal of Botany, British and Foreign* 37: 481–487.

Britten, J. 1905. Fraser's Catalogue, 1796. *Journal of Botany, British and Foreign* 48: 329–331.

Desmond, R. 1977. Victorian gardening magazines. *Garden History* 5 (3): 47–66.

Ewan, J., and N. Ewan. 1963. John Lyon, nurseryman and plant hunter, and his journal, 1799–1814. *Transactions of the American Philosophical Society* 53 (2): 1–69.

Fuchs, H. 1994. *Phlox: Stauden- und Polsterphloxe.* Stuttgart: Ulmer GmbH & Co.

Greene, E. L. 1890. Reprint of Fraser's Catalogue. *Pittonia* 2: 114–119.

Hamblin, S. F. 1928. Phlox, flower of flame: Half a hundred hardy species that are easily grown. *Garden & Home Builder* 47 (6): 531, 578, 589.

Hamblin, S. F. 1940. New types of moss phlox. *Horticulture* 18 (11): 4.

Harmer, J., and J. Elliott. 2001. *Phlox.* North East Lincolnshire [United Kingdom]: The Hardy Plant Society.

Harrison, M. 2000. Mary Gibson Henry, plantswoman extraordinaire. *Arnoldia* 60 (1): 2–12.

Harvey, J. H. 1974. *Early Nurserymen.* London: Phillimore.

Harvey, J. H. 1998. The English nursery flora, 1677–1723. *Garden History* 26 (1): 60–101.

Henry, M. G. 1943. A rock garden of natives. *Bulletin of the American Rock Garden Society* 1: 2–13.

Heriz-Smith, S. 1988. The Veitch Nurseries of Killerton and Exeter c. 1780 to 1863: part I. *Garden History* 16 (1): 41–57.

Hooker, W. J. 1836b. Biographical sketch of John Fraser, the botanical collector. *Companion to the Botanical Magazine* 2: 300–305.

Hunting, P. 2002. Isaac Rand and the Apothecaries' Physic Garden at Chelsea. *Garden History* 30 (1): 1–23.

Le Rougetel, H. 1986. Philip Miller/John Bartram botanical exchange. *Garden History* 14 (1): 32–39.

Loudon, J. C. 1836. A Notice of the Garden of Canonmills Cottage, the Residence of Patrick Neill. . . . *The Gardener's Magazine* 12: 333–341.

Pridham, A. M. S. 1928. Phlox. *Journal of the New York Botanical Garden* 29 (346): 249–259.

Pridham, A. M. S. 1931. Summer flowering phloxes. *Proceedings of the American Society for Horticultural Science* 28: 418–422.

Pridham, A. M. S. 1934. History, culture, and varieties of summer-flowering phloxes. *Cornell University Agricultural Experiment Station Bulletin* 588.

Rand, I. 1726. A catalog of the fifty plants from Chelsea-Garden, presented to the Royal Society, by the Company of Apothecaries, for the year 1725, pursuant to the direction of Sir Hans Sloane, Baronet, Pr. Coll. Med. S. P. V. Pr. By Mr. Isaac Rand, Apothecary. F. R. S. *Philosophical Transactions* 34: 125–127.

Rand, I. 1735. A catalog of the fifty plants, from Chelsea-Garden, presented to the Royal Society by the Company of Apothecaries, for the year 1733. Pursuant to the direction of Sir Hans Sloane, Bart, Med. Reg. Praes. Col. Reg. Med. & Soc. Reg. by Mr. Isaac Rand, Apothecary. F. R. S. Hort. Chel. Praes. Ac. Praelec. Botan. *Philosophical Transactions* 39: 1–4.

Robinson, G. W. 1938. New plants for the rock garden. *Journal of the Royal Horticultural Society* 63: 307–318.Stearn, W. T. 1972. Philip Miller and the plants from the Chelsea Physic Garden presented to the Royal Society of London, 1723–1796. *Transactions of the Botanical Society of Edinburgh* 41: 293–307.

Works Consulted: *Phlox* Systematics, Nomenclature, and Ecology

Cooperrider, T. S. 1986. The genus *Phlox* in Ohio. *Castanea* 51 (2): 145–148.

Crawford, D. J. 1971. IOPB Chromosome number reports XXXI: Polemoniaceae. *Taxon* 20 (1): 157–158.

Eater, J. W. 1967. A systematic study of subsection *Nanae* of the genus *Phlox*. MS Thesis, University of California at Santa Barbara.

Fehlberg, S. D., K. A. Ford, M. C. Unger, and C. J. Ferguson. 2008. Development, characterization, and transferability of microsatellite markers for the plant genus *Phlox* (Polemoniaceae). *Molecular Ecology Resources* 8: 116–118.

Grant, V. 1959. *Natural History of the Phlox Family*. The Hague: Martinus Nijhoff.

Grant, V. 2001. Nomenclature of the main subdivisions of *Phlox* (Polemoniaceae). *Lundellia* 4: 25–29.

Grant, V., and K. A. Grant. 1965. *Flower Pollination in the Phlox Family*. New York, New York: Columbia University Press.

Jennings, B. 1999. Species of *Phlox* in Colorado. Unpublished report prepared for the Colorado Native Plant Society, Fort Collins, Colorado.

Maslin, T. P. 1978. *Phlox nana* Nuttall. *Bulletin of the Alpine Garden Society of Great Britain*. 46: 163–167.

Mason, H. L. 1951. *Phlox* In *Illustrated Flora of the Pacific States, Volume III, Geraniaceae to Scrophulariaceae*. Ed. L. Abrams. Stanford, California: Stanford University Press. 408–413.

Peter, A. 1897. Polemoniaceae. *Die naturlichen Pflanzenfamilien* 4, 3a: 40–54. Leipzig: Wilhelm Engelmann.

Reveal, J. L. 1989. Proposal to conserve the names and types of three North American Linnaean species. *Taxon* 38 (3): 515–519.

Reveal, J. L. 1991. Typification of *Phlox glaberrima* Linnaeus (Polemoniaceae). *Phytologia* 71 (6): 475–478.

Reveal, J. L., C. R. Broome, M. L. Brown, and G. F. Frick. 1982. Comments on the typification of two Linnaean species of *Phlox* (Polemoniaceae). *Taxon* 31: 733–736.

St. John, H. 1936. The replicate species of *Phlox* of the Pacific Northwest. *Torreya* 36: 94–99.

Schassberger, L., and P. Achuff. 1991. Status review of *Phlox kelseyi* var. *missoulensis*, Lewis and Clark National Forest. Montana Natural Heritage Program, Helena, Montana.

Smith, D. M., and D. A. Levin. 1966. Preliminary reports on the flora of Wisconsin, no. 57: Polemoniaceae – Phlox family. *Transactions of the Wisconsin Academy of Sciences, Arts and Letters*. 55: 243–253.

Solecki, M. K. 2000. Sangamon phlox. *Illinois Steward* 9 (1): 22–23.

Taylor R. J., and C. E. Taylor. 1980. *Phlox oklahomensis* Status Report. Endangered Species Office, U.S. Fish and Wildlife Service, Albuquerque, New Mexico.

Taylor, R. J., and C. E. S. Taylor. 1981a. *Phlox pilosa* var. *longipilosa* (Waterfall) J. & C. Taylor comb. nov. (Polemoniaceae). *Sida* 9 (2): 183–184.

Taylor, R. J., and C. E. S. Taylor. 1981b. *Phlox pilosa* var. *longipilosa* Status Report. Unpublished report submitted to the U.S. Fish and Wildlife Service, Albuquerque, New Mexico.

U.S. Fish and Wildlife Service. 2000. Endangered and threatened wildlife and plants; Determination of endangered status for the plant Yreka phlox from Siskiyou County, California. Federal Register 65: 5268–5275. 03 February 2000.

U.S. Fish and Wildlife Service. 2004 Draft Recovery Plan for *Phlox hirsuta* (Yreka Phlox). Portland, Oregon.

Welsh, S. L. 1985. Revision of the *Phlox austromontana* (Polemoniaceae) complex in Utah. *Great Basin Naturalist* 45 (4): 791–792.

Works Consulted: Cultivation

Anon. 1931. Alpine phlox tried at Wisley. *Journal of the Royal Horticultural Society* 56: 256–257.

Cumming, R. W. and R. E. Lee. 1960. *Contemporary Perennials*. New York: The MacMillan Company.

Kelaidis, P. 1989. Fabulous phloxes. *Bulletin of the American Rock Garden Society* 47 (1): 13–16.

Lowzow, S. 1983. Don't discount Arizona. *Bulletin of the American Rock Garden Society* 41 (4): 185–188.

Lupp, R. 1999. Strategies for growing choice alpines. *Rock Garden Quarterly* 57 (4): 249–256.

Mineo. B. 1999. *Rock Garden Plants: A Color Encyclopedia*. Portland, Oregon: Timber Press.

Nicholls, G. 1998. *Phlox kelseyi* 'Lemhi Purple'. *Bulletin of the Alpine Garden Society* 66 (1): 22–23.

Nicholls, G. 2002. *Alpine Plants of North America: An Encyclopedia of Mountain Flowers from the Rockies to Alaska*. Portland, Oregon: Timber Press.

Sealy, J. R. 1946. *Phlox triovulata*. *Curtis's Botanical Magazine* 164: plate 9674.

Smith, J. R. and B. S. Smith. 1980. *The Prairie Garden: 70 Native Plants You Can Grow in Town or Country*. Madison, Wisconsin: University of Wisconsin Press.

Snyder, L. C. 1983. *Flowers for Northern Gardens*. Minneapolis, Minnesota: University of Minnesota Press.

Titchmarsh, C. C. 1916. Phloxes at Wisley, 1915. *Journal of the Royal Horticultural Society* 41: 250–264.

Index